# DESIGN FOR DYNAMIC LOADING
## THE USE OF MODEL ANALYSIS

# DESIGN FOR DYNAMIC LOADING
## THE USE OF MODEL ANALYSIS

Edited by G S T Armer and F K Garas

**Construction Press**
London and New York

Construction Press
Longman House
Burnt Mill, Harlow, Essex, UK

*A division of Longman Group Limited, London*

*Published in the United States of America
by Longman Inc., New York*

*First published 1982*

---

**British Library Cataloguing in Publication Data**

Dynamic modelling of structures
   1. Structural dynamics–Congresses
   2. Structural Engineering–Congresses
   I. Armer, G.S.T.    II. Garas, F.K.
   624.1'77    TA654
   ISBN 0-86095-706-3

**Library of Congress Cataloging in Publication Data**

Main entry under title:

Dynamic modelling of structures.

   "Based on the proceedings of an international seminar
organised by the Institution of Structural Engineers Informal
Study Group for 'Model Analysis as a Design
Tool,' . . . November 1981"–Pref.
   Includes bibliographical references and index.
   1. Structural dynamics–Congresses.   2. Structural
design–Congresses.   3. Engineering models–Congresses.
   I. Armer, G. S. T.   II. Garas, F. K.   III. Institution of
Structural Engineers (Great Britain). Informal Study Group
for "Model Analysis as a Design Tool."
   TA654.D94 1982    624.1'7'0724    82-7978
   ISBN 0-86095-706-3          AACR2

---

Printed in Great Britain at The Pitman Press, Bath

# Contents

# Preface

Designing for dynamic loads - those imposed by earthquakes, impact, vibration, wind and waves - has become an important requirement for many major civil engineering structures such as offshore oil and gas platforms, chemical and processing plants, power generators, shelters, tall buildings and bridges. Over the years engineers have been able to improve the construction and performance of such structures by using physical models to assess their response to dynamic loads. However, the involvement of physical model tests as part of the design process poses the many problems of model construction, test methods and interpretation of results.

This book is based on the proceedings of an international seminar organised by the Institution of Structural Engineers Informal Study Group for "Model Analysis as a Design Tool", the third meeting to be organised by the Group since its formation in 1977. This seminar took place at the UK Building Research Establishment in November 1981, and the 125 delegates represented 15 countries.

The object of the seminar was to provide an international forum for the exchange of knowledge between research and design engineers on the use of models and their role in the design of structures subject to dynamic loads.

Reproduced in this volume are 37 papers covering concrete structures, steel structures and foundations and the problems highlighted include those associated with limitations of physical models; correlation of experimental and analytical results; data analysis and interpretation; similitude requirements and model design; material properties and quality control; and loading and measuring techniques.

The discussion generated at the meeting is also reproduced and all together this volume represents an international compendium of knowledge which reflects the growing world-wide interest in the use of models for the design of dynamically loaded structures.

The editors wish to thank the authors for the work put into their papers and also Mrs Helen Stevenson, Miss Karen Bull, Mrs Patricia Rowley and Mrs Celia Belbin for their invaluable administrative help with the seminar.

G.S.T. Armer
F.K. Garas

# Opening Address

Dr Ivan Dunstan
*Director of the Building Research Establishment, UK*

Ladies and Gentlemen,

Dr. Garas has welcomed you on behalf of the Institution of Structural Engineers' Informal Study Group (on Model Analysis as a Design Tool) to this, the third International Seminar organised by the Group in recent years.

It is my agreeable task to thank Dr. Garas and the Study Group not only for all the preparatory work which has been done to bring us all together for this important event, but also for selecting the Building Research Establishment as the venue for the seminar.

BRE has been closely associated with the work of the Study Group since the Group's inception in 1977, and it has been our privilege to host both the previous international seminars held in 1978 and 1979.

We are particularly pleased to be closely associated with these seminars on two counts. First because they bring together at BRE powerful groups of professional researchers committed to programmes of common interest. That is good. And I can do no better than to quote the words of Prince Philip, Duke of Edinburgh, when he welcomed delegates to the 6th FIP Congress in Prague in June 1970. He said, and I quote:
"It is of course possible to practice any profession in determined isolation but regular opportunities to exchange knowledge, ideas, experience and expertise can greatly improve performance. Furthermore they make life more interesting".

I hope that this seminar will be such an occasion and it is with real pleasure that I welcome all our visitors both from the United Kingdom and from overseas. We are delighted that you have been able to attend - particularly at a time when pressure on travel funds must be very severe.

My second reason for applauding BRE's close association with these international seminars is that on each occasion and this event is no exception, the chosen theme has been very close to our own research programme interests. This means that we derive especial benefit from participation - not only by contributing papers, but also by learning about the work of other groups, and by joining in discussions both in formal session and informally.

This seminar takes as its theme "Dynamic Modelling of Structures".

The choice of subject reflects a current high level of interest in the importance of designing structures for dynamic loading and the behaviour of structures subjected to dynamic loading. Analytical approaches to the design problem are complicated - and this is not surprising bearing in mind both the range and variety of dynamic loads and the complexity of structural shapes. I quote from a recent article in the New Civil Engineer which described "the slender Priory Hall at Lanchester Polytechnic, Coventry as a good building for vibration testing. The projecting fins contribute to its lively behaviour, though it is by no means a bad building from the vibration sensitivity viewpoint".

Our own work at BRE and elsewhere, particularly in the United States, shows that valuable data can be obtained by carrying out tests on real structures, but if we are concerned with new concepts and untried designs then it is natural to turn to physical models.

Physical models have played a very important part in developing new methods of analysis and in checking or confirming the factors of safety of many important structures. And in recent years the value of physical models has been acknowledged in Codes of Practice in the United Kingdom, Europe and the United States. The provision for using physical models is usually couched in general terms and there is probably scope for more specific guidance as to the type of model which is considered suitable for a particular structure under given circumstances. I know that it is the hope of Dr. Garas and the Study Group that seminars like this will provide a forum for the exchange of knowledge between research and design engineers on the use of models and their role in the design of structures subject to dynamic loads.

Of course, the scope of the seminar is very wide. There will be opportunities to hear about vibration testing, aerodynamic studies, the effects of impact and earthquake loading - and the relevance of these studies to major undertakings such as the design and construction of offshore structures and nuclear installations. Throughout there is a recurring theme - the need to do research to validate analytical techniques describing the effects of dynamic loadings on structures - the need to achieve experimental validation of any calculation method.

This seminar provides a rich and varied menu. Dr. Garas, who is the Convener of the seminar, and his colleagues have worked hard to encourage a wide variety of interesting research papers. The response has been excellent. Now it is time for the authors to present those papers and for you, the delegates, to react to them. I hope that you will find these two days lively and rewarding, that the papers and discussion will lead to another valuable publication, and perhaps most important of all, that the new friendships you will be making will endure long after this seminar.

# Delegate List

| NAME | REPRESENTING (Country) |
|------|------------------------|
| Dr. D.P. Abrams | University of Colorado (USA) |
| G.R. Abrahamson | SRI International (USA) |
| K.F. Allbeson | National Nuclear Corporation Ltd (UK) |
| Dr. D. Anderson | University of Warwick (UK) |
| Dr. R.G. Anderson | R. Travers Morgan & Partners (UK) |
| T.H. Ang | Sheffield University (UK) |
| G.S.T. Armer | Building Research Station (BRE) (UK) |
| Prof. V. Askegaard | Technical University of Denmark (Denmark) |
| Dr. E.S. Awadalla | Teeside Polytechnic (UK) |
| N.K. Awan | Binnie and Partners (UK) |
| | |
| M.J. Baker | Imperial College London (UK) |
| P. Barr | United Kingdom Atomic Energy Authority (UK) |
| R. Bartley | National Nuclear Corporation (UK) |
| J.P. Batham | Central Electricity Research Laboratories (UK) |
| W.J. Beranek | TNO Delft (Holland) |
| A.J.K. Bisbrown | Tropical Products Institute (UK) |
| Dr. L.F. Boswell | The City University (UK) |
| H-P Brathaug | SINTEF (Norway) |
| I. Brown | Imperial College, London (UK) |
| Prof. H.A. Buchholdt | Polytechnic of Central London (UK) |
| H.L. Burrough | Central Electricity Research Laboratories (UK) |
| | |
| M. Casirati | ISMES (Italy) |
| P.S. Chana | Cement & Concrete Association (UK) |
| B.S. Choo | University of Glasgow (UK) |
| Dr. J.L. Clarke | Cement & Concrete Association (UK) |
| Dr. A.G. Collings | University College London (UK) |
| M.J. Cook | Bath University (UK) |
| G.M.E. Cook | Fire Research Station (BRE) (UK) |
| Dr. W.G. Corley | Portland Cement Association (USA) |
| W.H. Craig | University of Manchester (UK) |
| S. Creed | Queen Mary College, University of London (UK) |
| J. Cripps | Ove Arup and Partners (UK) |
| Dr. J.G.A. Croll | University College London (UK) |
| P. Cross | Ove Arup and Partners (UK) |
| | |
| J.T. Dale | Harris & Sutherland (UK) |
| Dr. D.N.E. D'Ath | SDRC Engineering Services Ltd (UK) |
| I.Ll. Davies | Taylor Woodrow Construction Ltd (UK) |
| Dr. R. Delpak | The Polytechnic of Wales (UK) |
| V.C.M. de Souza | Universidade Federal Fluminense (Brazil) |

| NAME | REPRESENTING (Country) |
|------|------------------------|
| Dr. R. Eatock Taylor | University College London (UK) |
| Prof. J. Eibl | University of Dortmund (W. Germany) |
| B. Ellis | Building Research Stations (BRE) (UK) |
| Dr. H.M. Emam | Cairo University (Egypt) |
| Dr. C.C. Fleischer | Taylor Woodrow Construction Ltd (UK) |
| Dr. F.K. Garas | Taylor Woodrow Construction Ltd (UK) |
| I. Gary | Laboratoire de Mechaniques des Solides (France) |
| Prof. J.E. Gibson | City University (UK) |
| Dr. M.H.R. Godley | Oxford Polytechnic (UK) |
| Dr. B. Goschy | Budapest |
| K. Handa | Chalmers University of Technology (Sweden) |
| S.O. Hansen | Danish Ship Research Laboratory (Denmark) |
| Prof. H.G. Harris | Drexel University (USA) |
| M.D. Hedges | DOE/PSA (UK) |
| S. Hetland | SINTEF (Norway) |
| B. Hobbs | Sheffield University (UK) |
| W.D. Howe | United Kingdom Atomic Energy Authority (UK) |
| G. Hughes | Cement & Concrete Association (UK) |
| A. Incecik | Glasgow University (UK) |
| F. Jarnott | Anthony Marchant & Associates (UK) |
| Dr. A. Jeary | Building Research Station (BRE) (UK) |
| Prof. D.J. Johns | University of Technology, Loughborough (UK) |
| J.T. Baldwin | BP Limited (UK) |
| Prof. N. Jones | University of Liverpool (UK) |
| J. Jowett | United Kingdom Atomic Energy Authority (UK) |
| V. Karthigeyan | Bechtel Great Britain Ltd (UK) |
| J. Knoop | Vickers Offshore (UK) |
| Dr. T. Krauthammer | Applied Research Associates Inc. (USA) |
| O. Kroggel | Technische Hogschule Darmstadt (W. Germany) |
| K.G. Kufuor | Imperial College of Science & Technology (UK) |
| Dr. B. Lee | Sheffield University (UK) |
| B. Leeming | CIRIA/UEG (UK) |
| Dipl.Ing.E. Maisel | Institutfuer Modellstatik der (W. Germany) Universitaet Stuttgart |
| D. Malam | Atkins R & D (UK) |
| J.D. McCann | The Queens University (UK) |
| A. McLeish | Taylor Woodrow Construction Ltd (UK) |
| R.D. McLeod | Ove Arup & Partners (UK) |
| Dr. J.B. Menzies | Building Research Station (BRE) (UK) |
| J. Miles | Ove Arup & Partners (UK) |
| Dr. J.H. Mills | Allott & Lomax (UK) |
| R.J.W. Milne | Institution of Structural Engineers (UK) |
| Dr. P.D. Moncarz | Failure Analysis Associates (USA) |
| P.L.T. Morgan | National Nuclear Corporation Ltd (UK) |
| Prof. Dr. Ing. R.K. Mueller | Institutfeur Modellstatik der (W. Germany) Universitaet Stuttgart |
| Dr. M.A. Murray | The Queens University (UK) |

| NAME | REPRESENTING (Country) |
|------|------------------------|
| P. Narzul | Institute Francais de Petrole (France) |
| D. Naylor | United Kingdom Atomic Energy Authority (UK) |
| J. Newell | Atkins R & D (UK) |
| A.J. Nielson | United Kingdom Atomic Energy Authority (UK) |
| E. O'Canainn | Building Design Partnership (UK) |
| A. Oetes | Technische Hochschule Darmstadt (W. Germany) |
| Dr. K.P. Osborne | A.E. Beer & Partners (UK) |
| H.H. Pearcey | National Maritime Institute (UK) |
| Dr. S.H. Perry | Imperial College London (UK) |
| M.W. Pinkney | Rendel Palmer & Tritton (UK) |
| I. Podhorsky | Gradevinski Institut Zagreb (Yugoslavia) |
| Dr. D.A. Reed | National Bureau of Standards (USA) |
| C.A. Robinson | Taylor Woodrow Construction Ltd (UK) |
| R.W. Robinson | The Queens University (UK) |
| C.M. Romer | DOE/PSA (UK) |
| Dr. B. Saravanos | National Nuclear Corporation Ltd (UK) |
| H. Schmidt | Riso National Laboratory (Denmark) |
| M.A. Silva | Cabinette Area des Sines (Portugal) |
| M. Simon | University of Manchester (UK) |
| C.R. Smith | National Nuclear Corporation Ltd (UK) |
| P.D. Smith | Royal Military College of Science (UK) |
| M. Sobaih | Cairo University (Egypt) |
| Dr. G. Somerville | Cement & Concrete Association (UK) |
| B-Q. Song | University of Liverpool (UK) |
| Dr. F. Stangenberg | Zerna Schnellenbach & Partners (W. Germany) |
| Dr. F.M. Thomas | University of Kansas (USA) |
| B.D. Threlfall | Cambridge University (UK) |
| T. Thuestad | SINTEF (Norway) |
| Dr. G.R. Tomlinson | University of Manchester (UK) |
| Dr. V.M. Trbojevic | Principia Mechnanica Ltd (UK) |
| Dr. B. Waine | Taylor Woodrow Construction Ltd (UK) |
| D.E. Walshe | National Maritime Institute (UK) |
| G.J. Wang | (USA) |
| J.G. Warren | Taylor Woodrow Construction Ltd (UK) |
| Dr. A.J. Watson | University of Sheffield (UK) |
| Dr. C. Williams | Plymouth Polytechnic (UK) |
| Dr. D. Williams | URS/John A Blume Associates (USA) |
| J.M. Wilson | University of Durham (UK) |
| V. Wesson | Seismology Research Institute (Australia) |
| A. Zelikson | Laboratoire de Mechanique des Solides (France) |

# Conflicts in dynamic modelling in geomechanics

W H Craig
*Simon Engineering Laboratories, University of Manchester, UK*

SUMMARY

When dynamic situations are to be modelled in the fields covered by the term geomechanics there are often processes involved which give rise to differing requirements when similarity conditions are studied. A number of instances are considered in which inertial loading similarity requirements are in conflict with those arising from viscous and fluid flow effects.

INTRODUCTION

In many engineering disciplines, particularly those involving fluid flows, eg. hydraulics, naval architecture and aerodynamics, model testing has long been almost a 'sine qua non'. This has not been the general situation in soil and rock mechanics and geology, although model work has made valuable contributions in some areas of study.

The technique of modelling in which small scale replicas of prototypes are subjected to centrifugal accelerations, in order to simulate increased gravitational accelerations and generate prototype self weight stresses, was first proposed in the last century, in relation to bridge structures. The first reports of actual tests appeared in the nineteen thirties – mining structures were investigated in the USA and soil mechanics problems in the USSR. The early history is documented elsewhere (1,2). Although static phenomena were investigated initially, time dependent and dynamic events were soon tackled. The present study is concerned with aspects of centrifugal modelling of such events in situations where self weight forces are important.

In general, to satisfy the requirements of equality of self weight stresses in model and prototype it is necessary to subject a model with identical material and linear scaling 1:N to an acceleration level of N times the gravitational acceleration 'g'. Alteration of material density is accounted for by suitable changes in applied acceleration when substitute materials are used.

The classic work in the geological literature is a study by Hubbert (3) dealing with the feasibility of models in a range of problems from the global level downwards. At the scale of endogenic processes in which a major role is played by gravity induced body forces, artificial model materials have to be used as substitutes for stronger, more viscous prototype materials which are treated as flowing continua. In soil and rock mechanics, when prototype dimensions are rather smaller, materials have to be modelled as particulate assemblies rather than continua and the use of substitute materials at reduced stress levels is rarely a satisfactory solution to modelling difficulties for two main reasons:
(i)   the impracticality of simultaneously meeting all the

requirements for material similarity of different phase
components and for the assemblies.

(ii) the dependence of many properties of such assemblies on the
absolute level of stresses.

## GEOLOGICAL MODELS

Hubbert (3) and Ramberg (4) recognised that in most geological
processes inertia plays no significant role. Rock flow can be
simulated at reasonable time scales if only self weight and viscous
forces are considered. With Reynolds numbers as low as $10^{-20}$ in the
prototype, artificial rock materials with comparatively low viscosity
must be used in small models subjected to very high accelerations.

When looking at the stability of mine structures, using field
materials Wright and Bucky (5) noted that prototype times to failure
could be assessed by taking the product of model time and the
acceleration factor N. This statement was not justified but would
appear to arise from consideration of similarity of gravitational and
inertial forces only, ie. Froude number.

## ENGINEERING MODELS

At a scale more readily conceived by engineers many structures have
significant dimensions in the range 1-100m and in principle can be
scaled down to manageable laboratory size, although simulation of
fine detail and significant field size imperfections may be a
limitation on minimum size. Engineering centrifuges have now been
developed in several countries with capability of subjecting models
with major dimensions of the order of 1m to acceleration levels
commonly of 100g and sometimes as high as 300g. This is in contrast
to the geological work where smaller models have been taken to
accelerations typically 1000-3000g.

In many situations the significant processes in models of soil and
rock obey a number of different time scaling relationships - leading
to complications in the modelling of dynamic phenomena. When field
materials are to be used in centrifuge models at prototype stress
levels, three broad categories of phenomena can be distinguished (6)
in models at a linear scale 1:N.

(i)     Creep and viscous effects. Time scaling for creep strains is
1:1 when model and prototype stress conditions are equal.
Viscosity must also be considered when material properties
(eg. strength) are strain rate dependent.

(ii)    Inertia - time scaling 1:N.

(iii)   Dissipative effects, eg. fluid flow - time scaling $1:N^2$.

These variations in time scaling can be seriously in conflict and due
attention must be paid when designing dynamic models.

### Earthquake and Blast Loading

When modelling dynamic behaviour in response to earthquake or blast
loading in a centrifuge the prime requirement is equality of Froude
number, ie. model acceleration levels and frequencies must all be N
times those in the field. Experiments of this type have recently
been performed by Schofield (2) and Zelikson (7). Under such imposed
conditions of 1:N time scaling, item (ii) above, possible

difficulties can arise because of the requirements (i) and (iii).

In saturated or nearly saturated soils, changes and reversals of shear stress associated with dynamic loading lead to an increase in pore pressures and reduction in effective stress levels with consequent loss of strength – in the extreme case a cohesionless soil may liquify. However the introduction of increases in pore pressure sets up hydraulic gradients which tend to dissipate by fluid flow – a process governed under (iii). In small models of highly permeable, natural soils with water as the pore fluid significant degrees of consolidation can occur even within the brief period of dynamic loading, which if unrecognised can lead to potentially dangerous conclusions being drawn. It is possible that in certain circumstances involving layered soils, dynamically induced pore pressures may spread through permeable horizons causing quasi-static failures after the period of dynamic loading has ended. In the latter case it may be acceptable to consider the inertial loading and fluid flow as two consecutive stages obeying separate time scaling laws without conflict, but in the former the concurrent nature of the two effects can only be correctly simulated by reducing model flow rates. Replacement of prototype pore fluid by one of greater viscosity is one possibility while inclusion of a small percentage of finer material to reduce the specific permeability of the solid matrix may be another.

At the other end of the spectrum of soil conditions dissipation effects can be ignored over short periods in cohesive soils of low permeability, but the essentially viscous nature of such materials leads to variation in other properties under increased rates of strain. Typical figures quoted are increases in undrained shear strength of the order of 10% per log.cycle of strain rate (8). If such figures can be relied upon then model material strengths may be reduced to allow for increased loading rates on the basis of series of conventional element tests under appropriate conditions, or alternatively loading levels can be increased.

## Projectile Penetration and Pile Driving

A range of problems which potentially involve all three time scaling requirements simultaneously are those involving rapid penetrations of small objects into soil. Simulation of inertial and gravitational loadings may be considered the dominant requirement in an impact study but the possibilities of performance being affected by consequent increases in strain rate and by simultaneous consolidation must be considered.

In cohesive soils the strain rate effects may be considered on the same basis as suggested above but care must be taken to ensure that element test data are obtained at appropriate strain rates – extrapolation of results may lead to erroneous conclusions. The strain rates in some zones of a model, eg. the thin annulus of soil being sheared adjacent to a circular pile shaft, may be higher than can generally be attained in laboratory elements.

While it may be possible to simulate correctly the inertial effects which seem to control the plugging or otherwise of open pipe piles under impact driving, with suitable adjustments for strain rate, fluid flow must still be considered. Dissipation of installation induced pore pressures during and after driving leads to 'set up' and significant time dependent increases in pile capacity. In clean sands the dissipation rates can be reduced as suggested above to maintain strict similarity and it is possible that in such soils drainage may be almost complete in the period between blows. In clays where the replacement technique is not practicable, partial drainage can occur in the overall period of driving if not between blows to an extent which may be difficult or impossible to simulate correctly.

## Gravity Structures Offshore

In modelling the wave loading on offshore structures, similarity of inertial loading may be secondary in importance to the need to apply adequate number of loading cycles to induce appropriate pore pressure changes over periods in which similarity of drainage must be approached. Practical difficulties (9) and considerations of viscous effects (10) have restricted the frequency of model loading to the point where similarity of inertial loading will be achieved. Pore fluid replacement has been used to satisfy the drainage requirements in sands. The presence of permeable fabric in clay deposits may allow some drainage to occur in the field even in the period of a single storm loading sequence. Use of remoulded clays with similar strength properties but lower overall permeability increases the possibility of achieving drainage similarity but in general model tests have been designed to remain essentially undrained (conservative), omitting simulation of very large numbers of low level loading cycles and concentrating on the fewer cycles at higher loading levels which induce the greatest part of the pore pressure changes.

## Silos

In models of the flow of granular materials in silos inertial loading and pore fluid flows may both be significant – in this case the pore fluid is compressible, generally air in the field. Nielsen (11) has considered the requirements in detail indicating the areas of conflict.

When radial flow occurs in the centrifuge in filling or emptying a silo the Coriolis acceleration associated with the radial velocity comes into play. Experiments (12) have shown that simulation of filling may be unrealistic where free fall of material is allowed, because the point of deposition is displaced and this may adversely affect the pressure distribution in the fill. Development of techniques for deposition down a moveable guide tube are a possible improvement.

There is some evidence (12) to suggest that achievement of similarity in the 'filled' condition by the alternative stress path of subjecting a prepared fill to increasing acceleration levels is an acceptable substitute for 'in flight' filling if a correct density can be achieved.

However model silos are filled, emptying in the centrifugal acceleration field again introduces the Coriolis effect. Material in free fall below the outlet moves in a curved path but flow patterns in the confines of the silo do not appear to be altered even when the radial flow velocity is high, as in the case of a narrow core flow.

## THE ROLE OF MODELS IN DESIGN

Models can be used to provide data for a specific project at the design stage or to investigate particular in-service or failure situations. They can also be used to carry out parametric studies to provide data to establish design rules of more general application. Alternatively they may be used for comparison with numerical studies as a means of calibrating the latter. Limitation to the last of these roles, as is sometimes advocated, trivialises physical modelling. Centrifuge models of geotechnical structures have indicated mechanisms of behaviour which have not been predicted by paper studies and which cannot always be reproduced by current analytical techniques (13) – the same is true in other disciplines. There are of course limitations to any model and field conditions of significance may be omitted whether the model is physical or analytical. If the omission is recognised it may be possible to make allowances, if it is not then modelling will be an unreliable guide.

The commissioning of model studies for design implies some credence in the results. Dynamic modelling is more difficult than static and it is important for the modeller to obtain feedback from the field in order to check and calibrate his work and to build up confidence. This is particularly important when conflicting requirements may involve compromising certain similarity conditions. The need for feedback may require considerable investments of time and money in the field but this must be recognised if model work is to progress.

REFERENCES

(1)    CRAIG, W H, 'Model studies of the stability of clay slopes', Ph.D Thesis, Univ. of Manchester (1974).

(2)    SCHOFIELD A N, 'Dynamic and earthquake geotechnical centrifuge modelling', Proc. Conf. on Recent Advances in Geotechnical Earthquake Engineering and Soil Dynamics, Missouri-Rolla (1981).

(3)    HUBBERT, M K, 'Theory of scale models as applied to the study of geologic structures', Bull. Geol. Soc. of America, Vol.48, pp.1459-1520 (1937).

(4)    RAMBERG, H, Gravity Deformation and the Earth's Crust, Academic Press, London (1967).

(5)    WRIGHT, F D and BUCKY, P B, 'Determination of room and pillar dimensions for the oil shale mine at Rifle, Colorado', Trans. AIMME, Vol.181, pp.352-359 (1949).

(6)    ROWE, P W et al, 'Dynamically loaded centrifugal model foundations', Proc. 9th ICSMFE, Tokyo, Vol.2, pp.359-364 (1977).

(7)    ZELIKSON, A et al, 'Scale modelling of a soil structure interaction during earthquakes using a programmed series of explosions during centrifugation', Proc. Conf. on Recent Advances in Geotechnical Earthquake Engineering and Soil Dynamics, Missouri-Rolla (1981).

(8)    BJERRUM, L, 'Problems of soil mechanics and construction on clays and structurally unstable soils', Proc. 8th ICSMFE, Moscow, Vol.3, pp.109-159 (1973).

(9)    CRAIG, W H, 'Cyclic loading equipment for offshore foundation models', in Offshore Structures: The use of Physical Models in their Design, Construction Press, Lancaster (1981).

(10)   ROWE, P W and CRAIG, W H, 'Application of models to the prediction of offshore gravity platform foundation performance', <u>Proc. Conf. on Offshore Site Investigation</u>, Graham & Trotman, London (1980).

(11)   NIELSEN, J, 'Model laws for grain materials with special reference to silo models', <u>Rpt No.49, Structural Research Lab, Technical Univ. of Denmark</u> (1974)

(12)   WRIGHT, A C S, 'Silos – model and field studies', Ph.D Thesis, Univ. of Manchester (1979).

(13)   Various Authors, 'The use of physical models in design', Discussion Session 9 in <u>Design Parameters in Geotechnical Engineering</u>, Vol.4, pp.315–360, British Geotechnical Society, London (1980).

# Adjustment of platforms' model using experimental results

P Lepert (SYMINEX)
P Narzul (IFP)

## INTRODUCTION

For some considerable time now designers have been very concerned about the structural integrity of buildings and above all those in which people must live and work. The methods used to ensure this safety, either during the design stage or once the structure is built generally involve four steps:

. the modelization of the structure
. the modelization of the maximum predictable loads to which this structure could be submitted during its life
. the simulation of the behaviour of the structure under these loads by response computation
. and, finally, the determination of the safety coefficients as ratio of expected maximum stresses versus acceptable stresses.

The structure is safe as long as these coefficients are larger than prescribed design values which take into account:

. the quality of the various materials used for the building
. the conditions of building
. the risks which are run when the structure is operating, particularly in terms of human life
. and, among several other considerations, the confidence people have in the methods used to determine safety coefficients: the most reliable the computed safety coefficients seem, the less their prescribed thresholds may be.

In most cases, the methods used for safety coefficients determination are considered as reliable as long as no dynamic effects are generating significant stresses. But, several experiments carried out over the last five years have proved that, in the offshore field, the classical methods give very bad results in both static and dynamic analysis. This led the offshore platform designers to accept very high values for their safety coefficients, or, in other terms, to design platforms with a high degree of redundancy. This policy, which was acceptable for small and middle water depth platforms, leads to prohibitive costs and technical difficulties for deeper platforms. This is why one of the main objectives of oilrig operators and designers is now to improve the methods used for the determination of safety coefficients in order to reduce the building cost of deep water rigs. Accidents which have occurred during recent years have made this tendency more pronounced.

The accurate modelization of offshore structures is of basic importance in the determination of these coefficients. This paper describes a study carried out by SYMINEX, ELF AQUITAINE, TOTAL CFP and the INSTITUT FRANCAIS DU PETROLE to develop and test a new modal analysis technique; it shows that it is possible to improve this modelization using experimental results.

## THE MODELIZATION OF OFFSHORE PLATFORMS

The modelization of offshore platforms at the design stage meets with at least four major problems :

. the stiffness of the foundations and, more generally speaking, the modelization of these foundations, is very difficult to deduce from the preliminary on-site soil mechanic tests. In particular, the soil behaviour has to be linearized which introduces supplementary errors in the modelization;

. the upper part of the platforms (decks) are always so complex that it is difficult to imagine its modelization in detail. The equivalent modelization (spring or beams, lumped masses, dampers etc.) is very unreliable, especially for dynamic computations;

. the distribution of masses on the various decks (machinery, modulus etc.) is not well-known at the design stage and can be changed by the operator at any time during the life of the platform;

. the operator sometimes adds to the platform itself a few less significant elements (oil riser, fire pump risers etc..) which in fact completely change local behaviours of this platform.

It clearly appears, therefore, that the modelization of a platform is inaccurate and unreliable, and that the operating platform is often slightly or even greatly different from the platform as built.

The main consequence of this observation is that the strains and stresses which actually appear in the platform are different from those predicted during the design. Thus the predicted safety co-efficients have in fact no meaning for the operating platform.

This conclusion was largely confirmed by permanent instrumentation of offshore platforms which has shown that (1) :

. The natural frequencies measured on operating platforms are largely different from those computed at the design stage;
. The same observation is applicable to measured strain and stresses.

Unfortunately, permanent instrumentation was unable to explain these differences and up to now all attempts to fit a model to experimental results led to limited success.

To overcome these problems, Modal Analysis was adapted to offshore structures and conditions. A special equipment was designed and realized, and the Modal Analysis of a steel jacket was performed in the Arabian Gulf. A Finite Element model of this structure was made, and it clearly appeared that theoretical results were completely different from experimental ones. The model was corrected using the four first modes measured on-site.

CORRECTING A MODEL USING MODAL ANALYSIS
---

Modal Analysis is now a well-known technique used to measure the modal parameters of a structure (frequencies, modes, shapes, modal masses and modal damping ratio). It has been successfully applied in the spatial and aeronautical fields as well as by automobile constructors or mechanical industries. In short, it consists of exciting the structure at a single point and measuring its responses in several locations by means of accelerometers. The resulting transfer functions are processed in order to extract the modal parameters of the structure.

Generally speaking, all the lower modes of a structure are used to fit a model. As higher frequency modes are concerned with local vibrations, they can be used for detailed correcting of the model. However, such detailed correcting requires more exacting processing methods and, in most cases, the use of computers. Thus, there are two ways to use Modal Analysis results to correct models :

. the correcting of main structural parameters (stiffness of links,
  lumped masses, equivalent stiffness and masses of substructures etc.)
  using only lower frequency modal parameters, by means of empirical
  methods;
. the detailed correcting of the model using higher frequency modal
  parameters, by either direct or iterative methods. (2)


## APPLICATION TO ABK LIVING QUARTER PLATFORMS

The ABK offshore oil field is located in the Arabian Gulf, in 30m
depth of water(3).It is a steel four legged jacket with three decks.
The total weight is around 1500 metric tons, of which 1100 tons are
the decks. (see figure 1)

The Modal Analysis was conducted during April 1980.   The equipment
was composed of three parts:

. an hydraulic shaker to excite the platform (see figure 2) which
  was clamped to a leg of the jacket at the boat landing level
. a triaxial waterproofed accelerometer (see figure 3) successively
  clamped on all the nodes of the jacket to measure the response.
  The response of each node was measured in turn, and divers were used to
  move the accelerometer from one location to the next
. an embarkable data acquisition unit (see figure 4) mainly composed
  of a Fourier analyser, to drive the excitation, and acquire the
  excitation (dynamic force) and response (accelerations) signals.
  The shaker was driven by the system's mini-computer in order to
  induce a proper excitation in the platform.   Several types of
  excitation were tested (random, swept sine, pseudo random etc...).

  The excitation and response signals were acquired, filtered, coded
  and recorded on line.   All the transfer functions were computed
  offshore so that bad measurements were immediately eliminated and
  repeated.

The platform was described by a 48 points network or, in other words,
48 response measurements were acquired.   This led to 144 transfer
functions in the 0.5 - 45 Hz frequency range.   The modal parameters
were extracted on shore : 39 modes were extracted between 0.85 Hz and
15 Hz.   Examples of these modes are shown on figure 5a to 5e.   The
main conclusions drawn from the qualitative observation of the mode
shapes were :

. foundations are so flexible that the low frequency motion of the
  jacket is almost a rigid body motion with deformations mainly in
  the piles
. the motion of the South face is larger than the North face one
. local vibrations (fire pump riser) induce unexpected fatique in some
  brackings of the third underwater level (-16,60m).


A Finite Element model of the platform was realised and the first four
modes measured on-site were used to adjust this model.


ADJUSTMENT OF FINITE ELEMENT MODEL
_____

The jacket of the ABK living quarter platform was modelized by a
576 dof (96 x 6) system using the finite elements method with only
beams.   The foundation of each leg is modelized with both an equivalent
pile and a system of spring at the mud line      (see figure 6).   The
machinery and modulus on the decks were considered as lumped masses.
A bridge between the platform and another one was described by a
special beam.

The natural frequencies obtained from the initial model are given in
Table N°1.   The first four measured natural frequencies of the real
platform are compared in this table.   On the basis of the observation
of the corresponding mode shapes, corrections were empirically made
on the model.   They mainly concerned :

. the stiffness of equivalent piles (foundation)
. the stiffness of mud-line springs (foundation)
. the masses on the upper deck.

Table N°2 gives the set of modifications which led to the best results,
from the global natural frequencies point of view.   The shape of the
first four modes of the corrected model were found to be very close
to the measured mode shapes.   Table N°3 gives the natural frequencies
of the corrected model.


CONCLUSION
_____

The experiment which was conducted in the Arabian Gulf during April 1980
clearly showed that it is very difficult to properly modelize a structure

as complex as an offshore platform. This observation was confirmed by the results of extensive campaigns of permanent data acquisition in the North Sea. It also appeared that it was actually possible to adjust the main structural element of a model of an offshore platform when the first modes of the platform are all available. Particularly, the interpretation of measured mode shapes are essential for success in this adjustment. The adjustment presented in this paper is only an example of what can be done and the comparison between the measured and computed frequencies is not perfect. However, it can be expected that the use of a greater number of modal parameters, measured in a higher frequency range, and the implementation of powerful computer methods will enable engineers to perform a detailed adjustment of theoretical models from experimental data. Such methods are now developed by SYMINEX, ELF AQUITAINE, TOTAL CFP and the INSTITUT FRANCAIS DU PETROLE and the first results will be issued during February 1982. The final objective is to increase the real safety of offshore platforms especially when installed in rough waters.

## REFERENCES

1. "Field Measurements of Correlation between Waves and Platform Response versus Significant Wave Height and Wave Direction"

   - P.F. ANSQUER and R.S. CARTON  OTC Paper N° 3797  1980

2. "Identification de Systèmes - Etude Bibliographique"

   - W. HEYLEN - P. VANHONACKER    LMS Report

3. "Vibrodetection applied to Offshore Platforms"

   - P. LEPERT, M. CHAY, P. NARZUL and J.Y. HEAS  OTC Paper N°3918 1980

TABLE N° 1

## NATURAL FREQUENCIES OF THE INITIAL MODEL

| Mode Number | Natural Frequency |
|:-----------:|:-----------------:|
| 1 | 0,424 Hz |
| 2 | 0,462 Hz |
| 3 | 0,695 Hz |
| 4 | 1,631 Hz |

TABLE N° 2

## MODIFICATION OF THE MODEL

| Structural part | modification |
|-----------------|--------------|
| Upper deck mass | divided by 1,5 |
| Section of piles | multiplied by 1,9 |
| Stiffness of foundation | multiplied by 1,7 |
| Stiffness of bridge between the platform and the next one | multiplied by 1,7 |

TABLE N° 3

## NATURAL FREQUENCIES OF CORRECTED MODEL

| Mode Number | Measured Natural Frequencies | Natural frequencies of corrected model |
|:-----------:|:----------------------------:|:--------------------------------------:|
| 1 | 0,857 Hz | 0,660 Hz |
| 2 | 0,922 Hz | 0,972 Hz |
| 3 | 1,638 Hz | 1,772 Hz |
| 4 | 1,803 Hz | 1,910 Hz |

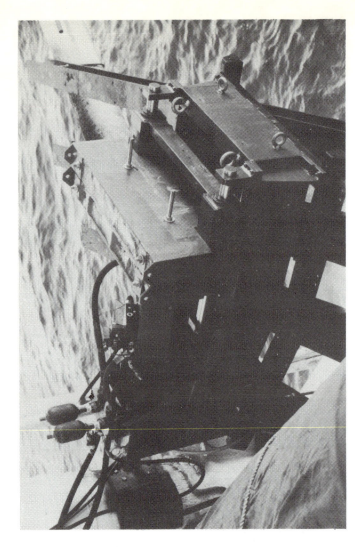

FIGURE N° 2 : HYDRAULIC SHAKER

FIGURE N° 1 : ABK PLATFORM

14

FIGURE N° 4 : DATA ACQUISITION UNIT

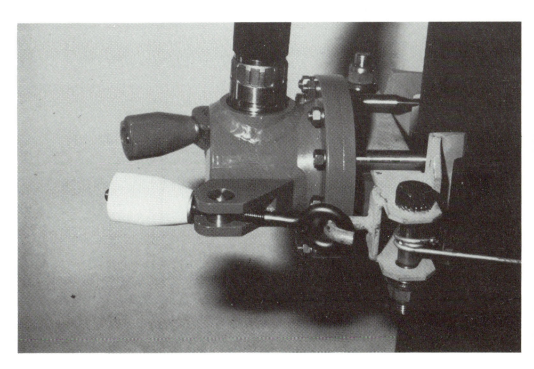

FIGURE N° 3 : ACCELEROMETERS

15

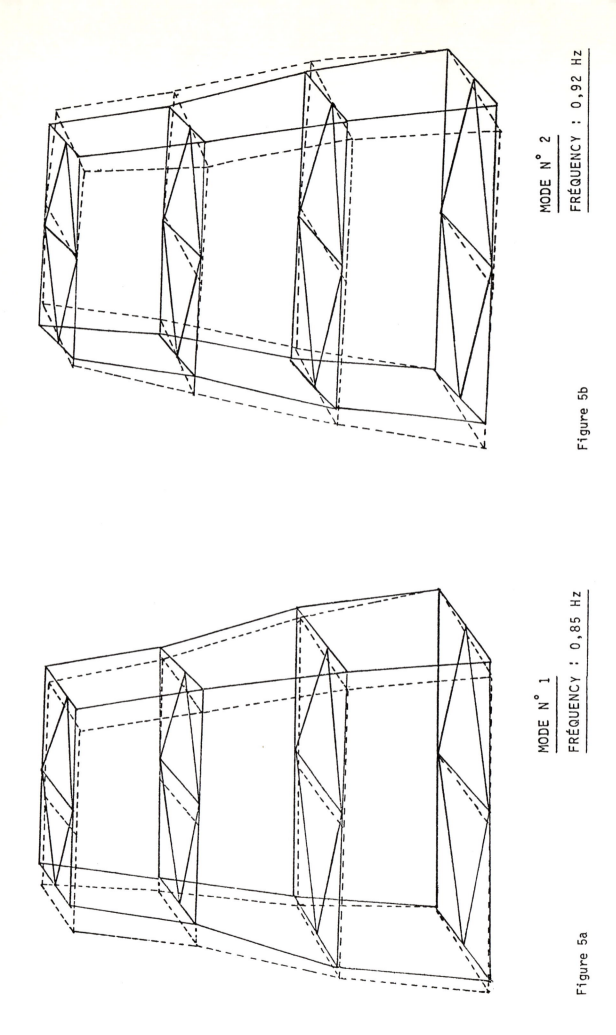

MODE N° 2

FRÉQUENCY : 0,92 Hz

Figure 5b

MODE N° 1

FRÉQUENCY : 0,85 Hz

Figure 5a

16

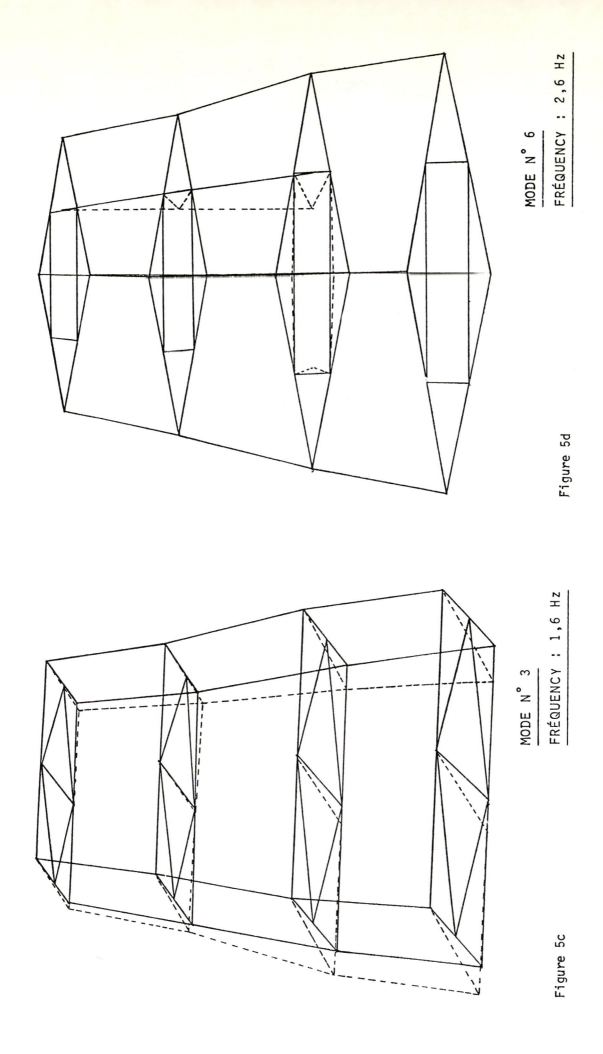

MODE N° 6

FRÉQUENCY : 2,6 Hz

Figure 5d

MODE N° 3

FRÉQUENCY : 1,6 Hz

Figure 5c

17

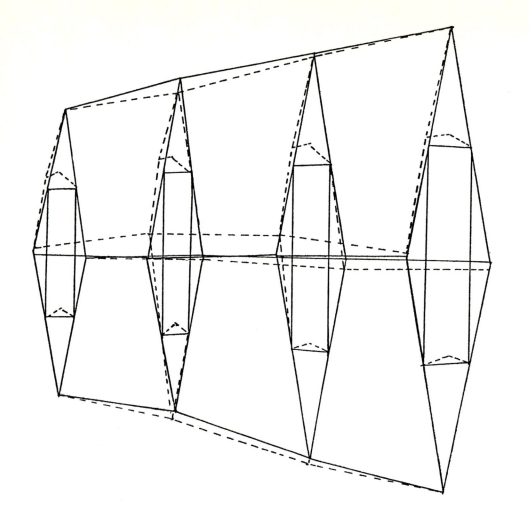

MODE N° 12

FRÉQUENCY : 3,9 Hz

Figure 5e

18

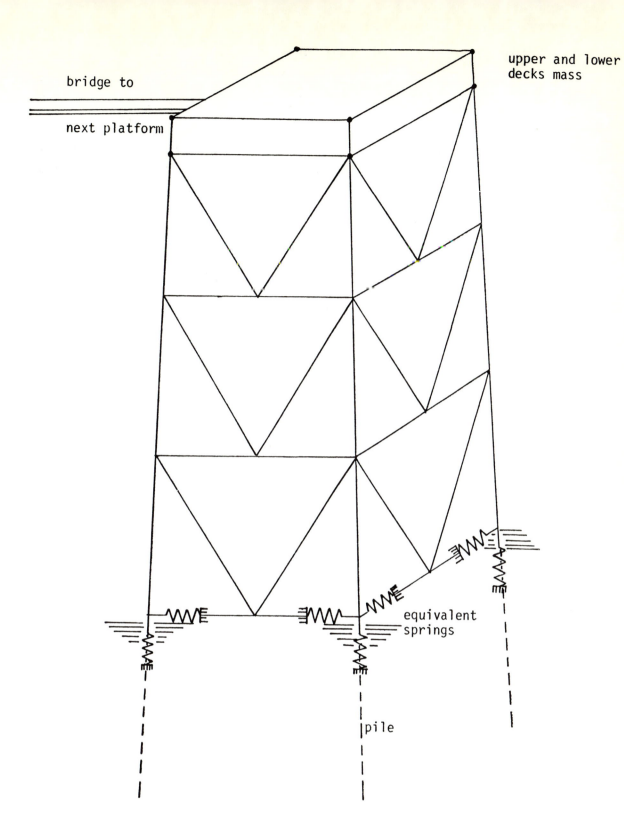

bridge to

next platform

upper and lower
decks mass

equivalent
springs

pile

Figure N° 6 : Sketch of the Jacket

19

# Contribution to discussion of Paper No. 2

J T Dale
*Harris and Sutherland, UK*

The authors are to be congratulated on their efforts at reconciling experiments and theoretical work on such a large structure.

The paper identified four areas of uncertainty in the modelling of offshore platforms:

(a)  the foundation stiffness,
(b)  the structural idealization of the upper platform,
(c)  the mass distribution, and
(d)  post-construction additions and alterations.

In order to obtain a measure of agreement between their finite element model and the measured response of the platform, the authors adjusted all four of the points (a) - (d). They report that an empirical reduction of one third of the upper deck mass of the model was required in order to match their observations. Was the original estimate of the mass based on a rigorous survey of the actual platform or was it based on the as-constructed drawings? As far as the writer is aware, the as-built masses of offshore structures tend to be significantly greater than assumed in design (ref 1). If this is so, could the authors suggest any physical reason why model masses should be <u>reduced</u>?

It would seem more logical to attribute the discrepancies between finite element models and real structures to points (a) to (d) above, especially (a). Finally, would the authors comment on the amount of "added mass" and damping applied to the platform legs that was found to give the best agreement between their model and the performance of the actual structure.

## Reference

1.  "Weight control of offshore structures"
    Offshore Research Focus, No. 25, June 1981.

## Authors Reply

The original estimate of the upper deck mass is merely based on the value used for the design of the jacket in 1974. It was certainly a very rough estimate, but no other data was available for our study. Likewise, it was not possible to obtain information about the as-built masses. Thus we felt free to modify this value in either way, as convenient. However, we agree that point (a) is the most important.

Added masses were not modified and we used throughout the values automatically computed by the program, the added mass coefficient being 1.

Finally, damping is not taken into account in the calculation of natural frequencies, as the program solves a classical eigenvalue problem.

# Dynamic modelling of the interaction between the modules and module support frame of an offshore oil production platform

L F Boswell and M L Taylor
*The City University, London, UK*

## SUMMARY

The problem of structural interaction between the modules and module support frame of an offshore platform has not received adequate attention and as a result design problems have occurred. Static interaction analyses which have been undertaken indicate that errors may occur if interaction is ignored. In this paper, a matrix method for the dynamic analysis of interaction is proposed. The method has direct analogies with the static analysis. The results of an example computer analysis are presented.

## INTRODUCTION

Clause 2.5.6 of API RP2A(1) states, 'Consideration should be given to the effect of deflections on the distributed load between the platform and supported rigid equipment packages'.

The implications of this clause would appear to have been overlooked in the design of a number of major platforms and as a result, serious strength and serviceability problems have occurred. Previous work (2),(3) which has involved case studies has shown that appreciable errors may occur in the calculation of the magnitude of the module reactions if the effect of interaction between the modules and module support frame is ignored. This work has been based upon a static analysis and ignores dynamic effects. In order to continue the investigation into the effect of module reactions on the local strength and fatigue life of the module support frame, a dynamic analysis is being undertaken using mathematical and physical models.

## INTERACTION ANALYSIS OF THE MODULE AND MODULE SUPPORT FRAME

### Equilibrium Equations

The mathematical model which has been used for the dynamic analysis is an extension of the static model. Reference will be made, therefore, to the static model at appropriate stages during the dynamic analysis since the associated structural idealisation and notation lead directly to dynamic considerations.

Figure 1 is a schematic arrangement of the loads acting on the modules and the platform of a typical offshore structure. The force vector P represents the dead and live loads applied to the modules both from the operating equipment it supports and from the various environmental forces acting upon it, whereas the force vector Q is due to the reaction of the module support frame on the module. The module loading on the module support frame is obviously −Q, whilst all other loading on the platform is represented by the force vector W.

The undamped equations of motion for a single module are:

$$\begin{Bmatrix} p \\ q \end{Bmatrix} = \begin{bmatrix} k_{11} & k_{12} \\ k_{21} & k_{22} \end{bmatrix} \begin{Bmatrix} u \\ v \end{Bmatrix} + \begin{bmatrix} m_{11} & m_{12} \\ m_{21} & m_{22} \end{bmatrix} \begin{Bmatrix} \ddot{u} \\ \ddot{v} \end{Bmatrix} \tag{1}$$

where $\{q\}$ is the module reaction vector and $\{v\}$ are the corresponding displacements.

$\{p\}$ represents all other module forces and $\{u\}$ are the corresponding displacements.

For sinusoidal motion, $\ddot{u} = -u\omega^2$ and $\ddot{v} = -v\omega^2$. Equation (1) may be re-written as

$$\begin{Bmatrix} p \\ q \end{Bmatrix} = \left[ \begin{bmatrix} k_{11} & k_{12} \\ k_{21} & k_{22} \end{bmatrix} - \omega^2 \begin{bmatrix} m_{11} & m_{12} \\ m_{21} & m_{22} \end{bmatrix} \right] \begin{Bmatrix} u \\ v \end{Bmatrix} \tag{2}$$

For fixed supports, $v = 0$ and

$$\{p\} = \left[ k_{11} - \omega^2 m_{11} \right] \{u\} \tag{3}$$

or

$$\{u\} = \left[ k_{11} - \omega^2 m_{11} \right]^{-1} \{p\} \tag{3a}$$

and

$$\{q\} = \left[ k_{21} - \omega^2 m_{21} \right] \{u\} \tag{4}$$

hence

$$\{q\} = \left[ k_{21} - \omega^2 m_{21} \right] \left[ k_{11} - \omega^2 m_{11} \right]^{-1} \{p\} \tag{4a}$$

By expanding equations (2) and eliminating u, the unknown module support reactions q may be written in terms of the module forces p and the support deflections v, thus

$$\{q\} = \left[ k_{21} - \omega^2 m_{21} \right] \left[ k_{11} - \omega^2 m_{11} \right]^{-1} \{p\} + \left[ \left[ k_{22} - \omega^2 m_{22} \right] \left[ k_{21} - \omega^2 m_{21} \right] \left[ k_{11} - \omega^2 m_{11} \right]^{-1} \left[ k_{12} - \omega^2 m_{12} \right] \right] \{v\} \tag{5}$$

or

$$\{q\} = \{p^*\} + \left[ k^* \right] \{v\} \tag{5a}$$

where $\{p^*\} = \left[ k_{21} - \omega^2 m_{21} \right] \left[ k_{11} - \omega^2 m_{11} \right]^{-1} \{p\}$ is the vector of module reactions when the support points are fixed, equation (4a).
$\left[ k^* \right] = \left[ k_{22} - \omega^2 m_{22} \right] - \left[ k_{21} \omega^2 m_{21} \right] \left[ k_{11} - \omega^2 m_{11} \right]^{-1} \left[ k_{12} - \omega^2 m_{12} \right]$ is a frequency dependent stiffness matrix corresponding to the module support points. $k^*ij$ is the amplitude of reaction at support i when support j is subjected to a displacement amplitude of unity.

From equation (5a) the system of equations for n modules becomes,

$$\{Q\} = \{P^*\} + \left[ K^* \right] \{v^m\} \tag{6}$$

where $\{Q\} = \{q_1, q_2 \ldots \ldots q_n\}^T$

$\{P^*\} = \{P_1^*, P_2^* \ldots \ldots P_n^*\}^T$

$$[K^*] = \begin{bmatrix} k_1^*, & 0 \ldots \ldots 0 \\ 0 & k_2^* & \\ & \vdots & \\ & \vdots & \\ 0 & & k_n^* \end{bmatrix}$$

$\{V^m\} = \{v_1, v_2 \ldots \ldots v_n\}^T$ and corresponds to the top of the module bearing plates.

Equations (6) reduces to the static equilibrium equation (4) when $\omega^2 = 0$.

The static equilibrium equations for the module support frame are derived in the same way as those for the modules (4) and by direct analogy with equation (6) correspond to the dynamic case when $\omega^2 = 0$. The dynamic equations for the module support frame may be written as

$$-[F^*]\{Q - Q_E\} = \{V^d - V_I\} \tag{7}$$

where $[F^*]$ is the frequency dependent module support frame flexibility matrix.
$\{Q_E\}$ is a vector of estimated module reactions.
$\{V^d\}$ is a vector of module support frame deflections corresponding to the bottom of the module bearing plates.
$\{V_I\}$ is a vector of module support frame deflections obtained from an analysis of the platform under the action of environmental forces W and estimated module reactions $Q_E$.

Compatibility Condition

Equations (6) and (7) contain three sets of unknowns and these are $Q$, $V^m$ and $V^d$. One further equation is required and is given by the compatibility condition which relates $V^m$ and $V^d$.

In practice, the modules are not fixed to the module support frame but rest on support plates and separation is, therefore, possible. Also, in order to account for fabrication errors it is usual to introduce shims between the upper and lower bearing plates. The equation of compatibility between the modules and module support frame may be written as

$$\{V^d - V^m\} = [C]\{Q\} - \{T\} \tag{8}$$

where $[C]\{Q\}$ is the compression in the bearing plates due to a force $\{Q\}$ and $\{T\}$ is the shim thickness. C may be used to account for the uplift or separation condition. For the ith bearing plate, if $C_{ii} = 0$ then $V_i^d - V_i^m = -T_i$, i.e. no uplift or if $C_{ii}$ tends to infinity then $Q_i$ tends to zero, i.e. uplift has occurred and the reaction has become zero.

## Final Equations

Equations (6), (7) and (8) may be combined to give the following set of simultaneous equations

$$\left[ K^* \, (F^*+C) + I \right] \{Q\} = \left\{ K^* \, (T + V_I + F^* Q_E) + P^* \right\} \qquad (9)$$

which are solved for the reactions $\{Q\}$. The module support frame deflections are then solved using equation (7). Finally, equations (6) and (8) may be used in combination to determine the module deflections.

## COMPUTATIONAL PROCEDURE

The program performs a Fourier type analysis on $P^*$ and $V_I$ to determine the components having the same frequency and phase;values of $Q$, $V^m$ and $V^d$ are computed for each of these components and the contributions of each are summed at appropriate time intervals. To account for uplift at a bearing, $C_{ii}$ is set to a large positive number and the analysis repeated.

## NUMERICAL EXAMPLE

Figure 2 shows a simple structure containing the necessary features for an interaction problem. The data for the problem are given as follows:

### Static components of load

$$\{P^*\} = \{0.3125, 1.375, 0.3125\}^T \quad V_I = \{-0.01432, -0.02083, -0.01432\}^T$$

$$\left[ F^* \right] = \begin{bmatrix} 0.01172 & 0.01432 & 0.009115 \\ 0.01432 & 0.02083 & 0.01432 \\ 0.009115 & 0.01432 & 0.01172 \end{bmatrix} \quad \left[ K^* \right] = \begin{bmatrix} 96.0 & -192.0 & 96.0 \\ -192.0 & 384.0 & -192.0 \\ 96.0 & -192.0 & 96.0 \end{bmatrix}$$

$$\{Q_E\} = \{0.0, 0.0, 0.0\}^T \qquad \{T\} = \{0.0, 0.0, 0.0\}^T$$

### Dynamic components of load

Period = 5.0 s
Start time = 0.0 s

$$\{P^*\} = \{0.15625, 0.6875, 0.15625\}^T \quad \{V_I\} = \{-0.02864, -0.04166, -0.02864\}^T$$

$$\left[ F^* \right] = \begin{bmatrix} 0.02344 & 0.02864 & 0.01823 \\ 0.02864 & 0.04166 & 0.02864 \\ 0.01823 & 0.02864 & 0.02344 \end{bmatrix} \quad \left[ K^* \right] = \begin{bmatrix} 96.0 & -192.0 & 96.0 \\ -192.0 & 384.0 & -192.0 \\ 96.0 & -192.0 & 96.0 \end{bmatrix}$$

$$\{Q_E\} = \{0.0, 0.0, 0.0\}^T \qquad \{T\} = \{0.0, 0.0, 0.0\}^T.$$

In order to determine that the mathematical model and hence the computational procedure adequately represent the dynamic interaction problem, two cases have been considered. Firstly uplift has been prevented and secondly uplift has been permitted. The results of the two cases are given below.

24

|       | Uplift prevented | | | Uplift permitted | | |
| TIME | $Q_1$ | $Q_2$ | $Q_3$ | $Q_1$ | $Q_2$ | $Q_3$ |
| --- | --- | --- | --- | --- | --- | --- |
| 0.000 | 2.0312 | −2.0624 | 2.0312 | 1.0000 | 0.0000 | 1.0000 |
| 0.625 | 3.2465 | −3.7859 | 3.2465 | 1.3536 | 0.0000 | 1.3536 |
| 1.250 | 3.7499 | −4.4998 | 3.7499 | 1.5000 | 0.0000 | 1.5000 |
| 1.875 | 3.2465 | −3.7859 | 3.2465 | 1.3536 | 0.0000 | 1.3536 |
| 2.500 | 2.0312 | −2.0624 | 2.0312 | 1.0000 | 0.0000 | 1.0000 |
| 3.125 | 0.8159 | −0.3389 | 0.8159 | 0.6464 | 0.0000 | 0.6464 |
| 3.750 | 0.3125 | 0.3750 | 0.3125 | 0.3125 | 0.3750 | 0.3125 |
| 4.375 | 0.8159 | −0.3389 | 0.8159 | 0.6464 | 0.0000 | 0.6464 |
| 5.000 | 2.0312 | −2.0624 | 2.0312 | 1.0000 | 0.0000 | 1.0000 |

It can be seen that at each time step the two sets of results give the same total reaction, which is to be expected, and symmetry is maintained. An independent hand calculation has also been made and the results are identical. Whilst the example is of no practical significance, it demonstrates the validity of the method of analysis.

CONCLUSIONS

A mathematical model representing the dynamic interaction between the modules and module support frame of an offshore structure has been proposed and a simple example studied. Further work will involve a comparison between the mathematical model and a physical model of an offshore platform.

REFERENCES

1.   American Petroleum Institute, "Recommended practice for planning, designing and constructing fixed offshore platforms", Eleventh Edition, Dallas (1980).

2.   Bunce, J.W., Boswell, L.F. and Taylor, M.L., "Calculation of module reactions in accordance with API RP2A Clause 2.5.6, OTC paper 3970, Thirteenth Annual Offshore Technology Conference, Houston (1981).

3.   Bunce, J.W., Boswell, L.F. and Taylor, M.L., "The effects of differential deflections and fabrication errors on module support frame loadings. The Structural Engineer. To be published.

4.   Bunce, J.W., "Interaction between the topside facilities modules and the deck of an offshore platform", Supplementary Papers, The Royal Institution of Naval Architects, Vol. 122, London (1980).

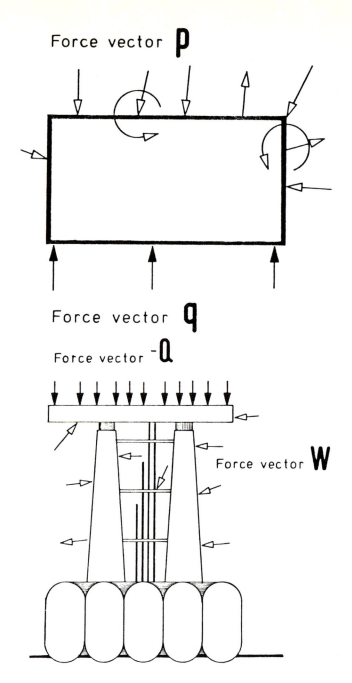

Figure 1.  Schematic arrangement of the loads acting on the modules and the platform of a typical offshore structure

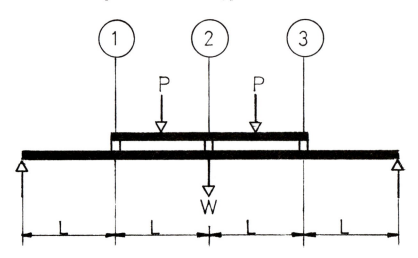

Figure 2.  Simple structure for dynamic interaction analysis

26

# Multi-degree of freedom testing techniques for analysing the hydrodynamic behaviour of offshore structures

## 4

R W Robinson, M A Murray, J D McCann
*Queens University, Belfast, UK*

SUMMARY

Testing and analysis techniques are described for determining the hydrodynamic properties of an oscillating water column wave energy device. In some instances these techniques could be more generally applied to other forms of offshore structure. Added mass and radiation damping coefficients are found using transient, monochromatic, and multifrequency testing methods. Measurement procedures are described, including the use of a remote tracking system for monitoring body motions, as well as a versatile data acquisition and analysis program.

## 1. DEVELOPMENT OF A MODEL FOR AN OSCILLATING WATER COLUMN DEVICE

The analysis treated the water column as a distinct mass of fluid contained within a hollow chamber, the fluid being excited by a wave at the column entrance. The forces acting on the fluid mass are similar to those experienced by any large structure placed in a wavefield, viz an inertia due to the fluid mass ($M\ddot{x}$), a stiffness due to the tendency of the water column to return to mean water level ($Kx$), and a damping load caused in this instance by the power take-off mechanism ($D_a\dot{x}$). The column when excited by an incident wave, generates a radiated wave by its motion, and in so doing experiences additional forces. These forces can be expressed as a radiation damping force ($D_a\dot{x}$), due to the radiated wave, and an additional inertia ($M_a\ddot{x}$) due to the entrainment of fluid at the column's entrance. Flow into the column experiences large rotations which generate vortices at the entrance lips, and these effects can be treated as an additional damping term ($D_1\dot{x}$). Although it is non-linear (i.e., velocity dependant) the term was never sufficiently large as to invalidate the equation of motion that was developed i.e.;

$$(M+M_a)\ \ddot{x}\ +\ (D_r+D_1+D_a)\ \dot{x}\ +\ Kx\ =\ F_w(t) \qquad (1)$$

The damping and added mass terms in the above equation are related to both wave frequency and column geometry, and an experimental program was conducted to identify the geometric wavefield influences on these coefficients.

Applying the equation of motion more generally it is possible to construct the equations for each of the six degrees of freedom for any body which is of sufficient size to behave in a similar manner. The equations describing the three translational degrees of freedom - heave, surge and sway - have a similar form to (1). The effect of moorings can be included by an additional stiffness term (the inertia and damping effects tend to be negligible). The rotational degrees of freedom - pitch, roll and yaw - are written in terms of angular rotation but also have a similar form to (1). The complete dynamic system can be obtained from evaluating the matrices of mass, damping and stiffness coefficients, plus those of the wave forces and body motions.

It is possible to determine the radiation damping and added mass coefficients for simplified structures using boundary integral or finite element methods, they can also be obtained by physical modelling.

The experimental analysis of a body with all six degrees of freedom is a daunting task, however the problem can be eased by dealing with each body freedom in turn. Most floating structures have a preferred body motion determined by the wave forcing available to, and the natural frequency of, that body motion. In the case of an oscillating water column the major contributor to wave energy absorption is not the motion of the containing structure but that of the water column itself. Hence the first stage in the experimental programme was to fix rigidly the containing structure, leaving the water column free to respond to the wave. Further tests were carried out which allowed restricted body freedoms in order to find the sensitivity of the body to these. Final release of the body will eventually lead to a total understanding of the complicated behaviour of water column devices.

## 2. MEASUREMENTS

### 2.1 Wave Measurements

Wave heights can be measured quite accurately by pre-calibrated twin wire resistance probes. In wide tank flumes care is required when measuring monochromatic waves since parasitic transverse standing waves can set up tank resonances. To minimise their effect it is best to employ several wave probes at discrete intervals determined by the node positions of the transverse wave, so that a true assessment of the incident waveheight can be made. Also the presence of reflected waves from the model under test can cause longtitudinal standing waves in the wave tank. However the amplitudes of the incident and reflected waves can be measured by spacing two racks of wave probes at a quarter wavelength seperation on a travelling trolley. The probe output from each rack is displayed as a Lissajous figure on an oscilliscope. The trolley is moved until the Lissajous figure achieves maximum eccentricity i.e., one probe measures the sum, while the other measures the difference, of the wave amplitudes.

### 2.2 Damping Measurements

The amount of damping applied externally in the case of a wave energy absorber can be derived from the power take-off mechanism. In the model tests of the water-column the damping was applied using a variable rectangular orifice plate. The applied damping coefficient ($D_a$) was defined as the ratio of the rms pressure force in the column to the rms velocity of the free surface.

### 2.3 Body Motion Measurements

Although in the initial tests of the wave energy absorber the only motions required were those inside the water column, subsequent tests required the measurement of the containing structure motions. The problem of a contacting measuring system influencing the body motion, at the model scale, led to the development of a relatively low cost, remote sensing device (McClean (1)). This was based on a pair of video cameras and a microprocessor one of whose functions was to digitise the T.V. image into a 256x256 dot matrix. Two pairs of light sources were mounted orthogonally on the body (fig.1) in the same plane as each camera, in order to isolate the six degrees of freedom. Voltage levels from each camera were passed through a comparator which gave either a high or a low level value to appropriate elements in the 256x256 matrix in the microprocessor's memory. By presetting the voltage levels within the comparator it could be made to respond only to the light sources on the model. The memory image was updated at 50Hz allowing ample time for the processor to scan the previous one. This provided analogue voltage outputs as cartesian co-ordinates of the location of the light sources in the memory image. The device was calibrated by placing the cameras at sufficient distance to allow the maximum motions of the body to remain within the monitor screen limits. The voltage difference between two displaced light sources provided a correlation with their linear displacement. An accuracy of ½% was established and any depth -of-field errors in the plane of the camera could be adjusted using the second camera.

## 3  DETERMINING THE HYDRODYNAMIC COEFFICIENTS

As already discussed in section 1 the physical model was designed so that the relationships between the geometric-wavefield parameters, (i.e., entrance depth, width and orientation in both the vertical and horizontal planes), and the added mass and radiation damping could be established, as well as isolating and determining the viscous losses at the column entrance.

### 3.1  Calculating The Added Mass

The added mass can be determined from the resonant frequency, provided that the total damping term is not excessive. By definition resonance occurs when the reactance terms of stiffness and inertia balance. Provided that the stiffness and body mass are known (as obtained from the body geometry), the added mass can be determined. Hence at resonance

$$f_r = 2\pi\sqrt{(K/(M+M_a))} \qquad (2)$$

### 3.2  Transient Tests

If a body has an initial displacement, and thereafter no further external force is applied, it will tend to oscillate at its natural frequency. This oscillation will continue ad infinitum when there are no damping losses, however in practice a decay in amplitude will occur with time due to the presence of damping forces. Both of these features allow the radiation, and loss damping, and the added mass to be determined.

The tests were carried out in a large tank, which had wave absorbing material placed around its perimeter (to prevent the reflection of waves generated by the model from returning to the model site, and exciting it after its initial displacement). The wave probe, mounted inside the chamber on the model centreline, had its output displayed on an ultra violet recorder as a time trace. As the column was displaced the time trace was started and the amplitude was recorded over several cycles (fig.2). The resonant frequency was obtained by dividing the number of cycles by the recording time. The damping was derived from the relationship developed by Morrison (2) i.e.,

$$D_{tot} = \frac{\rho g a \ln(a_1/a_n+1)}{\omega n^2 T} \qquad (3)$$

During the tests the orifice plate was removed so that the total damping present was caused by radiation and losses, however the two terms can be separated due to the non-linearity of the losses. The tests were repeated for decreasing values of initial displacement, and due to the dependance of the loss term on velocity the total

29

damping tended towards the asympotic value of radiation damping. Thus having obtained the radiation term it could be removed from the total damping allowing the relationship between column velocity and viscous loss damping to be established, (fig.3).

## 3.3 Monochromatic Frequency Tests

This technique relied on repeated testing of the water column using sinusoidal waves within the tank frequency range (0.6-1.8Hz). The amplitudes were chosen using a wave steepness criteria, since amplitudes are naturally limited by wavelength. A wave steepness of 0.025 was chosen since this has the most common occurrence at the proposed full scale location of wave energy devices. The resonant frequency of the column (with no applied damping) was measured using alternative methods.

The first procedure measured the phase relationship between the column and wave displacements using the resistance probes. Since at resonance, as previously mentioned, the reactance term disappears, then from (1) the column velocity must be in phase with the external wave. Hence the column displacement will lag the wave by ninety degrees. By repeatedly measuring the phase angle between the column and wave displacements over the range it was possible to plot the phase angle locus (fig.4) and obtain the frequency at which quadrature occurred. The analysis was carried out using the program ANYS on the PDP1103 tank computer. This program scanned the specified channels and obtained the discrete number of wavelengths n in each channel. The phase angle was calculated using the following relationship:

$$\text{Cos } \phi + \frac{\sum\limits_{i=i}^{n} a_i \, b_i}{\sqrt{\sum\limits_{i=i}^{n} a_i^2} \, \sqrt{\sum\limits_{i=i}^{n} b_i^2}} \quad (4)$$

The second procedure for determining resonance involved measuring the magnification factor, defined as the ratio of the column to the wave amplitude. Repeated testing of the model for several frequencies provided sufficient points in order to plot magnification factor against frequency (fig.4), from which the frequency corresponding to maximum magnification factor was defined as resonance. Since both the magnification factor and phase angle could be obtained simultaneously it was normal procedure to use both to determine the resonant frequency.

An equation for power absorption can be derived from equation (1) viz:

$$P_o = \frac{\frac{1}{2}D_a \, \omega^2 |F_w|^2}{(K-(M+M_a)\omega)^2 + (D_r+D_a+D_l)^2\omega^2} \quad (5)$$

and differentiating this with respect to the applied damping gives a relationship for maximum power at resonance;

$$D_a \text{ opt} = D_r + D_l \quad (6)$$

By measuring the power and applied damping at resonance (defined as the product and ratio of pressure and velocity respectively), a locus was obtained, from which the damping corresponding to maximum power could be derived (fig.5). The tests were repeated for decreasing wave amplitudes to separate the radiation from the loss damping as already described.

A further method for determining the radiation and loss damping terms is obtained from the following relationship:

$$D_r + D_1 = F_w / x_{max} \qquad (7)$$

This method involved testing the model under two conditions. Firstly measuring the force experienced by the water column when it was held fixed (by sealing the orifice), and secondly measuring the column velocity when the orifice was removed. From the ratio of maximum pressure and velocity for any given wave amplitude the damping values could be determined. These are separated in the manner described previously.

## 3.4  Multi-frequency Testing

The major weakness with a monochromatic test approach is that the process is slow and laborious. This led to the development of a multi-frequency approach. A special program RASEA was developed which could generate a two-dimensional sea of twenty one frequencies. The choice of frequencies and amplitudes could be decided by the user or weighted as a Pierson-Moskowitz, or a Jonswap  spectrum. The problem of obtaining a good range of frequencies led the authors to opt for a specially designed spectrum suited to the wave tank. This spectrum used a constant orbital velocity weighting, i.e., $a\omega$ was kept constant. This spectrum guaranteed a reasonable range of stable frequencies.

The second stage was to develop a versatile software package to analyse multi-frequency seas. The data acquisition was separated from the analysis unlike the monochromatic program ANYS. A data acquisition program ACQ scanned the specified channels, sorting them into channel blocks in engineering units which could be displayed graphically.

The data was processed in the analysis program RANYS which could recall any channel block. The processing was carried out in several stages. The first of these involved adjusting the time series e.g., differentiation. This was followed by a statistical routine which provided mean rms and cross-over period analysis of the signal. From this a decision could be made as to whether the channel was worth further processing. The last stage involved transforming the data into the frequency domain. The data was adjusted with a Hamming window function to reduce leakage prior to fast Fourier transform analysis (Otnes (3)). The data was readjusted after the transform to restore any energy lost due to windowing. The transform produced real and imaginary components from which the amplitude and phase angle for each frequency component could be established for that channel. The computer core was limited so that only four channels could be in core at any one time and a processing sequence was developed so that minimum over-write of the core occurred. When sufficient channels were processed and stored in core the user could perform several post transform analysis. The first of these was a phase analysis between two signals. By comparing the difference between the phase relationships for each frequency component of the two channels a phase angle - frequency locus could be achieved (fig.4). Similarly a magnification factor was calculated by examining the ratio of the channel amplitudes for each frequency component giving a magnification factor - frequency plot (fig.4). Another routine calculated an efficiency operator by comparing the energy absorbed by the device with that available in the wave for each discrete frequency (fig.6). The same routine calculated a damping operator which estimated the pressure-velocity ratio for the same discrete frequencies.

The program was originally developed in order to assess the performance of an oscillating water column device, however another application was found

31

in the slow drift oscillatory forces on the moorings of the wave energy device.

## 4. CONCLUSIONS

### 4.1 Determining The Resonance

The authors opted for the multi-frequency approach in order to determine the resonant frequency. This was because the transient method was inaccurate due to tank effects as was the monochromatic testing which was also rather slow. Both phase angles and magnification factors were used in the analysis in order to double check the resonant frequency.

### 4.2 Finding Radiation and Loss Damping

Since these tests were best suited to resonance testing the analysis was carried out using monochromatic waves. The time taken to carry out lengthy analysis using RANYS made a multi-frequency approach unattractive, while the tanks effects made transient methods difficult.

### 4.3 Estimating the Efficiency Bandwidth Response

The multi-frequency approach was the most often used since a bandwidth response could be achieved in one test. The major limitation of the single frequency approach, apart from the time taken to build up a frequency response, was that single frequencies excite tank resonances which made the measurement of wave amplitudes inaccurate. Multi-frequency testing had the advantages that the effect of tank resonances were reduced due to the averaging of the wave over several cycles and the tendency for the parasitic effects to cancel each other.

### 4.4 Applications to Other Off-shore Structures

The method outlined in the latter part of section 3.3 may be of interest in other offshore applications although it may not be as easy to apply a damping to the structure. By testing a structure in a rigidly held frame and measuring the forces experienced by the model and subsequently releasing it and measuring the maximum velocities it may be possible to identify the damping coefficients.

The development of a remote scanning system using video cameras and a microprocessor proved a successful method in measuring free-body responses. The low cost of development of the unit may prove of interest to the other establishments.

ACKNOWLEDGEMENTS

The authors would like to thank the Department of Energy for financial support, and the co-operation of both W. McClean and A. Thompson in the development of the remote scanning system. We also acknowledge the support of the wave energy team, and the staff of the Department of Civil Engineering, The Queen's University of Belfast.

REFERENCES

(1)  McCLEAN, W   'A Microprocessor Based Remote Measuring Instrument', Microprocessor Applications, Vol. 4, No.1, (1980).

(2)  MORRISON, J L M, CROSSLAND, B,  An Introduction To The Mechanics of Machines, Longmans, (1964).

(3)  OTNES, R K, ENOCHSON, L,  Digital Time Series Analysis, John Wiley and Sons, New York, (1972).

Figure 1

Model device in wide tank

Figure 2

Transient test time trace

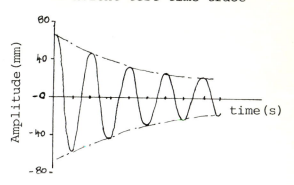

Figure 3

Damping - velocity relationship

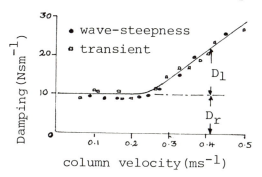

Figure 4

Determination of resonance

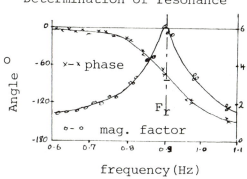

Figure 5

Determination of optimum damping

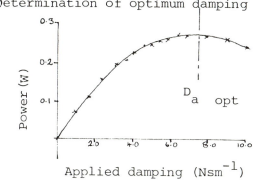

Figure 6

Device efficiency - frequency

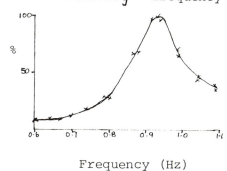

# Analytical and experimental studies on ship–ship and ship–platform collisions

## 5

A Incecik and E Samouilidis
*Department of Naval Architecture and Ocean Engineering,*
*University of Glasgow, UK*

## SUMMARY

This paper gives an introduction to collision problems and discusses mathematical and experimental models for the understanding and determination of the various parameters involved in collision dynamics. The ultimate aim of this study is to develop design methods which will be directly available to the designers for assessing the extent of damage to ships or offshore platforms, and predicting the post-damage behaviour of these structures.

## INTRODUCTION

Between 1970 and 1980 25% of the total accidents involving ships were due to ship-ship collisions, and there were 76 collisions involving fixed and offshore structures. These accidents often result in extensive material damage and in loss of production time as well as, less frequently, in loss of life. Ship and offshore structure designers should be provided with the analytical tools so that they can predict the extent of damage for low energy collision, and also make structural arrangements which may minimise the damage.

In the field of collision research and development, because of the complexity of the phenomena, each design tool developed should be tested with model experiments before being presented to the designers. The extensive literature survey on ship-ship collision is given in reference (1). On the other hand, ship-platform collisions have not been studied in great depth, although some areas in the ship-ship collision research field can be linked to ship-offshore platform collision problems.

## 1. DESCRIPTION OF THE COLLISION PROBLEMS

### 1.1 Ship-Ship Collision

In general, collision problems cover the following research areas:

a) Hydrodynamics
b) Structural response
c) Structural performance
d) Materials
e) Navigation
f) Pollution

Initially, existing knowledge concerning ship collision problems may be integrated to develop an analysis procedure. During this process shortcomings in existing knowledge from the point of view of collision mechanics should be investigated analytically as well as experimentally. Once these shortcomings are clearly defined, more fundamental studies may be needed to analyse the problem more accurately.

Since the collision consequences will depend on the types of ship involved, one needs to specify these types. The following classifications may be suggested:

(a)   Merchant Ships
      1.   Ships with two or more continuous longitudinal bulkheads (tankers, LNG carriers, containerships, the larger passenger ferries)

      2.   Ships without internal longitudinal bulkheads (dry cargo ships, bulk cargo carriers, Ro-Ro ships, etc.)

(b)   Navy Ships

In this study only ships in group (a) will be considered.

The total energy of the striking ship will be absorbed after the collision as follows:

1.   Energy absorption due to the transverse rigid body acceleration of the struck and striking ships

2.   Energy absorption due to the rotational acceleration of the struck and stiking ships

3.   Energy absorption due to the acceleration of the fluid which surrounds the struck and the striking ships in transitional modes

4.   Energy absorption due to the acceleration of the fluid which surrounds the struck and striking ships in rotational modes

5.   Energy absorption due to the overall elastic deformation of the struck ship

6.   Energy absorption due to elasto-plastic deformation of the structural members around the impact region of the struck ship

7.   Energy absorption due to the crack or rupture of the structural elements of the struck ship

8.   Energy absorption due to the elasto-plastic deformation of the structural members around the impact region of the striking ship

9.   Energy absorption due to the crack or rupture of the structural elements of the striking ship

10.   Energy absorption due to the overall elastic deformations of the striking ship

11.   Energy absorbed by cargo in damaged region

The energy conservation equation may be written as follows:

$$E_T = \sum_{i=i}^{11} E_i \qquad (1)$$

where $E_T$: Total energy of the striking ship and defined as:

$$E_T = \frac{1}{2}(M_S + M''_S)V_S2 \qquad (2)$$

$M_S$       :    Mass of the ship

$M''_S$     :    Added - mass of the ship

$E_i$       :    The energy absorption components as summarised above

It may be worthwhile to classify the energy absorption parameters as follows:

a)     Outer (or external) collision parameters which are
$E_1 - E_5, E_{10}$

b)     Inner collision parameters
$E_6 - E_9, E_{11}$

The kinetic energy and the momentum relations in respect of the transverse and the rotational rigid body acceleration of the struck ship and striking ships were calculated in Reference (2).

Similar relations were derived for the energy absorption due to the acceleration of the fluid particles which surround the ship. In these expressions added-mass and the added-moment of inertia terms were used. The added-mass and added-moment of inertia concepts for sinosoidally oscillating ships are discussed in detail in References (3,4). On the other hand, added-mass and added-moment of inertia values should be determined under the impact loading for collision studies.

In Reference (5) a theoretical and an experimental study was carried out to define the added-mass values under step or ramp type of impact force as a function of added-mass and damping quantities in the frequency domain as well as the collision duration time. One can theoretically determine the added mass and the damping values for any given ship geometry in the frequency domain. (See for example Reference 6). On the other hand, duration of the collision time will be a function of the local and the overall stiffness characteristics of the struck and striking ships.

Theoretically, assuming the strip theory approach, the added-moment of inertia can be obtained by taking the second moment of added-mass values in the sway mode about the assumed centre of rotation. However, one also needs to take into account the effect of three dimensionality in the calculation of the added-moment of inertia. Experimental work being carried out at Glasgow University will throw some light on the effect of this three-dimensionality, as well as providing a correcting factor for the calculated values to be used in the strip theory approach. (See also section 2).

So far external collision parameters have been summarised in terms of their energy absorption values. If the masses and the speeds of the striking and struck ships are known, those values can be approximated. Since the shape and the magnitude of the collision force is not known, determination of the added mass as a function of the duration of the collision time will only be approximate.

In the energy approach structural damage (inner energy absorption) versus input energy should be known so that the extent of the damage can be predicted. In Reference (6) Minorksy collected the information on 50 damaged ships and determined inner-energy absorption values. He also gives an analytical expression to relate absorbed inner energy to the volume of the damaged elements of the ship. In this study, energy absorption due to elastic deformation of the whole ship beam and the local structural elements is included within the energy absorption of the damaged elements, and no allowance is made to take into account strength variation between the striking and the struck ships. Minorsky's method is valid only for existing conventional ships.

In Reference (9), extensive model testing has been carried out with various bows and side structures and a formulation of the absorbed internal energy based on these experimental results has been given. Experimental results and the semi-empirical results given in Reference (9) are in the form of absorbed energy versus penetration and load versus penetration. The energy and penetration values are the total value for the struck and striking ships. Further results have been presented to obtain the energy absorption ratio between the struck and the striking ships. A semi-empirical formulation of energy absorption versus penetration, and of load versus penetration, is made using the statical behaviour of the structural elements. Experiments were performed under static and dynamic loading. It is found that energy absorption during the dynamic tests was about double that during the static test. Such effects can only be determined by model tests, but even then the effects of scaling have to be carefully considered. These results also show that there is a need for a thorough investigation into the dynamic behaviour of the main structural elements used in ship or offshore structures. These investigations should be of both a theoretical and an experimental nature.

In the literature, some theoretical and experimental work exists to determine the behaviour of plate and shell elements under impact loading for certain geometries as well as for certain loading and boundary conditions. These studies were reviewed in Reference (10). There is a definite need to continue this theoretical and experimental research with certain structural elements, boundary conditions and loading, which are experienced with ship and offshore structures during the collision. So far the energy method has been summarised to predict the extent of damage in both the striking and the struck ship. This method requires full scale or experimental data to determine the extent of damage as a function of inner enegy absorption. The accuracy of the prediction of outer energy absorption is sensitive to the added mass characteristics which are being investigated with a series of experiments. Using this energy approach there is also no immediate need to determine the variations or the magnitude of load or the duration of collision time. However, if, during the collision study of a ship or offshore platform, there is not enough collision information on the particular design, the energy method cannot predict the structure's response to collision accurately. In that case we may carry out an iterative process to determine the outer and inner collision responses of a ship with varying forcing function with the proposed method discussed in Section 1.2.

$$\Delta E = \tfrac{1}{2}(U_{ss} - U_s)^2 M_{ss} \tag{11}$$

where $U_{ss}$  :  Velocity of struck ship

$U_s$  :  Velocity of striking ship

$M_{ss}$  :  Mass of struck ship

energy $\Delta E$ must be absorbed by the deformation of the structural members around the impact region under increased load $[f(\tau) + \Delta f]$.

b)    Triangular Force Impulse:

The total energy absorption will be calculated. If the input energy is greater than the total absorbed energy the impact force should be increased as in the following equation:

$$f = \frac{2f_o}{\tau_i}(n\,\Delta t_i + 1) \quad \text{for } n\,\Delta t_i + 1 \le \tau_i/2 \tag{12}$$

$$f = 2f_o\left(1 - \frac{n\,\Delta t_i + 1}{\tau_i}\right) \quad \text{for } n\,\Delta t_i + 1 > \tau_i/2 \tag{13}$$

The calculation steps will be repeated starting from step C and the process should be continued until the input energy is equal to the absorption energy.

The velocity checks for the struck and striking ships, as well as the energy balancing if necessary, can be done as described in the step input case.

If absorbed energy is greater than the total input energy, the maximum load, $f_o$, should be reduced and computation may be repeated until the input and the absorbed energies are equal. The velocity checks and necessary further corrections should be performed as described earlier.

c)    Parabolic Force Impulse:

The calculation procedure is exactly the same as in the triangular force impulse case, except that the triangular force-time variation curve is replaced with the parabolic force-time representation.

d)    Rectangular Force Impulse:

The calculation process should be carried out by increasing the time steps until the velocities of the struck and the striking ships become equal. At the end of the process, if the input and the absorbed energies are not identical, the magnitude of collision force should be changed and the calculations should be repeated, starting with step C.

(I)    If the collision results in rupture of the ship's hull plating, post-damage behaviour of the ship in waves should be checked from the point of view of motion response.

This method can also make direct use of the available theoretical or experimental information on the elasto-plastic behaviour of the main structural elements under static or dynamic loading. In References (9) and (11) the structural behaviour of some simple ship strength members under static load and, in References (9, 10, 12, 13, 14 and 15), similar members' behaviour under dynamic loading has been discussed. It should be noted here again that the available information on the dynamic behaviour of ship structural elements is very limited, so that it can only approximately determine the collision behaviour of a ship under impact loading.

### 1.2 A Method to Predict Ship Response and Extent of Damage During a Ship-Ship Collision Using the Best Available Information

(A) Choose an initial value for the impact force, say, $f_0$. It will also be necessary to define the shape of the force time variation curve, i.e. step input force, parabolic input, triangular input, rectangular input, and so on. At the start of the analysis available records on a similar type of design's collision variables can help to predict the initial force value and to give some idea of the duration of collision time, say, $\tau$. The duration time can be divided into equal amounts of time spacing so that the impact force $f$ can be increased at the end of each time interval, $\Delta\tau$ for triangular or parabolic input cases.

(B) Calculate the striking ship's input energy from Equation (2).

(C) Determine the rigid body displacements from Equations (14) and (15) of Reference (2). These two equations give motion in only two degrees of freedom, and will only be valid for the small amplitude, uncoupled, linear, rigid body displacements. Depending on the collision point the motion of a ship may well be in the large displacement range during the collision stage. Then non-linear coupled equations will be needed instead of Equations (14) and (15) of Reference (2). A general form of motion response equation has been devised by the author so that the non-linear coupled motion of a ship structure can be determined in a six degrees of freedom system Reference (16). At the end of the first time step the absorbed energy due to the rigid body motion, and the distribution of rigid body motion induced loading, can be determined.

(D) Using the existing theoretical or experimental information to predict the behaviour of the structural members under impact loading, the response of the member around the impact region under collision load $f(t)$ and the rigid body induced loading will be predicted. At the end of this time step the energy absorbed due to the local deformations of the impact region will be calculated.

(E) Assuming that the ship collision will occur when the ship is travelling amongst waves, the ship's new position in the waves will be altered using the new displaced value from C at the end of the first time step and the new structural rigidity will be calculated after the damage has occurred. With these new parameters of geometry and the stiffness of the ship beam, the structural response in waves will be calculated to determine overall structural integrity. If structural integrity does not exist, the analysis procedure will be terminated after a check between input and the absorbed energies, i.e. $B < C + D$.

(F)     Assuming that the ship can be represented by a free-free beam, elastic deformations of this beam under impact load f(t) and the rigid body motion induced loading will be calculated to obtain absorbed energy due to the elastic deformations.

(G)     At this stage we can check the duration of the first collision with the parameters calculated in the previous sections, and modify the load time variation curve as necessary. The striking ship's lateral velocity at the beginning of each time increment can be calculated from the rigid body equation of motion (Equation 14 of Reference 2). It will also be assumed that velocity will remain constant during each time increment. The maximum penetration of the striking ship at the end of the time increment is denoted as $W(t_i)$. The following relations can be written between $\Delta \tau$, $W(t_i)$ and the striking ship's velocity as:

$$\Delta \tau_i = \frac{W(t_i)}{V(t_i - 1)} \tag{3}$$

If this value is found to be different from that initially assumed, $\Delta \tau_0 = \frac{\tau}{n}$ (where : total duration time of collision and n: number of time intervals), the shape of the force-time curve will be changed as follows:

a)     Triangular force impulse

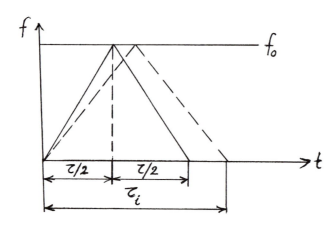

Figure 1

$$\frac{df}{dt} = \frac{f_0}{\tau_i/2} \tag{4}$$

or     $$\frac{\Delta f}{\Delta \tau_i} = \frac{2f_0}{\tau_i} \tag{4-A}$$

$$\tau_i = 2f_0 \frac{\Delta \tau_i}{\Delta f} \tag{5}$$

and the modified time increment becomes:

$$\Delta \tau_i + 1 = \frac{\tau_i}{n} \tag{6}$$

b)    Parabolic force impulse

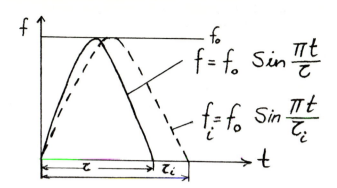

<u>Figure 2</u>

$$\frac{df}{dt} = f_o \quad \frac{\pi}{\tau_i} \quad \cos \quad \frac{\pi t}{\tau} \tag{7}$$

or $$\quad \frac{\Delta f}{\Delta \tau_i} = f_o \quad \frac{\pi}{\tau} \quad \cos \quad (\frac{\pi \Delta \tau_i}{\tau i}) \tag{8}$$

$$\tau_i = \frac{f_o}{\Delta f} \Delta \tau \, \pi \, \cos \, (\frac{\pi \, \Delta \tau_i}{\tau_i}) \tag{9}$$

The modifed time duration can be obtained from Equation (9) by iteration and the new increment becomes:

$$\Delta \tau_i + 1 \quad = \frac{\tau i}{n} \tag{10}$$

(H)   During this step various controls will be carried out to continue the process of calculation, as follows:

a)    <u>Step Force Impulse:</u>

If the $\Delta \tau_i$ value calculated in Equation (3) is greater than a certain predefined amount, then the collision load may not be assumed to be a step input. If $\Delta \tau_i$ is small enough, the absorbed energies calculated in stages C, D and F will be summed up and cmpared with the input energy of the striking ship calculated in step B. When the total absorbed energy is small than the input energy the collision load $f_o$ should be increased and the calculation process should be repeated starting from step C until both the input and the total absorbed energies are equal. At the end of the process, when energy balancing is met, a check should be made to ensure that both the striking and the struck ship have reached the same speed.

If the striking ship's velocity is found to be greater than that of the struck ship, this indicates that the inner collision energy absorption was overestimated.

Since overestimation of the absorbed inner energy will predict more damage than will probably occur, such calculations are acceptable. If the struck ship's velocity is higher than the striking ship's velocity, similarly this will show underestimation of the absorbed inner energy. If we defined the energy difference as:

41

## 1.3 Ship-Offshore Structure Collision

In order to study ship-offshore structure collision problems, first of all offshore structures will be classified as either fixed or floating platforms.

a)    Fixed Platforms:

Jacket, jack-up, hybrid and gravity platforms.
Guyed tower and articulated column type structures will also be included in this category for the collision studies.

b)    Floating Platforms:

Semi-submersibles, tension-leg platforms.

The most likely element of any type of offshore structure to be involved in a collision is a circular cylindrical member. The geometrical size and the boundary conditions of these circular members vary depending on the type of offshore platform. There is a large amount of literature available to predict the behaviour of these circular cylinders with various stiffener arrangements under static loading (17). Some theoretical and experimental studies also exist on the stability and the strength of these tubular members under dynamic loading (18, 19, 20, 21). The present knowledge on the behaviour of cylindrical elements under dynamic loading is inadequate for the application of offshore platform assessment studies. However, collision and post-collision assessment of offshore structures may be carried out with the following suggested method, which makes use of both dynamic and quasi-static analysis techniques. The collision damage assessment predicted with this method should be checked and improved with model tesing. Model testing would also indicate any significant difference in the behaviour of struck elements due to quasi-static calculations.

### 1.4.1 A Method to Predict Fixed Platform Response and Extent of Platform - Ship Damage During a Collision Using the Best Available Information

The general form of this method will be similar to that suggested for the ship-ship collision. The method can be summarised with the following steps:

a)    Space frame representation of the fixed offshore platform and beam elements. The mass of each beam will be concentrated in the middle of these beam elements as lumped masses. The beam elements may be chosen as the structural members in between the physical joints of the structure. The impact region of the platform may be modelled with the increased number of beam elements for greater accuracy in calculations.

b)    Determination of cross-sectional areas, second moment of inertia values for each beam element.

c)    Calculation of the kinetic energy of the striking ship.

d)    Initial predictions of collision load in terms of magnitude and its variation with time and the prediction of duration of collision load.

e)   The non-linear structural response of fixed platforms under $f = f_0$ (t), $0 < t < \tau$ collision is determined. The method of calculation is given in detail by the author in Reference (22). Similarly, the response of a striking ship is also calculated using the method suggested in Section 1.3. During the calculation of the structural response of a fixed structure, two types of structural behaviour will be experienced.

   i)   Non-linear, inelastic or plastic structural response of the members which are in the impact region. Calculations to predict structural behaviour will be carried out using static load-deflection characteristics. That is where future research is needed to determine these load-deflection characteristics under dynamic loading.

   Geometrical non-linearities obtained from quasi-static analysis for the next load increment, but the members' behaviour with those geometrical non-linearities under the next increment of impact load will again be unknown until further experimental investigations are carried out. This problem has recently been studied for the static loading case with the tubular members (23,24).

   ii)   The linear elastic structural response of the members which are not in the impact region can be obtained under impact loading with more confidence. However, the decision on the ultimate strength of the members which are experiencing the structural load around their limit values should also incorporate dynamic behaviour.

   The absorbed energies of the platform and the ship will be calculated.

f)   The structural response of the fixed platform under wave loading will be calculated (see Ref. 25). In the calculations the new displaced members' geometrical and structural load due to impact and to the wave forces will be obtained to check the limit state for each member. Since the members performances in the limit states are obtained from the static considerations, the results in the dynamic case may be inaccurate. If the failure of some members has been predicted in this step, the integrity of the structure will be recalculated under wave loading. If the input energy is smaller than the absorbed energies and integrity does not exist, the analysis will be terminated.

g)   Similarly to steps G and H of Section 1.3, the collision duration time will be checked and the variation of the collision force will be carried out until the energy balancing between the input energy of the striking ship and the energy absorption by the struck platform and the striking ship is satisfied.

## 1.4.2   A Method to Predict Floating Platform Response and Extent of Damage During a Collison using the best available information

The method of calculation will be similar to the fixed platform case, but with additional steps. These steps may be summarised as follows:

1)    Determination of rigid body displacements under impact loading. The general method of calculation to predict rigid body motion of floating structures is six degrees of freedom system is derived by the author in Reference (15). The absorbed energies due to the rigid body displacements will be calculated. Similarly to the ship hydrodynamic energy absorption parameters, added mass and damping values should be checked experimentally during floating platform collision to validate the existing theroetical applications.

2)    The structural response should be predicted under $f = f_0$ (t) collison force as well as rigid body motion induced loading. Detailed analysis methods for determining the structural response of floating platforms have been reported by the author in Reference (26). In the same reference the increases in the structural loads due to the failure of the struck members are also illustrated for a typical semi-submersible platform (See Figures 3-4).

## 2.    EXPERIMENTAL STUDIES

In order to verify and improve various analytical formulations for the prediction of overall and local structural response under impact loading, a series of model tests are being carried out at the Hydrodynamics Laboratory of Glasgow University. Tests are being concentrated on simplified small scale tanker models at present and will be continued with models of offshore structures. In order to measure various energy absorption parameters with the minimum interaction, tests are being carried out by fixing the model on a rigid test rig and applying impact loads to various compartments of the tanker model with various rigid bow geometries. During these set of tests, inner collision parameters, as well as overall structural response of the model and the effect of water in the various compartments from the energy absorption point of view, will be studied. During the experiments local and the overall deformations of the model, as well as the dynamic impact load-time variation and the total input energy, are measured.

The tests which are being carried out on the fixed test rig will be repeated in the towing tank in order to measure various outer energy absorption parameters, i.e. rigid-body acceleration and velocity, as well as added-mass and added-moment of inertia values.

These test results will also provide a comparison between the level of damage caused by static and dynamic loads for any given level of equivalent energy.

## 3.    CONCLUSION

In this paper a summary of the research being carried out at Glasgow University on ship-ship and ship-offshore platform collisions has been discussed. This research aims at developing design tools for assessing the extent of damage and minimising the damage under unexpected loading. As work has progressed it has become increasingly clear that there are many aspects of the problem which cannot be resolved entirely analytically, and that extensive model testing will be required to provide the basic data for analytical tools to be able to predict behaviour in a specific case. The development of the design tools involves the integration of the results of current research at Glasgow University's Naval Architecture and Ocean Engineering Department, with contributions from other institutions working, or having worked, on related fields elsewhere.

## ACKNOWLEDGEMENT

This research is being carried out under the sponsorship of the Marine Technology Directorate of the Scientific and Engineering Research Council, U.K.

The help received from the academic and technical staff at the Department of Naval Architecture and Ocean Engineering, in particular Mr. N.S. Miller and Dr. P.A. Frieze, is warmly acknowledged. Considerable assistance given to the Department by Det norske Veritas is gratefully acknowledged. The authors are also grateful to Miss I. Campbell for typing the manuscript.

## REFERENCES

1. JONES, N.: "A Literature Survey on the Collision and Grounding Protection of Ships", Ship Structure Committee Report, SSC-283, U.S. Coastguard, Washington, 1979.

2. INCECIK, A.: "Research into Ship-Ship and Ship-Platform Collisions", Department of Naval Architecture and Ocean Engineering Report No. NAOE-HL-81-07, Glasgow University, 1981.

3. FRANK, W.: "Oscillation of Cylinders In or Below the Free Surface of Deep Fluids", David Taylor Model Basin Report No. 2375, Washington DC., 1967.

4. VUGTS, J.H.: "The Hydrodynamic Coefficients for Swaying, Heaving and Rolling Cylinders in a Free Surface", Netherlands Ship Research Centre TNO Report No. 112S, 1968.

5. MOTORA, S., FUJINO, M., SUGIURA, M.: "Equivalent Added Mass of Ships in Collisions", Jnl of the Society of Naval Architects of Japan, 118, 1965.

6. ATLAR, M.: "Frank-Close fit Computer Programme for the Calculation of Added Mass and Damping Coefficients of Oscillating Cylinders", Department of Naval Architecture and Ocean Engineering Report No. NAOE-HL-81-09, Glasgow University, 1981.

7. MINORSKY, V.U.: "An Analysis of Ship Collision with Reference to Protection of Nuclear Power Plants", Jnl of Ship Research, Vol. 3, No. 2, October 1959.

8. PETERSEN, M.J.: "Dynamics of Ship Collisions", The Danish Centre for Applied Mathematics and Mechanics Report No. 182, 1980.

9. AKITA, Y., ANDO, N., FUJITA, Y., and KITAMURA, K.: "Studies on Collision Protective Structures in Nuclear Powered Ships", Nuclear Engineering and Design, Vol. 19, 1972.

10. HERRMANN, G.: "Dynamic Stability of Structures", Proc. of an Intl. Conference held at Northwestern University, Illinois, Pergamon Press, 1965.

11. McDERMOTT, J.F., KLINE, R.G., JONES, E.L., MANIAR, N.M., CHIANG, W.P: "Tanker Structural Analysis for Minor Collisions", Trans. SNAME, 82, 382-414, 1974.

12. JONES, N.: "A Theoretical Study of the Dynamic Plastic Behaviour of Beams and Plates with Finite Deflections", Intl Jnl of Solids and Structures, Vol. 7, 1971.

13. JONES, N., URAN, T.O. and TEKIN, S.A.: "The Plastic Behaviour of Fully Clamped Rectangular Plates", Intl Jnl of Solids and Structures, Vol. 6, 1970.

14. JONES, N., DOS REIS, H.L.M.: "On the Dynamic Buckling of a Simple Elastic-Plastic Model", Intl Jnl of Solids and Structures, Vol. 16, 1980.

15. WITMER, E.A., BALMER, H.A., LEECH, J.W. and PIAN, T.H.H.: "Large Dynamic Deformations of Beams, Rings, Plates and Shells", AIAA Jnl, Vol. 1, 184-8, 1963.

16. INCECIK, A.: "Motion Response of Floating Offshore Structures", Department of Naval Architecture and Ocean Engineering Report No. NAOE-HL-80-29, Glasgow University, 1980.

17. "Collected Papers on Instability of Shell Structures", National Aeronautics and Space Administration Report No. NASA TN D-1510, 1962.

18. BUDIANSKY, B.: "Dynamic Buckling of Elastic Structures: Criteria and Estimates", Proc. of an Intl Conference on Dynamic Stability of Structures, Edited by G. HERRMANN, Pergamon Press, 1965.

19. BUDIANSKY, B. and HUTCHINSON, J.W.: "Dynamic Buckling of Imperfection Sensitive Structures", Proc. of the Eleventh International Congress of Applied Mechanics, Munich, 1964.

20. FLORENCE, A.L. and VAUGHAN, H.: "Dynamic Plastic Flow Buckling of Short Cylindrical Shells due to Impulsive Loading", Intl Jnl of Solids and Structures, Vol.4, pp 741-756, 1968.

21. JONES, N. and OKAWA, D.W.: "Dynamic Plastic Buckling of Rings and Cylindrical Shells", Nuclear Engineering and Design, 37, 1976.

22. INCECIK, A.: "Modelling of a Guyed Tower Problem for Structural Response", Department of Naval Architecture and Ocean Engineering Report No. NAOE-HL-81-06, Glasgow University, 1981.

23. SMITH, C.S., KIRKWOOD, W. and SWAN, J.W.: "Buckling Strength and Post-Collapse Behaviour of Tubular Bracing Members including Damage Effects", Proc. of the Second International Conference on the Behaviour of Offshore Structures, England, 1979.

24. TABY, J., MOAN, T. and RASHED, S.M.H.: "Theoretical and Experimental Study of the Behaviour of Damaged Tubular Members in Offshore Structures", Norwegian Maritime Research, No. 2, 1981.

25. INCECIK, A.: "A General Method and a Computer Program to Calculate Wave Loading on the Circular Cylindrical Members of Fixed and of Floating Offshore Structures", Department of Naval Architecture and Ocean Engineering Report No. NAOE-HL-80-19, Glasgow University, 1980.

26. INCECIK, A.: "Structural Response of Floating Offshore Structures", Department of Naval Architecture and Ocean Engineering Report No. NAOE-HL-81-03, Glasgow University, 1981.

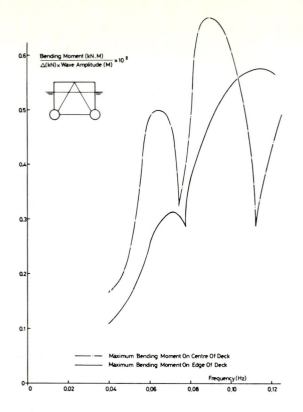

Figure 3 - a

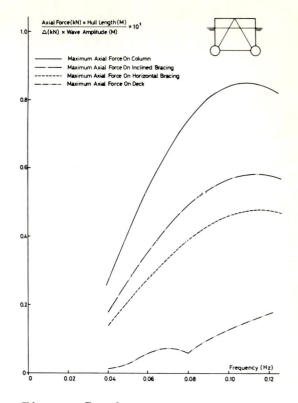

Figure 3 - b

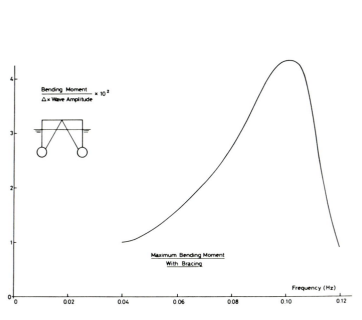

Figure 4 - a

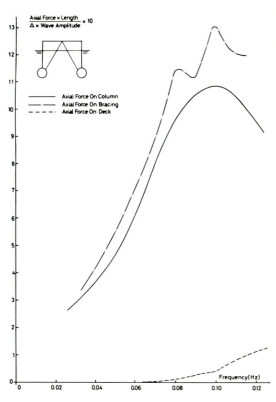

Figure 4 - b

# The accuracy of mathematical models of structural dynamics

6

A P Jeary and B R Ellis
*Building Research Establishment, UK*

## 1. INTRODUCTION

The purpose of producing models of dynamic structural behaviour is to allow the assessment in the laboratory or in the design office, of the probable response of a structure to a prescribed loading. A designer is often interested in the response to extreme or rare events and needs to predict his structure's likely behaviour in these circumstances. The model chosen to represent the dynamic behaviour may be physical or mathematical, but it is obviously of paramount importance to construct these models to reflect reality as accurately as possible. The problems with this are that the real dynamic behaviour of structures is imperfectly known, and that there are always differences between models and reality.

In this paper the use of mathematical models of structural dynamics is considered since a rigorous analysis of the implications of an imperfect knowledge of real structural dynamics can be made for these cases. In particular the implications for the confidence associated with response predictions are investigated. This analysis of confidence intervals suggests areas where research effort should be directed. Additionally, cases in which measurements on an existing structure have been made are considered. In these circumstances there still remain significant differences between prediction and actual response and the reasons for this are considered.

## 2. DYNAMIC MODELLING PROCEDURES

In the mathematical treatment of vibration it is often assumed that the response in each mode of vibration can be described by a linear 2nd order differential equation, and that the total response can be derived by superposing the response in each mode of vibration. In practice it is seldom necessary to consider more than a small number of modes (eg 3) for the calculation of response to be as accurate as is attainable. Most currently used dynamic design methods use this technique.

### 2.A. The basic equations

It is often assumed that the equation which best describes motion in a single mode of vibration takes the form (1):

$$\ddot{X}_r + 4\pi f_r \zeta_r \dot{X}_r + 4\pi^2 f_r^2 X_r = \frac{F_r(t)}{M_r} \tag{1}$$

in which $X_r$ is the response for mode 'r' where

   $f_r$ is the natural frequency

   $\zeta_r$ is the modal damping ratio

   $M_r$ is the modal mass

and

   $F_r(t)$ is the time varying modal force.

Additionally, a mode shape $\phi(v,n)$ is usually defined such that it has a value of 1 at the greatest displacement.

If equations, of the same form as equation (1) are derived for a complete multi-degree-of-freedom system(2), it is found that an equivalent or modal mass can be defined such that:

$$M_r = \sum_{v=1}^{j} M_v \phi^2_{vn} \qquad (2)$$

and an equivalent or modal force can be defined as:

$$F_r = \sum_{v=1}^{j} F_v \phi_{vn} \qquad (3)$$

for a system having j masses and n normal modes.

It can be seen that equations (1), (2) and (3) can be used to give a prediction of the motion of a structure for any particular forcing function $F_r$ (and for $F_r(t)$ if $F_r$ is time varying), provided that information about the following parameters exists:

1)   Natural frequency   $(f_r)$

2)   Damping ratio       $(\zeta_r)$

3)   Modal mass          $(M_r)$

4)   Mode shape          $(\phi_r)$

5)   Mass distribution   $(M_v)$

6)   Force distribution   $(F_v)$

As an alternative to information about $M_r$, the modal stiffness $K_r$, defined as $4\pi^2 f_r^2 M_r$, may be used. Additionally, if information exists about items 4) and 5) then the modal mass may be calculated directly.

2.B.   Forms of forcing function

There are three different categories into which the forcing function may be placed and equation (1) can be modified to handle these cases.

These three categories are now considered in turn.

(i)   Deterministic

A deterministic forcing function is one which can be described by a unique mathematical function (such as a sine wave). There are very few natural sources which can be described by such a function, but one example in this category is machine induced vibrations. Equation (1) is often used directly to obtain an estimate of response caused by a deterministic forcing function.

## (ii) Long-term random

Long-term random forcing functions are those which can only be described using statistical techniques. Furthermore the 'long-term' part implies that the statistical quantities are invariant with time (within prescribed limits) and a measurement made during one time interval would be reproduced if made at a different time. Data that satisfy this requirement are termed 'stationary'. Those forces caused by wave, wind and traffic are often considered to be in this category and whilst it is not strictly true for long periods (trends often appear), it represents a pragmatic approach to the design process.

It is normal practice, nowadays, to deal with vibration in this category by using spectrally based techniques. Equation (1) can be re-formulated into the frequency domain as:

$$H(f) . F_r(f) = X_r(f) \tag{4}$$

where $H(f)$ represents the structure (or system) and is called the 'complex frequency response function'. $H(f)$ takes the form:

$$H(f) = \frac{1/(2\pi f_r{}^2)M_r}{1 - \left(\dfrac{f}{f_r}\right)^2 + j\ 2\ \zeta_r\ \dfrac{f}{f_r}} \tag{5}$$

where $j = \sqrt{-1}$

By combining equations (4) and (5) we have:

$$\frac{F_r(f)}{4\pi^2 f_r{}^2 m_r} \left[ 1 - \left(\frac{f}{f_r}\right)^2 + j\ 2\ \zeta_r\left(\frac{f}{f_r}\right) \right]^{-1} = X_r(f) \tag{6}$$

Each of the parameters in equation (6) must remain stationary for the spectrally based predictors to produce a precise estimate of response.

It is worth noting that equation (6) can be transformed back into the time domain and can be used with deterministic data. Equation (6) can also be used to derive an equation for statics if zero frequency is considered. The resulting equation is:

$$X_r(t)_{f=0} = \frac{F_r(t)_{f=0}}{K_r} \tag{7}$$

In words, equation (7) states that the force is the product of the stiffness and the displacement. The consideration of different frequencies leads to different equations. At resonance $f = f_r$ and equation (6) reduces to:

$$X_r(t) = \frac{F_r(t)_{f=f_r}}{K_r\ j\ 2\ \zeta_f} \tag{8}$$

Equation (8) shows that the response at resonance is larger than the statis response to a similar magnitude force by a factor of $1/2\zeta_r$. This quantity is often called the dynamic magnification factor.

## (iii) Short-term random

Short-term random signals do not last long enough for stationarity to obtain, and accordingly they can be treated only as time varying quantities. Forces such as those caused by earthquakes and occasional traffic loads are in this category.

The Duhamel integral approach is often used to describe the response of a system to a short-term random loading. For this purpose, the forcing function is idealised as a series of short impulses and the response of the structure to each impulse is calculation, for discrete times, and then all the resulting displacements are superposed to produce a total response. Equation (1) is used with a forcing function which describes an impulse. The resulting equation for the ensuing motion is:

$$x(t) = \frac{1}{M_r 2 \pi f_r} \int_0^t F(\tau) e^{-\zeta \omega (t-\tau)} \sin \omega_r (t-\tau) d\tau \qquad (9)$$

This is solved at discrete values of $\tau$ for each impulse and a numerical summation technique is used to evaluate the total response.

It is often assumed that:

$$F(\tau) = M_r \ddot{V} \qquad (10)$$

where $\ddot{V}$ is the acceleration of the ground. The forcing function can then be replaced by the ground acceleration. This is particularly useful in earthquake engineering as records of ground acceleration exist for certain earthquakes.

## 2.C.  The finite element method

The finite element method is often used to develop equations of motion for complete structures. It is used in conjunction with forces in any of the three categories discussed. This method seeks to characterise a complete structure by assembling idealisations of small elements. The resulting matrix equation is in the form:

$$\overline{M}\ddot{X} + \overline{C}\dot{X} + \overline{K}X = F(t) \qquad (11)$$

and is directly analogous to equation (1).

The amount of detail that goes into forming equation (11) tempts the user to suppose that it is inherently more accurate than equation (1), but this is not necessarily so. The model produced, albeit complex, will be only as good as the information used in its formation. Very often this information just does not exist (this is particularly troublesome when trying to describe interconnections between elements in a structure), and in these cases no mathematical model can be fully justified.

## 3.  ERRORS INVOLVED IN ESTIMATING DYNAMIC PROPERTIES

In this section the parameters in equation (1) are examined in turn, in order to estimate the confidence with which we can predict them. The experience gained in many studies throughout the world has been used to arrive at these figures.

## 3.A.  Natural frequencies

In a recent study (3), Ellis compared theoretically produced estimates of natural frequency with those actually measured (and published) for 17 buildings, and went on to use a sample of 163 buildings to derive a 'best fit' formula. The comparison of theory with practice showed that very simple formulae gave estimates of natural frequency that were better correlated with measured natural frequencies than estimates from computer based methods. The best fit formula is:

$$f_1 = 46/H \hspace{4cm} (12)$$

where H is the height of the building in metres.

Errors in excess of 50% were common whatever predictive method was used, no matter whether they were the very simple estimators or the computer-based method.

## 3.B. Damping ratios

There is a great deal of confusion, at present, over what value of damping should be used in any specific instance. The authors have suggested previously (4) that this confusion stems in part from measurements being made using poor techniques and in part from being made at different amplitudes. In recent publications changes of damping with amplitude have been detailed for various buildings (4,5).

Values of damping used in codification or design are, at present, guessed and have very wide tolerances on them (at best ± 100%).

## 3.C. Forces

There is only one instance of forces on a building being compared directly with those predicted (6). This recent work has shown that in the field of wind engineering tolerances on estimates may be relatively small (of the order 10-20%). The problems associated with directionality for wind and wave engineering and of spectral content for earthquake engineering add significant uncertainties. Because there are fewer data about wave loading and earthquake loading the tolerances for the magnitude of forces are larger in these fields than for wind engineering.

## 3.D. Mass

Although at first sight a simple matter, it is not particularly easy to make a good estimate of the dead load of a building. This is because setting-out and construction are often not precise, and occupancy loads are uncertain.

An attempt, in a recent test at Leicester University (7), to measure a well controlled difference in mass by means of vibration testing was not successful. A tank full of water at the top of the building represented about 10% of the modal mass of the first translational mode of the building and vibration tests were conducted on the building both with the tank full and with it empty. Since vibration testing is the only way presently available for measuring the mass of an existing building, the failure of these tests to resolve a 10% difference accurately implies that the error bound for mass for all buildings must be at least 10%.

Additionally, some tests have been performed (3) in which the modal mass, for first translational mode activity, has been calculated at different amplitudes of vibration. Variations of up to 20% (with no trend) have been noted in these cases.

In view of these factors it seems likely that the error bound for estimating the mass of a building should be considered to be 20%.

## 4. THE IMPLICATIONS OF PARAMETER ERROR BOUNDS FOR RESPONSE CALCULATION

Equation (6) is the basis of all predictive methods currently used. It is used directly in wind and wave engineering, in the response spectrum approach to earthquake engineering, and is the basic assumption on which the Duhamel Integral approach is based. With a knowledge, or a guess, of all the error bounds involved, in the parameters that go to make up equation (6), it is possible to calculate the error bounds for resultant response prediction.

In the following, the error bounds are calculated for two different situations. The first considers the design stage where no measurements have been made and the designers must rely on past experience and 'engineering judgement'. The second is for an existing structure which has been tested to give a priori knowledge (albeit imperfect because of measurement error) of the dynamic parameters.

Since the major part of the response of engineering structures occurs at resonance, equation (6) is considered for this case (ie $f = f_r$) and the error bound in the response is given as:

$$\frac{\delta x_r(f)}{X_r(f)} = \frac{\delta F_r(f)}{F_r(f)} - \frac{\delta M_r}{M_r} - \frac{\delta \zeta_r}{\zeta_r} - \frac{2\delta f_r}{f_r} \qquad (13)$$

This equation is now applied to the two different situations.

### 4.A. The design stage

The error bounds discussed in section 3 suggest that the error bounds of individual parameters may be considered to be:

$$f_r \quad \pm \quad 50\%$$
$$\zeta_r \quad \pm \quad 100\%$$
$$M_r \quad \pm \quad 20\%$$
$$F_r(f) \quad \pm \quad 20\%$$

The error bound for $d_r(f)$ is, therefore:

$$\pm 0.20 \quad \pm 0.20 \quad \pm 1.0 \quad = \quad \pm 2.40$$
$$\text{or} \quad \pm 240\%.$$

### 4.B. Based on test results from an existing structure

The individual error bounds now become:

$$f_r \quad \pm \quad 0.1\%$$
$$\zeta_r \quad \pm \quad 10\%$$
$$M_r \quad \pm \quad 20\%$$
$$F_r(f) \quad \pm \quad 20\%$$

and so the error bound for $X_r(f)$ becomes:

$$\pm 0.20 \quad \pm 0.10 \quad \pm 0.20 \quad \pm 0.002 \quad = \quad 0.502$$

or approximately 50%.

These two scenarios consider the case where the best possible information is used. In practice even this is not the case - the present generation of wind design guides do not consider that turbulence decreases with increasing height for instance. The two error bound figures must, then, be considered as lower bound solutions for the confidence interval.

It can be appreciated from the foregoing that a precise knowledge of natural frequency alone reduces the error bound by approximately 100%. It can also be appreciated that an error bound of 240% is not merely unacceptable; it makes the calculations virtually worthless. There is, therefore, a great temptation to ignore the implications of our imperfect knowledge.

## 5. OTHER SOURCES OF ERROR IN RESPONSE PREDICTION

As has already been pointed out, the error bounds calculated in section 4 are lower bound solutions. There are many occasions when basic data are badly handled, to give a poorer estimate of one of the parameters involved, or not all of the available knowledge or techniques available are used. Some particular examples are considered here.

### 5.A. Wind data

The use of wind statistics to estimate probable forces which will act on a structure is taken for granted by many designers, and there is a certain complacency that, at least in Britain, the positioning of isopleths is reasonably accurate. Unfortunately the necessary corrections to the raw anemometer data are not always applied (8) by the Meteorological Office. Furthermore the location of some sites where wind force estimates are required may be so far away from Meteorological Stations that the precision of the estimates of wind forces may be downgraded.

### 5.B. Codes of Practice

Codes of Practice normally follow a long way behind the best theoretical methods. The error bounds, for calculation of response, must, therefore, be larger than the lower bound ones of section 4. It should be noted that since the conventional static equations are merely a special case of equation (6), similar error bounds apply to conventional quasi-static design procedures.

### 5.C. Earthquake data

The statistics used to calculate an earthquake regime are similar to those used for wind engineering. The difference is that there are fewer data about earthquake forces and, accordingly the error bound for force must be larger than that considered in section 4.

### 5.D. Offshore force data

Data about loading caused by ocean waves are scanty, and consequently the error bounds, for any particular location, are wide. The problem is compounded for offshore engineering because earthquake and wind data need to be considered as well.

### 5.E. Non-linear effects

The analysis of confidence limits shown in section 4 are based on a visco-elastic system. Unfortunately most buildings do not conform to this model over the whole practical range of responses because of non-linearities of frequency and damping as a function of amplitude.

The use of very precise eccentric mass vibrators capable of delivering various force levels at the same frequency has allowed the study of the changes of frequency and damping with amplitude in the current BRE research programme. To date, however, insufficient data have been collected for these changes to be predicted.

For the case of wind engineering these effects are not normally significant, but they can be of paramount importance for earthquake and offshore engineering.

## 5.F.  Dynamic-design guides for wind engineering

The BRE dynamics research programme has provided full-scale test data which have subsequently been used in the calibration of design guides. Response to wind loading has been measured and compared with values calculated from design guides for corresponding conditions for five different buildings. The most advanced design guides have been used for this purpose(9,10,11). These design guides often gave a prediction of response which was much larger than the measured response and the average overestimate for the five buildings was 1470%(5). Deaves and Harris(12) have suggested alterations to the calculations for wind forces in these design guides. The nature of the alteration is such as to reduce the calculated forces, and this may go some way to reducing the error.

Because the information about the dynamic properties of the buildings was precise, this exercise has been a calibration of the design guide only, and has served to show that there are defects in them as they are presently constituted.

## 6    SOME IMPLICATIONS OF WIDE CONFIDENCE LIMITS FOR RESPONSE CALCULATION

From the foregoing it is clear that much practical research is needed if we are to make dynamic models and predictive procedures reliable and accurate. More information is urgently required about the forces that act on structures and overall structural behaviour (including information about natural frequencies, damping, mass); but above all the comparison of theory with practice is vital if models of dynamic behaviour are to be refined sufficiently to give good estimates of response.

The use of full-scale vibration tests can reduce confidence limits to such an extent that it may be possible to give assurances about an existing structure's likely response during its lifetime.

The response of owners of some structures (particularly offshore structures) to a lack of knowledge about likely forces or real behaviour, is to monitor the behaviour of the structure for changes of response which will indicate changes in the structure itself. This process is variously termed integrity monitoring, finger-printing or footprinting. If no a priori knowledge of all the dynamic parameters and variables noted in sections 4 and 5 exist, then any changes of response noted may not be attributable to changes in the structure. However, only if a structure is calibrated in a forced vibration test and the response at times when the forcing is at a similar level is monitored so as to accumulate sufficient 'stationary' data, is there a chance that these techniques will be successful.

ACKNOWLEDGMENT

The work described has been carried out as part of the research programme of the Building Research Establishment of the Department of the Environment and this paper is published by permission of the Director.

REFERENCES

(1)  JEARY A P and SPARKS P R.  Some observations on the dynamic sway characteristics of concrete structures.  ACI Symposium on vibrations in concrete structures, New Orleans (October 1977). Available as BRE CP 7/78.

(2)  BIGGS J M.  Introduction to structural dynamics.  McGraw Hill (1964).

(3)  ELLIS B R.  An assessment of the accuracy of predicting the fundamental natural frequencies of buildings and the implications concerning the dynamic analysis of structures.  Proc Instn Civ Engrs, Part 2 (1980), 69, p 763-776.

(4)  ELLIS B R and JEARY A P.  Recent work on the dynamic behaviour of tall buildings at various amplitudes.  Proc 7th World Conference on Earthquake Engineering, Instanbul (1980).

(5)  JEARY A P and ELLIS B R.  Recent experience of induced vibration of structures at varied amplitudes.  Proc ASCE/EMD Conference on Dynamic Response of Structures, Atlanta, GA (Jan 1981).

(6)  EVANS R A and LEE B E.  The assessment of dynamic wind loads on a tall building:  A comparison of model and full scale results. 4th US National Conf on Wind Engineering Research, Paper IV-5, Seattle (July 1981).

(7)  JEARY A P, BEAK M, ELLIS B R and LITTLER J D.  Vibration tests at Leicester University engineering tower.  BRE N 40/80 (April 1980).

(8)  LEE B E.  The problems of anemometer exposure in urban areas. A wind tunnel study.  (July 1981), Meteorological Magazine.

(9)  Engineering Sciences Data Unit.  Data Item 76001.  ESDU, 251-259 Regent Street, London (September 1976.  Amended June 1977).

(10)  National Building Code of Canada, (1975) and commentaries on Part 4.  National Research Council of Canada, Ottawa.

(11)  WYATT T A.  The calculation of structural response, Paper 6, Proc of the Seminar on Modern Design of Wind Sensitive Structures, CIRIA (1976).

(12)  DEAVES D M and HARRIS R J.  A mathematical model of the structure of strong winds.  CIRIA Report 76 (May 1978).

## 7

# Dynamic instability of a freely floating platform in waves

R Eatock Taylor and J Knoop
*Department of Mechanical Engineering, University College London, UK*

## SUMMARY

The possibility of heave instability of a large volume floating body in waves is investigated by mathematical and physical models. Responses in regular and random waves are briefly reviewed, leading to tentative conclusions concerning subharmonic responses of full scale platforms.

## INTRODUCTION

Potential wave induced instabilities of tethered buoyant platforms in lateral modes have recently attracted considerable research interest [e.g. (1)], as have roll instabilities in ships and other freely floating bodies. Less familiar, however, is the instability associated with heave of a large freely floating structure such as a monolithic drilling, production or storage facility in waves. Rainey (2) established the theoretical basis for such an instability and also provided a convincing experimental demonstration in regular waves. The present paper describes an attempt to investigate this essentially non-linear phenomenon in greater detail, by extending the study to include random wave excitation and the comparison of experimental data with theoretical predictions.

The experimental model is a simple derivative of a 500ml round bottom flask, which was tested in regular and random waves in the 2.2m wide by 1m deep tank at University College London. Two theoretical approaches were adopted to describe the vertical motions of the flask. The first is based on three dimensional linear potential flow and use of a source-sink idealisation of the flask to solve the wave diffraction and radiation problems, and hence the linear motion response. The second theoretical approach is based on the long wave (small body) approximation. The latter provides a somewhat crude simplification of this particular problem, but it may readily be extended to incorporate a simple non-linear mechanism leading to heave instability of the type observed experimentally.

## THEORY

Relative to an inertial frame of reference the equation of motion y in the upwards vertical direction is

$$\rho V \ddot{y} = P(t) \tag{1}$$

where V is the displaced volume of the floating body, $\rho$ is the fluid density and P is the total hydrodynamic force. If we make the assumption of ideal flow, the linearised first order hydrodynamic problem may be solved in terms of wave diffraction and radiation potentials [e.g. Newman (3)], leading to the following expression for vertical force in a sinusoidal wave of frequency $\omega$

$$P = - A\ddot{y} - B\dot{y} - Cy + F . \tag{2}$$

The added mass and damping terms, A and B respectively, are generally functions of $\omega$; the hydrodynamic stiffness for a body having water-plane area $S_o$ is

$$C = \rho g S_o ;$$

and the body is assumed to have symmetry so that heave is uncoupled from the other motions. F is the wave force on the fixed body, resulting from the pressure gradient in the undisturbed incident wave plus the effect of wave scattering.

In the long wave approximation, the following simple expressions are obtained:

$$A = \rho V a_m, \quad B \approx 0, \quad F = \rho V(1 + a_m) \dot{v} + \rho g S_o \eta .$$

Here $a_m$ is the added mass coefficient for vertical motions of the body, the undisturbed wave elevation is $\eta$, and the characteristic undisturbed vertical wave particle acceleration is $\dot{v}$, at the equilibrium position of the body. (We note that whereas it would be consistent with this approximation (3) to neglect $a_m$ for a slender body such as a spar buoy, this term must be retained for a body having large displaced volume but small water plane area). The equation of motion may therefore be written

$$M\ddot{y} + Cy = F = M\dot{v} + C\eta \tag{3}$$

where

$$M = \rho V (1 + a_m) .$$

Equation (3) provides only a crude description of the system behaviour, but it gives useful insight into a possible instability mechanism. Let us assume that $\dot{v}$ is evaluated at depth h. Then in deep water waves

$$F = (- \omega^2 M e^{-\frac{\omega^2 h}{g}} + C) \eta \tag{4}$$

and two features about the wave force emerge. Firstly we see that, according to this simple approximation F is zero at a frequency slightly higher than the heave natural frequency defined by $\omega_1^2 = (C/M)$. Secondly, we may consider the second order effect of the motions y on the force F if we replace h by (h-y) in eq.(4). Expanding the exponential involving y as a series and retaining only the first two terms (up to the term linear in y), we obtain a modified equation of motion which may be written

$$\ddot{y} + \omega_1^2 (1 + r \cos \omega t) y = \frac{F}{M} . \tag{5}$$

We have defined

$$r = \frac{\bar{\eta} S_o}{(1 + a_m) V} \left(\frac{\omega}{\omega_1}\right)^4 e^{-\frac{\omega^2 h}{g}} \tag{6}$$

for a wave elevation at the position of the body given by

$$\eta = \bar{\eta} \cos \omega t \quad .$$

The homogeneous form of eq(5) is of course the Mathieu equation, solutions to which grow exponentially at certain frequencies. We have therefore a mechanism by which eq (3) can lead to dynamic instabilities, particularly in the region of practical significance $\omega \simeq 2\omega_1$.

The limitations of this theoretical model are readily apparent. A more rigorous linear model would be based on the values of A, B and F obtained by numerical solution of the wave diffraction-radiation boundary value problem. We have computed results for the flask model described below using the axisymmetric source distribution technique developed by Eatock Taylor and Dolla (4). The resulting vertical force and heave response amplitudes, per unit wave amplitude, are plotted against frequency in Figs 1 and 2 respectively. The wave force cancellation just above the predicted heave resonant frequency $\omega_1$ = 3.92 rad/s is readily apparent. It is also seen that at zero frequency F tends to a value of about 14.5 N per metre wave elevation: since for this body C = 14.4 N/m, this value of F is consistent with the low frequency limit of eq (4).

On this basis it seems plausible that eq (5) could form the starting point for investigating stability of the flask. Linear and quadratic damping were incorporated into the equation, and responses to regular and pseudo random waves were simulated by numerical integration of the non-linear equation of motion. Figure 3 shows a typical computed time history of unstable response to a regular wave. Analogue simulation of the damped equation of motion was also employed, and the full lines of Fig. 4 show stability boundaries obtained by this technique for regular wave excitation around twice the system natural frequency, for different values of damping ratio $\mu$.

PRELIMINARY MODEL TESTS

After some initial trials with different models, the arrangement shown in Fig. 5 was chosen, both to emphasize the instability and to minimise external damping. The latter was associated both with the simple instrumentation used to measure displacement (a rotary potentiometer) and the constraints imposed to restrict motion to one degree of freedom. The glass flask was attached at one end of a slender aluminium arm, the other end of which incorporated a pivot: on one side was a stainless steel shaft rotating in a nylon bush, while the potentiometer was mounted on the other side. The neck of the flask was enlarged in diameter and length by incorporation of a concentric piece of plastic tubing. This served both to optimise the underwater geometry in accordance with maximum instability predictions based on eq (6), and also to bring the natural frequency close to one half of the frequencies of the wave spectra peaks generated in the UCL random wave tank.

When the rig was mounted as a pendulum in air, it oscillated after an initial small displacement for over 100 cycles: friction in the pivot was very small. With the rig mounted in the tank, the total damping was observed from transient decay tests to be linear, corresponding to a damping ratio of 0.02. The natural frequency was found to be 3.84 rad/s, giving an added mass coefficient $a_m$ = 0.43.

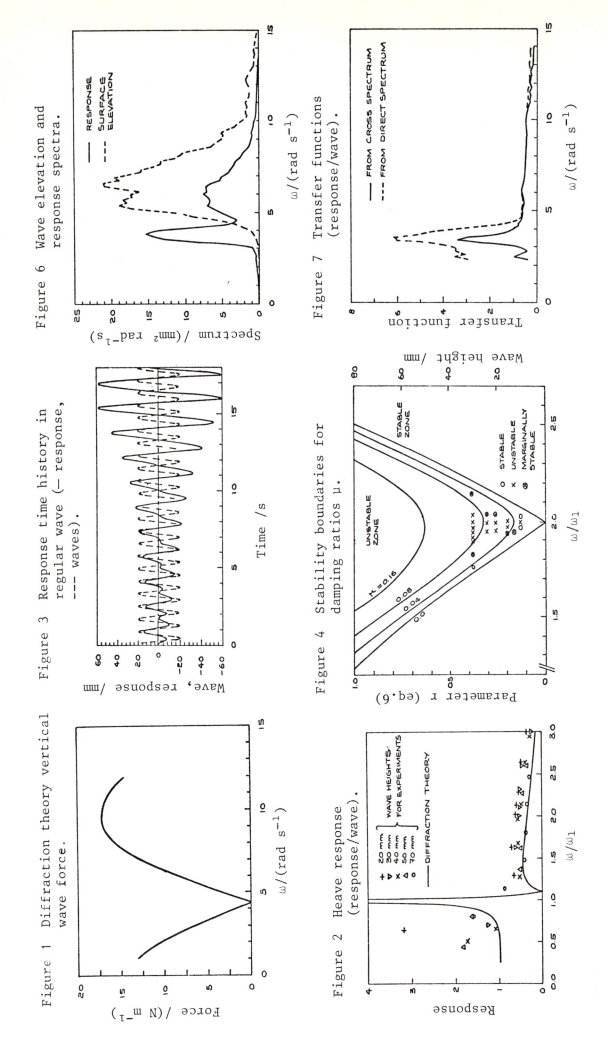

Figure 6 Wave elevation and response spectra.

Figure 7 Transfer functions (response/wave).

Figure 3 Response time history in regular wave (— response, --- waves).

Figure 4 Stability boundaries for damping ratios μ.

Figure 1 Diffraction theory vertical wave force.

Figure 2 Heave response (response/wave).

60

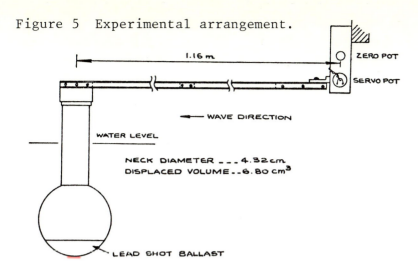

Figure 5  Experimental arrangement.

The added mass obtained from the source distribution analysis is almost constant at 0.265 kg over the frequency range 2 to 10 rad/s, corresponding to $a_m = 0.39$. The predicted radiation damping at resonance corresponds to a damping ratio of less than $10^{-4}$, although this ratio increases to 0.03 at 12 rad/s. It is clear that viscous damping must be important at resonance.

Regular wave tests were conducted over a range of wave heights from 20 to 120mm. Wave height and flask displacement were sampled at 25 Hz, leading to the experimental points on the transfer function shown in Fig. 2. The behaviour near resonance was,not unexpectedly, rather difficult to define, but at higher frequencies systematic trends could be discerned. Increasing wave height tended to decrease the post resonant response, but the reason for this is not readily apparent: the numerical simulations suggest that the effect of quadratic drag would be negligible throughout most of this frequency range.

The preceding experimental results were obtained during quasi-steady state oscillations, when there was no evidence of instability. Further regular wave tests were then conducted near a wave frequency of twice the natural frequency, and unstable responses were clearly defined. As described by Rainey (2), the vertical displacement of the flask increased in amplitude until limited by emergence of the bulb of the flask above the water surface. Time histories very similar to Fig. 3 were logged, and the stability boundary briefly investigated as in Ref. (2). Experimental points roughly indicating this boundary are plotted in Fig. 4.

RANDOM WAVE EXPERIMENTS

Tests were conducted for several long crested seas whose spectra had characteristic periods in the range 0.58 to 1.3s and significant wave heights between 30 and 40mm. No instability as such was clearly defined, but large responses at the resonant frequency were measured, even in waves having no apparent energy at that frequency. Figures 6 and 7 show typical results from a test in the mid frequency range; the characteristic period was 0.81s and the significant wave height was 30mm. Estimates of wave elevation and response spectra are given in Fig. 6, and estimated transfer functions in Fig. 7. The transfer function obtained from the wave and response direct spectra (broken line) differs significantly near resonance from that evaluated (solid line) from the cross spectrum; but over the rest of the frequency range agreement is very close, in this as in all of the

random wave tests. It is suggested that the discrepancy near resonance could be at least partly due to the same non-linear mechanism
that gives rise to instability in the regular wave tests. Above
5 rad/s, however, the transfer functions obtained from the random
wave experiments are similar to those predicted in Fig. 2 by the
linear diffraction-radiation analysis.

DISCUSSION

In an attempt to provide further insight into the behaviour of this
system in random waves, several digital simulations were performed.
These were based on the simple theoretical model described above,
and numerical integration of the equation of motion. Response spectral estimates from the resulting theoretical time series were compared with the experimental data for corresponding pseudo random
wave trains. Theoretical responses at resonance were again excited
by waves having no energy at the resonant frequency. The non-
linearity in the simple model provided thé appropriate sub harmonic
response, but the magnitudes of the simulated resonant responses
were considerably smaller than those obtained experimentally. This
is thought to be due partly to the imperfect wave force model used
in the simulation, and partly to the difficulty of estimating the
damping associated with motions in random waves.

Certain conclusions are nevertheless corroborated by both the physical and the mathematical models. The cause of the instability
appears to be the dependence of wave force on the position of the
body in the wave. In a regular wave at twice the natural frequency
of the system, transients become unstable and the response grows
exponentially at the natural frequency. The sensitivity to such an
instability depends on the draft of the body and the ratio of displaced volume to waterplane area [eq.(6)]. Viscous drag appears to
have a rôle in limiting subharmonic motions. Although the evidence
is far from conclusive, it seems unlikely that full scale floating
platforms would exhibit an actual instability of this form in real
seas. But their response spectra may exhibit significant energy
near resonance that is not predicted by linear theory. It is important to note that this could happen in spite of a platform being
designed to have such long natural periods that wave induced resonance is thought (following a linear analysis) to be precluded.

REFERENCES

(1)  ROWE, S J and JACKSON, G E, An experimental investigation of
     Mathieu instabilities on tethered buoyant platform models,
     NMI Report R73, (1980).

(2)  RAINEY, R C T, Parasitic motions of offshore structures,
     RINA Paper W3, (1980).

(3)  NEWMAN, J N, Marine Hydrodynamics, MIT Press, (1977).

(4)  EATOCK TAYLOR, R and DOLLA, J P, Hydrodynamic loads on vertical
     bodies of revolution, Trans RINA, Vol. 122, (1980).

D A Reed
*Department for Building Technology, National Bureau of Standards, Washington, DC 20234, USA*

## SUMMARY

Full-scale time series pressure-difference data collected on two hyperbolic cooling towers were analyzed by ARIMA time series methodology. ARIMA representation of the pressure process in the frequency domain was shown to be adequate. Transfer function models in the frequency domain using ARIMA expressions were found to provide a good characterization of the relationship between input wind velocity and output pressure-difference.

## INTRODUCTION

Certain full-scale data were analyzed by the ARIMA time series methodology [1] and by a spectral approach traditionally employed by engineers. The two approaches are shown to be consistent. Results for fluctuating pressure loadings as well as transfer function models in the frequency domain relating input wind velocity to output wind pressure-difference are discussed.

## FULL-SCALE DATA SOURCES

The time series considered in this paper consist of wind pressure-difference data recorded around the circumference of two reinforced hyperbolic cooling towers, Martin's Creek [3, 5, 7], located in the United States, and Schmehausen, located in West Germany [5]. A diagram of the Martin's Creek tower appears in Figure 1. The tower is of hyperboloid form, 127 m high, with transducer placement at 95.4 m. These transducers, 16 in all, are approximately equally spaced around the throat level. A diagram of the Schmehausen tower appears in Figure 2. The transducers were located at a height of 55.3 m on the 122.2 m-high tower. Only 12 of the total 16 transducers were operating on this tower.

## TIME DOMAIN ANALYSIS

### Pressure Loadings: Stagnation Point

Previous ARIMA analyses of the records for a single storm of the Martin's Creek data, and for the Schmehausen data, have shown that windward fluctuating pressure loadings on reinforced concrete cooling towers are AR(2) processes [3, 5, 7]. This paper summarizes the analysis of eight additional storm records for the Martin's Creek tower (last eight entries in Table 1). The results confirm the findings of References 3, 5 and 7. In addition it was found that the AR(2) model parameters are independent of the mean, $\bar{u}$, of the oncoming flow. Table 1 contains the AR(2) model parameters. The ratio of the residual variance, $\sigma_a^2$, to the pressure variance, $\sigma_p^2$, is in the range of 5 to 24 percent.

The coefficients are similar for all data sets. The average AR(2) model for fluctuating pressure, $p_t$, at the stagnation point, is given below:

$$p_t = 1.2181p_{t-1} - .2946p_{t-2} + a_t \qquad (1)$$

where $p_t$ = fluctuating component of the pressure loading at time t
$a_t$ = random shock at time t, "white noise."

Equation 1 may be rewritten in terms of the linear difference operator, $\nabla$, as [1, 7]:

$$(\nabla^2 + 2.134\nabla + .259)p_t = f_t \qquad (2)$$

where $\nabla^2$, $\nabla$ = the linear difference operators of order 2 and 1, respectively
$f_t$ = the forcing function, $f_t = 3.39a_t$.

If the forcing function $f_t = 0$, equation 2 represents the difference equation form of the free vibration of an overdamped single-degree-of-freedom linear oscillator with natural frequency, $\omega_n = 0.51$ rad/sec and damping ratio, $\zeta = 2.09$.

<u>Pressure Loadings: Transition and Rear Zones</u>

The ARIMA parameters for the pressure loadings display the same circumferential pattern for all of the recorded storm data. The transition zone, $110° \leq \theta \leq 150°$, average models are:

AR(2)    $\theta \approx 110°$ from windward

$$p_t = .4639p_{t-1} + .2863p_{t-2} + a_t \qquad (3)$$

AR(1)    $\theta \approx 132°$ from windward

$$p_t = .6496p_{t-1} + a_t. \qquad (4)$$

The AR(2) model with positive coefficients marks the beginning of the transition zone. This model was not apparent in previous studies; it is similar to the following average AR(2) model of the fluctuating wind velocity, $u_t$:

$$u_t = .6586u_{t-1} + .1454u_{t-2} + a_t. \qquad (5)$$

The AR(1) model of equation 4, frequently labeled as a "Markov" process, is consistent with previous results [5, 7]. These models reflect a transition of the flow from one type of process to another; physically, this region corresponds to a region of flow separation. The linear difference form of equation 4 is:

$$(1 + 1.85\nabla)p_t = 2.85 a_t. \qquad (6)$$

This model describes a first-order dynamic system with steady state gain equal to 2.85.

The average AR(2) model for the rear zone is

$$p_t = .9744p_{t-1} - .2348p_{t-2} + a_t. \qquad (7)$$

This AR(2) model is similar to that for the stagnation point, (Equation 1). The linear difference form of equation 7 is

$$(\nabla^2 + 2.154\nabla + 1.1090)p_t = f_t \qquad (8)$$

where $f_t = 4.26\, a_t$.

With $f_t = 0$, Equation 7 corresponds to the free vibration of an approximately critically damped single-degree-of-freedom linear oscillator with $\omega_n = 1.053$ rad/sec and $\zeta = 1.023$. In comparison with the windward system, $\omega_{n,\ rear} > \omega_{n,\ windward}$ and $\zeta_{windward} > \zeta_{rear}$.

## Transfer Function ARIMA Models

At the stagnation point, the ARIMA model relating input wind velocity, $u_t$, and output wind pressure-difference, $p_t$, is

$$p_t = \frac{\omega_o}{(1-\delta_1 B)} u_t + n_t \qquad (9)$$

where $\omega_o$ = input lag parameter
   $\delta_1$ = output lag parameter
   $n_t$ = noise component of the system
   $B$ = backwards difference operator; $Bu_t = u_{t-1}$.

Table 2 contains parameter values for the various data traces. In the absence of noise, the average model was found to be

$$p_t = \frac{.0231}{(1 - .7635B)} u_t. \qquad (10)$$

The linear difference form of equation 10 is

$$(1 + 5.48\nabla)p_t = .1158 u_t. \qquad (11)$$

This model reflects a first-order dynamic system [reference 1, Chapter 10]. ARIMA transfer function models in the time domain could only be obtained at the stagnation point. The Q values [1] for the models fit for the transition and rear zones indicated that the models were inadequate.

## FREQUENCY ANALYSIS

### RMS Pressures

The average RMS pressure ratio, $\sigma_p(\theta)/\sigma_p$, max, was plotted vs. $\theta$ in Figure 3. It also includes plots of the average $\phi_1$ and $\phi_2$ values. The figure shows a noticeable interdependence between $\phi_1$ and $\phi_2$; i.e., if $\phi_1$ increases, $\phi_2$ decreases, and vice-versa. The average $\phi_2$ value follows the trends of the average $\sigma_p$. Consistent with the findings of Hashish [4] and Sageau [6], the peak RMS value for most of the data traces occurred at $\theta \approx 88°$.

The RMS pressures were nondimensionalized using the method of Durbin and Hunt [2] and plotted in Figure 4. The nondimensionalized RMS pressure is defined as [2]:

$$\sigma_p^* = \sigma_p / \rho\, \bar{u}\, \sigma_u$$

where $\rho$ = the density of air

   $\bar{u}$ = mean of the upstream velocity

   $\sigma_u$ = variance of the upstream velocity

The turbulence scale, $\varepsilon$, is defined as

$$\varepsilon = Lx/a$$

where Lx = the upward integral scale of turbulence

   a = the radius of the structure

For the cooling tower data, $\varepsilon$, is in the range of intermediate turbulence and the intermediate scale values of $\sigma_p^*/\varepsilon^{1/2}$ fall inbetween those for large ($\varepsilon = 100$) and small ($\varepsilon \to 0$) scale turbulence. Figure 4 suggests that the expressions used to describe fluctuating pressures for small scale turbulence could be extrapolated to describe those for intermediate scale results.

## Spectral Expressions

The normalized spectral density functions, $g_{pp}(n)$, defined below, were obtained for all three regions:

$$g_{pp}(n) = S_{pp}(n)/\sigma_p^2, \quad n = \text{frequency [Hz]} \tag{12}$$

where $S_{pp}(n)$ and $\sigma_p^2$ = the spectrum and variance of the pressure, respectively.

In Figure 5, $ng_{pp}(n)$ for several storms vs. $na/\bar{u}$ (a = radius) at the stagnation point was plotted. It can be seen that the average ARIMA curve is consistent with the average of the spectral values calculated directly from the data. The best fit curve, constant x $(na/\bar{u})^{-1.33}$, closely follows the ARIMA curve. Similar plots for the transition and rear zones appear in Figures 6 and 7. Though it somewhat underestimates the average values, the AR(1) curve in Figure 6 broadly follows their general trend. The AR(2) curve closely follows the average of the spectral values obtained directly from the data through the Fourier transform for the rear zone. The correspondence between the ARIMA and the Fourier spectra is adequate.

## Transfer Function Models

The transfer function, $|h(n)|^2$, is defined by

$$|h(n)|^2 = g_{pp}/g_{uu}. \tag{13}$$

Expressions for $|h(n)|^2$ were determined for the three zones using the average ARIMA models (Equations 1, 4, and 7). The results are plotted in Figure 8 along with the quasi-static prediction, $|h(n)|^2 = 1.0$. Note that each curve must be mutiplied by the appropriate ($\sigma_p^2/\sigma_u^2$) to obtain the aerodynamic admittance. The actual transfer function at the stagnation point for the Martin's Creek storm with maximum mean velocity, u = 18.6 m/sec, is plotted for comparison. The close correspondence at the stagnation point with the ARIMA curve is noteworthy. It is seen in Figure 8 that (1) the quasi-static prediction is conservative for high frequencies, (2) an approximately constant relationship between input velocity and output pressure exists in the transition zone for high frequencies, and (3) the character of the transfer function is different for each zone.

## CONCLUSIONS

It has been shown that the ARIMA models describing full-scale fluctuating pressure loadings on cooling towers adequately characterize the data in both frequency and time domains. The fit of the transfer function model is best in the time domain at the stagnation point for the present data. The ARIMA frequency representation of

$|h(n)|^2$ offers a better approach than the ARIMA time domain representations to the modeling of the input velocity-output pressure loading system for the three tower zones.

REFERENCES

1.  BOX, G P and JENKINS, G M, Time Series Analysis: <u>Forecasting and Control</u>, Holden-Day, San Francisco, California (1976).

2.  DURBIN, P A and HUNT, J C R, "Fluctuating Surface Pressures on Bluff Structures in Turbulent Winds: Further Theory and Comparison with Experiment," <u>Proceedings of the 5th International Conference on Wind Engineering</u>, July 1979, Vol. 1, pp. 491-509.

3.  FORTIER, L J and SCANLAN, R H, <u>A Cooling Tower Wind Loading Model Based on Full-Scale Data</u>, Proceedings of the Fifth International Conference on Wind Engineering, Colorada State University. Fort Collins, Colorado, (J.E. Cermak, Ed.) Vol. 2, pp. 1217-1226 (July 8-14, 1979(.

4.  HASHISH, M G, <u>Wind Response of Hyperbolic Cooling Towers</u>, University of Western Ontario, Ph.D Thesis (September 1978).

5.  REED, D A and SCANLAN, R H, <u>Cooling Tower Wind Loading</u>, to be presented at the Fourth U.S. National Conference on Wind Engineering Research, Seattle, Washington (July 26-29, 1981).

6.  SAGEAU, J F, <u>In-Situ Measurement of the Mean and Fluctuating Pressure Fields Around A 122 Metres, Smooth, Isolated Cooling Tower</u>, Direction des Etudes et Recherches, Electricite de France (1979).

7.  SCANLAN, R H and FORTIER, L J, <u>Turbulent Wind and Pressure Effects Around a Rough Circular Cylinder at High Reynolds Number</u>, Research Report, Department of Civil Engineering, Princeton University, Princeton, New Jersey.

Table 1   Stagnation point pressure: AR(2) comparison

Model: $P_t = \phi_1 P_{t-1} + \phi_2 P_{t-2} + a_t$

| Storm | $\bar{u}$ (m/sec) | $\phi_1$ | $\phi_2$ | $\sigma_a^2 / \sigma_p^2$ |
|---|---|---|---|---|
| Schmehausen | 31.2 | 1.0070 | −.1683 | .24 |
| Martin's Creek | | | | |
| Reference 6 | 17.6 | 1.2800 | −.3410 | .17 |
| 4076 EV4 | 18.6 | 1.0592 | −.1653 | .08 |
| 4076 EV6 | 18.3 | 1.3020 | −.3393 | .05 |
| 4064 | 15.9 | 1.3303 | −.3933 | .07 |
| 4092 | 13.9 | 1.2835 | −.3900 | .12 |
| 4027 | 11.9 | 1.0880 | −.1394 | .09 |
| 4087 | 9.8 | 1.4353 | −.4831 | .05 |
| 4095 | 7.4 | 1.0710 | −.1479 | .13 |
| 4050 | 3.9 | 1.3248 | −.3779 | .07 |
| AVERAGE | | 1.2181 | −.2946 | .11 |

Table 2   Transfer function models at the stagnation point

| Storm | $\bar{u}$ (m/sec) | $\omega_o$ | $\delta_1$ |
|---|---|---|---|
| Schmehausen | 31.2 | .0273 | .8769 |
| Martin's Creek | | | |
| 4076 EV4 | 18.6 | .0317 | .4744 |
| 4076 EV6 | 17.6 | .0395 | .7394 |
| 4064 | 15.9 | .0235 | .6863 |
| 4092 | 13.9 | .0089 | .9269 |
| 4027 | 11.9 | .0079 | .8769 |
| AVERAGE | 18.1 | .0231 | .7635 |

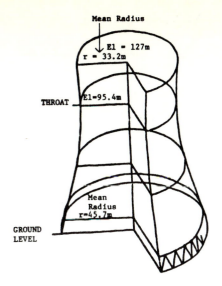

Figure 1.   Tower geometry,
Martin's Creek Tower.

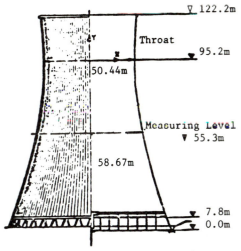

n = 144 (number of ribs)
k = 3.0 cm (height of each rib)

Figure 2.   Tower geometry,
Schmehausen Tower.

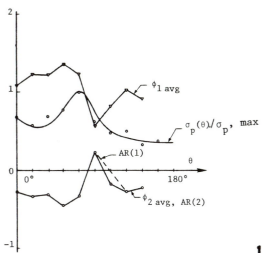

Figure 3.
Parameter comparison.

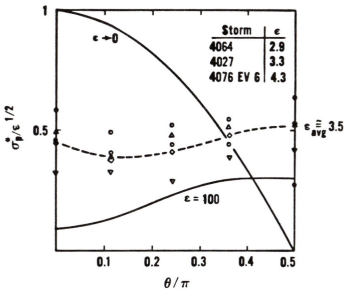

Figure 4.

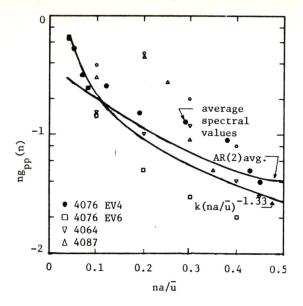

Figure 5. Spectral density comparison at the stagnation point.

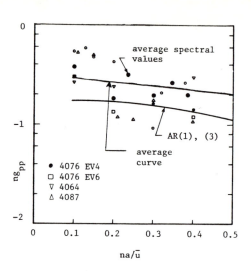

Figure 6. Spectral density comparison in the transition zone.

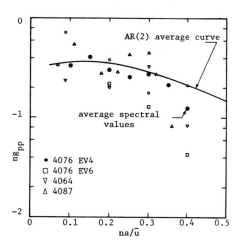

Figure 7. Spectral density comparison in the rear.

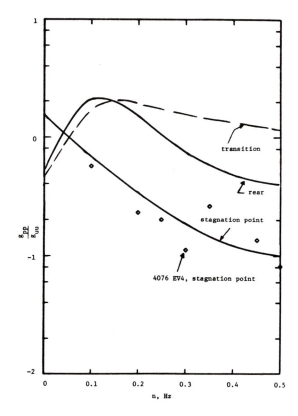

Figure 8. ARIMA transfer function model,
$$|h(n)|^2 = g_{pp}/g_{uu}$$
for the three flow zones.

70

# The use of small scale models to measure the dynamic wind forces on buildings

9

R A Evans and B E Lee
*Department of Building Science, Sheffield University, UK*

## INTRODUCTION

From time to time the technical press contains reports of serviceability problems associated with excessive building movement caused
by wind excitation. Experience has shown that only a few cases get
reported and that problems related to building response, including
occupant discomfort, noise, damage to partitions and lack of weather-
tightness are quite common.

The relationship between wind characteristics and building response
is a complex one involving an aerodynamic transfer function relating
the gust structure to the load function, and a mechanical transfer
function relating the load function to the response. Full aeroelastic modelling techniques have been evolved which enable the
designer to reproduce all the elements of this process in a wind
tunnel in order to assess the response of a proposed structure to the
local wind environment of its site and have been shown to produce
acceptably accurate predictions for both the along wind and the cross
wind response cases when compared with the available full scale data.
However the cost of this full modelling procedure for any particular
building has tended to be prohibitively high in all but exceptional
cases and the occurrence of unacceptably large dynamic responses in
modern lightweight buildings is sometimes the penalty paid for such
savings in the design process.

The other consequence of the full aeroelastic modelling method is
that it only enables the response for a particular set of structural
characteristics to be determined and that the intermediate step in
the transfer process, i.e. the load function, is not usually
characterized. If the load function defined in an appropriate manner,
were itself capable of independent determination as a function of
only the wind structure and the aerodynamic shape of the building,
then this would greatly assist our understanding of the nature and
origins of fluctuating wind loads. Furthermore, if a large body of
suitable loading function data were available and covered a wide
range of wind characteristics, building shape and common site interference problems the expensive requirement for full aeroelastic model
tests on specific buildings would diminish, provided appropriate
mechanical transfer functions could be deduced for a particular
design. In order to make some progress towards achieving this
objective, programmes of full scale and model dynamic wind load
assessment have been carried out at Sheffield University over the
past 6 years.

# FULL SCALE DYNAMIC WIND LOADS ASSESSMENT

The full scale measurement programme concerns the determination of the modal wind loads experienced by the 20 storey Arts Tower at Sheffield University. The methods used to evaluate these forces, including a full scale forced vibration calibration method, have been previously reported by Jeary, Lee and Sparks (1), where full details of the building's structural characteristics are given. The building is approximately 78m high, 36m wide and 20m deep and stands on sloping ground with its N - S axis, wide face, at -20° to true north. Structurally it consists of a cast in situ reinforced concrete core with deep floor slabs spanning between the core and external reinforced concrete columns. Non structural blockwork partitions are on each floor and the cladding is of both glass and lightweight panelling. The building is founded on piles driven into shale.

The full scale results are presented as values of modal force for sets of values of wind speed and wind direction, and are given in their original form in Table 1. It has been considered both necessary for comparative purposes and more informative, to alter the form in which these results are presented and the modified format is given in Table 2. An extensive survey of the wind structure in the area around the Arts Tower has been carried out, Evans and Lee (2), and indicates that it is not possible to define a representative mean wind speed at an effective height of 10m in terrain conditions where the obstruction size itself is of the order of 10m. Thus the wind speed data is re-presented as that measured at 84m, corrected for building - induced directional effects only.

In Table 1 the full scale force data are presented as "modal forces", where the definition of this term has been taken from Sparks and Crist (3), who state, "The force, $F_r^2$, exists for each normal mode and is a function of the frequency of that mode. It is in fact a measure of the wind force power spectral density function at the frequency of that mode". The values of the modal forces given in Table 1 are directly related to the peak values of the accelerometer response spectra multiplied by the dynamic building calibration factor determined from forced vibration test. The modified form of data presentation given in Table 2 presents the force measurements now as equivalent R.M.S. forces. The conversion of power spectral density function modal force values to those of equivalent R.M.S. force is given by Hurty and Rubinstein (4). For a lightly damped single degree of freedom system subjected to random excitation, it is found that

$$\overline{\sigma_r^2} = \frac{\overline{F_r^2} \cdot \pi \cdot f_r}{4 \zeta_r} \qquad \text{(Hurty and Rubinstein Eqn. 11.58)}$$

where  $\overline{\sigma_r^2}$  = modal mean square response, $(N^2)$

$\overline{F_r^2}$  = modal power spectral density function, $(N^2/Hz)$

$f_r$  = modal frequency (Hz)

$\zeta_r$  = Modal damping

In terms of the structural properties of the Arts Tower building this relationship becomes, for the N - S mode

$$\overline{\sigma_{N-S}^2} = \overline{F_{N-S}^2} \times 62.0$$

and for the E - W mode

$$\overline{\sigma_{E-W}^2} = \overline{F_{E-W}^2} \times 78.5$$

## WIND TUNNEL MODELLING SYSTEM

The model experiments have been carried out in the Sheffield University 1.2 x 1.2m Boundary Layer Wind Tunnel. The wind tunnel and its atmospheric boundary layer wind simulation system are described by Lee (5).

A rigid body modelling technique has been evolved by which the modal loads may be determined. The one-component balance system is designed to operate at specific model building resonances and consists of a series of strain gauged transducers which, in conjunction with an appropriate model building, resonate at predetermined frequencies. Additionally the system contains an integral variable damping mechanism thus enabling both model frequency and model damping to be varied independently. The model building is constructed from 3mm carbon fibre reinforced plastic side panels with dural end plates and is filled with polyurethane foam to inhibit panel vibration.

## WIND TUNNEL MODEL RESULTS

In order that a direct comparison can be made between equivalent R.M.S. forces measured on both model and full scale buildings the values of damping chosen for the model and building are nominally identical. The full procedure by which model forces measured under a variety of conditions of frequency, damping, bandwidth and wind speed can be normalized to represent those acting on a geometrically similar full scale building having arbitrary dynamic characteristics are currently under investigation, results to date being very encouraging. Such a normalizing procedure will be necessary in order to present dynamic wind forces in the form of design guide information.

In order to determine appropriate values for the frequencies of the model/balance transducer configuration in both N-S and E-W bending modes the model to full scale length ratio has been assumed to be 1:350 and the corresponding velocity ratio has been assumed to be 1:3. For the model results presented, this scaling ratio when applied to the wind tunnel speed of 9.4 m/s results in a corresponding full scale wind speed of 28.2 m/s at 84m, since the model and full scale modal frequencies are related by the relationship:

$$f_m = f_f \times \frac{U_m}{U_f} \times \frac{D_f}{D_m}$$

where    $f_m$ is the model modal frequency

$f_f$ is the full scale modal frequency

73

$\dfrac{U_m}{U_f}$ is the model to full scale velocity ratio

$\dfrac{D_m}{D_f}$ is the model to full scale length ratio

The joint choice of these scaling ratios implies that the N-S mode whose frequency is 0.68 Hz in full scale can be modelled using an 80 Hz balance transducer and the E-W mode, 0.86 Hz full scale, modelled by a 100 Hz balance transducer.

The model results for the isolated building in the simulated atmospheric boundary layer are shown in Figure 2 which depicts the variation of R.M.S. force with wind direction for the E-W mode, strong axis, and the N-S mode, weak axis. The interference effects caused by the surrounding buildings can be seen in Figure 3 where the variation with wind direction of the R.M.S. force for the N-S mode, weak axis, is shown both with and without a site model. The values of R.M.S. force presented in these figures are those measured directly by models. In order to convert these measurements to the appropriate full scale values the following relationship may be used,

$$\frac{\text{Model Force}}{\text{Full Scale Force}} = \left(\frac{U_m}{U_f}\right)^2 \times \left(\frac{D_m}{D_f}\right)^2$$

## DISCUSSION

At present the full scale data available for comparison is sparse and does not adequately cover the range of wind speeds and directions. From the wind tunnel results presented in Figure 2 it can be seen that the R.M.S. force values for the weak axis, N-S mode, vary from 0.0025 N at 360° through a minimum of 0.0015 N at 310° to 0.0020 N at 270°. After scaling up the equivalent full scale values are 2.75 KN, 1.65 KN and 2.2 KN respectively for an 84m wind speed of 28.2 m/s. The corresponding values for the E-W mode, strong axis, are 1.3 KN at both 360° and 270° and a minimum of 1.0 KN at 310°. Whilst a comparison of these figures with the nearest corresponding data in Table 2 indicates that the wind tunnel results are on the high side, it may be fairer to say that the scatter in the available full scale results is too great for a proper comparison to be made, other than that the model technique is certainly yielding results of the correct order. This conclusion alone is considered to be most encouraging particularly since some of the higher wind speed full scale data points have low confidence limits since they are based on a small number of wind storm records. So far as the variation of modal force with wind direction is concerned the model results depict maxima when the wind is normal to the faces and minima when the wind is on the corner.

The layout of the university site is given by Evans and Lee (2) who show the Arts Tower lies to the north east of the site surrounded by numerous low-rise buildings with two seven storey buildings to the south at distances of ∿ 100m and ∿ 200m. The effects of aerodynamic interference from the surrounding buildings may be seen in Figure 3. The more noticeable interference effects occur when the wind is to the south and west of the Arts Tower, where the results indicate an increase in the R.M.S. force thought to be caused by the increased buffeting on the Arts Tower from the wakes of the adjacent low rise buildings. Since this buffeting is restricted to the lower levels of the Arts Tower the large moment arm necessary to produce a more significant influence on the dynamic load is avoided.

## ACKNOWLEDGEMENT

This work has been performed under contract from the U.K. Building Research Establishment and is reproduced by their kind permission.

## REFERENCES

(1)  JEARY, A, LEE, B E and SPARKS, P R, 'The determination of modal wind loads from full scale building response measurements', Proc. 5th Int. Conf. Wind Engineering, Colorado State University Publ. by Pergamen Press, (1980).

(2)  EVANS, R A, and LEE, B E, 'The problems of anemometer exposure in urban areas - a wind tunnel study', Meteorological Magazine, (July 1981).

(3)  SPARKS, P R, and CRIST, R A, 'Determination of the response of tall buildings to wind loading. ASCE-EMD Speciality Conf. Dynamic Response of Structures, UCLA, (1976).

(4)  HURTY, W C, and RUBINSTEIN, M F, (1964), Dynamics of structures Prentice Hall Inc., New Jersey. (1979).

(5)  LEE, B E, 'The simulation of atmospheric boundary layers in the Sheffield University 1.2 x 1.2m boundary layer wind tunnel', Department of Building Science Report No. BS 38, Sheffield University (1977).

Figure 1

The Model – Balance

Transducer System

Table 1. Full Scale Dynamic Wind Loads - Original Presentation

| Mean Wind Direction (Building North) | Mean Wind Speed (10m height) m/s | Modal Force, $\overline{Fr^2}$ | |
|---|---|---|---|
| | | N-S$_1$ Mode kN$^2$/Hz | E-W$_1$ Mode kN$^2$/Hz |
| 254 | 9.3 | 6.3 | 2.2 |
| 254 | 11.1 | 8.5 | 1.7 |
| 275 | 8.5 | 3.4 | 1.8 |
| 275 | 11.8 | 35.2 | 6.5 |
| 295 | 9.1 | 8.3 | 2.8 |
| 295 | 11.8 | 24.1 | 6.5 |
| 305 | 7.0 | 3.4 | 1.5 |
| 305 | 9.6 | 15.2 | 4.3 |
| 305 | 11.4 | 23.4 | 6.1 |
| 305 | 13.3 | 27.0 | 6.7 |
| 338 | 10.0 | 3.4 | 0.3 |
| 338 | 11.9 | 6.7 | 1.1 |
| 338 | 13.3 | 8.4 | 1.7 |
| 350 | 8.1 | 5.8 | 0.6 |

Table 2. Full Scle Dynamic Wind Loads - Revised Presentation

| Mean Wind Direction (Building North) | Mean Wind Speed (84m height) m/s | Equivalent R.M.S. Force | |
|---|---|---|---|
| | | N-S$_1$ Mode (Weak Axis) KN | E-W$_1$ Mode (Strong Axis) KN |
| 254 | 20.2 | 0.62 | 0.42 |
| 254 | 24.1 | 0.73 | 0.36 |
| 275 | 18.9 | 0.46 | 0.38 |
| 275 | 26.3 | 1.48 | 0.71 |
| 295 | 22.9 | 0.72 | 0.47 |
| 295 | 29.7 | 1.22 | 0.71 |
| 305 | 18.3 | 0.46 | 0.34 |
| 305 | 25.0 | 0.97 | 0.58 |
| 305 | 29.7 | 1.20 | 0.69 |
| 305 | 34.6 | 1.29 | 0.72 |
| 338 | 22.9 | 0.46 | 0.15 |
| 338 | 27.2 | 0.64 | 0.21 |
| 338 | 30.4 | 0.72 | 0.36 |
| 350 | 20.5 | 0.60 | 0.30 |

## Figure 2   Variation of RMS force with wind direction, no site model

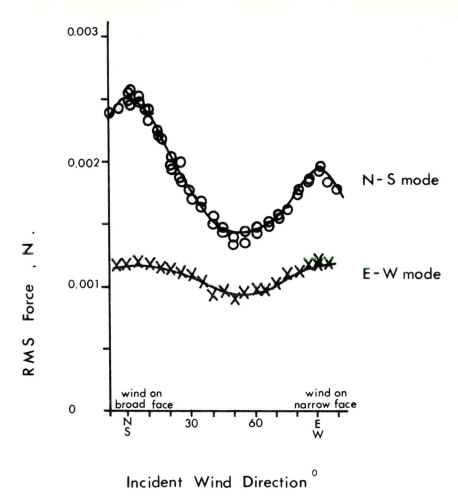

## Figure 3   Effect of site model on variation of RMS force with wind direction

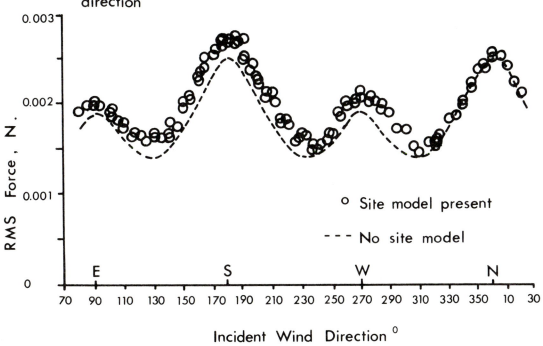

# Dynamic loading of a guyed mast

## 10

Claës Dyrbye
*Department of Structural Engineering, Technical University of Denmark*
Ejgil M Jensen
*Research Assistant of the Danish Electricity Supply Undertakings*
Mette Thiel Nielsen and Peter Nittegaard-Nielsen
*Department of Structural Engineering, Technical University of Denmark*

SUMMARY

A 3.7 m high model of a guyed radiomast has been constructed in stainless steel. The mast consists of 4 flanges
connected with diagonal bars. The mast is simply supported
at its base and further it is supported by two sets of
guys.

Figure 1  Main dimensions of model. Lengths are in mm.

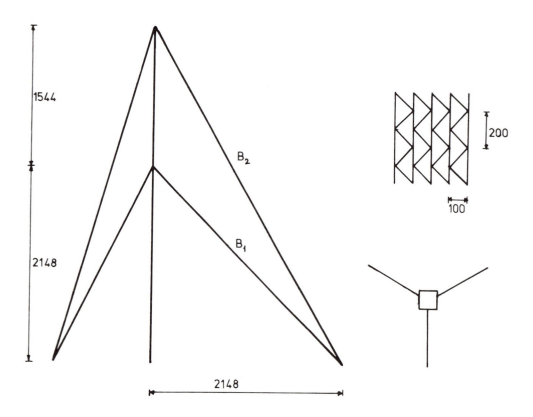

Two different problems have so far been examined by the model. The first problem was an investigation of the dynamic actions caused by a sudden rupture of a guy. The other was observations of deflections and cable forces due to a harmonic lateral force on the mast.

A main reason for model tests instead of calculations, or as a supplement to calculations, is that guyed masts are non-linear systems and therefore it is difficult to perform adequate mathematical modelling and calculations. The non-linearity, however, also gives rise to considerable difficulties in applying harmonic forces in the model tests. A harmonic force will introduce subharmonic and superharmonic responses, and so it was unavoidable that the force consisted of more than a single harmonics.

Figure 2    Photograph of the model. The shaker is to the left of the mast.

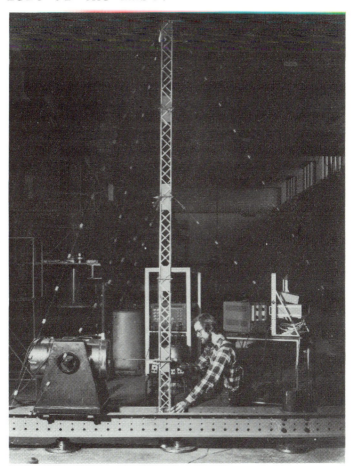

DESCRIPTION OF THE MODEL

The investigations were carried out as part of master theses in civil engineering. This caused limitations both with respect to duration and with respect to costs of the project. Thus it was necessary to construct the model from materials which could be provided without delay.

The model represents a Danish radiomast in geometric scale 1:15. The main dimensions of the model are shown in Fig. 1. The mast is designed as a four-sided truss in stainless steel. The flanges are angular sections and the diagonals are flat bars, connected to the flanges by cold rivets. The cables were 1.75$\phi$ mm circular steel wires.

In order to fulfil the model laws, both the mast and the guys were provided with additional masses as shown in Fig. 2.

## RUPTURE OF A GUY

Before testing the effects of a sudden rupture, the guy in question was replaced by a wire of less diameter and without additional masses. This wire was stretched such that the mast returned to the vertical position. The guy exposed to rupture was one in the upper set.

After the rupture, the structure has a new state of equilibrium. The structure performs free vibrations with the initial conditions that the velocity is zero and the deflections are the difference between the original and the new equilibrium shapes. During the vibrations, the sag of the cables will change, thus causing non-linear restoring forces. Calculations are further complicated from the fact that a combined in-plane and out-of-plane motion of the guys takes place.

Calculations were made by a finite-element method, here the mast was divided into 5 elements and each of the guys

Figure 3  Measurements after rupture of a guy
1: Deflections at lower point of guy attachment.
2: Deflections at the top.
3 and 5: Forces in symmetrically lower guys.
4: Force in lower guy in plane of the ruptured guy.
6 and 7: Forces in upper guys.

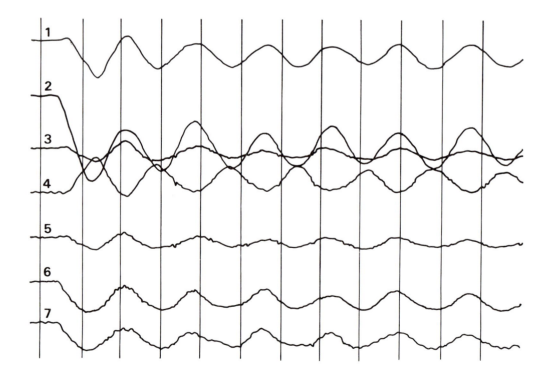

divided into 3 elements. This choice was motivated from
the limitations in time and economy, but it is recommended
to use a much more refined system for calculations,
especially for the guys. However, the calculations made so
far have proved useful as they demonstrate the ability of
the finite-element method to the problem in question.

It was considered most interesting to measure the cable
forces after a rupture. Therefore, strain-gages were glued
to all the guys and used as force-transducers. Each of the
transducers consisted of two BT-4-120 gages in series.
Special care was necessary when mounting the gages on the
rather thin guys.

Further, the deflections of the mast at the levels where
the guys are attached, were measured by means of deflection transducers, type Hewlett-Packard 7DCDT-1000.

Fig. 3 gives an impression of the dynamic effects following
immediately after the rupture. The deflections at the top
of the mast has a maximum value ab. 0.1 sec after the
rupture has taken place, and the motion ressembles a
damped vibration about a new state of equilibrium. The top
never returns to its original position.

Measurements of deflections at the lower point of attachment for guys show a different picture, as this point
roughly speaken performs a vibration with the original
position as the one extreme.

Figure 4   Force spectrum.   $f_0$ = 10.4 Hz, Amplitude 20 N.

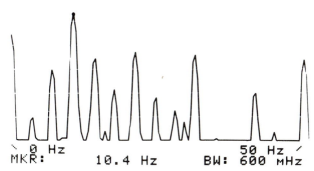

Figure 5   Force variation in a short guy. Generator
frequency 4.0 Hz. Figure shows predominant
frequency 8.0 Hz and beating.

## FORCED, HARMONIC VIBRATIONS

A Brüel & Kjaer shaker, type 4802/4 818 was placed in
horizontal position 72 cm above the ground support of the
mast, see fig. 2. A forcetransducer, B & K 8200, was
screwed to the mast and a horizontal bar connected the
shaker table with the force transducer. At the same level,
an accelerometer, B & K 4338, was attached to the mast.

Figure 6    Acceleration spectra at the top of the mast.
            Forcing frequency 13.6 Hz corresponds to
            maximum peaks. Upper picture: force amplitude
            10 N, lower picture: force amplitude 80 N.

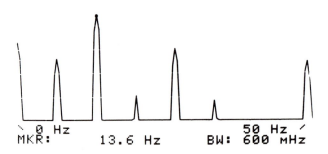

MKR:    0 Hz        13.6 Hz        BW:  50 Hz 600 mHz

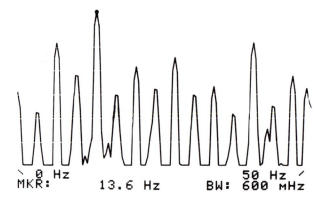

MKR:    0 Hz        13.6 Hz        BW:  50 Hz 600 mHz

The signals from these two transducers were amplified and
conditioned in B & K preamplifiers 2628 and 2626 resp.,
and taken to a signal generator B & K, 1026. An output-
signal from the generator goes to a power amplifier B & K
2708 which in turn gives power to the exciter 4802. This
circuit allows for harmonic forces with a predetermined
force amplitude.

Further instrumentation used at these tests were two B & K
accelerometers, 4338 and 4366 resp. with preamplifiers
B & K 2626, and the same strain gages as in the guy rupture
tests attached to the guys. Signals were analysed with a
spectrum analyser Hewlett Packard, type 3582 A.

The non-linearity of the restoring forces gave clear
indications at the tests. Thus, at some frequencies, it
became impossible to get a pure harmonic force from the
shaker. As an example, Fig. 4 shows a force spectrum,
where the frequency at the generator was $f_0$ = 10.4 Hz and
the force amplitude 20 N. While the predominating frequency
of the spectrum is $f_0$, there is clear indications of
spectral content at $2f_0/3$, $4f_0/3$, $2f_0$ and $3f_0$ so that
both subharmonic and superharmonic vibrations take place.

Fig. 5 shows a 5 sec record of the force in a short guy,
when the frequency generator gave $f_0$ = 4,0 Hz and force
amplitude 60 N. Apart from the beating, the dominating
frequency of the guy force is $2f_0$ = 8 Hz.

Fig. 6 shows spectra of the acceleration at the top of the mast when the frequency generator was tuned to $f_0 = 13.6$ Hz, and the force amplitude was 10 N or 80 N. At 10 N, subharmonic vibrations at $f_0/2$ and superharmonic at $2f_0$ are seen. At 80 N there is a remarkable superharmonic at $3f_0$ and besides, vibrations at multiples of $f_0/4$ are easily recognized.

REFERENCES

(1)   NIELSEN, M T & P NITTEGAARD-NIELSEN, 'Guyed Masts, Vibrations at Guy Rupture' (In Danish), Master thesis, Dept. of Struct. Engng., Techn. Univ. of Denmark, (1980).

(2)   JENSEN, E M, 'Guyed Masts, Forced Vibrations' (In Danish), Master thesis, Dept. of Struct. Engng., Techn. Univ. of Denmark, (1981).

# Chairman's summing up

H H Pearcey
*Head of Research, National Maritime Institute, UK*

This is an extremely interesting group of papers covering a range of subjects. The interest lies almost as much in what they convey about the whole science and application of modelling as in the information they give on their specific subjects.

Dr. Craig's paper starts us off with a very salutory reminder that the whole art and science of using scaled-model testing have to be continually worked on with a sharp eye on the fundamental processes involved. He identifies four distinct roles for model testing, namely, to predit the full-scale behaviour of some specific design, to study a specific failure in design or application, to conduct parametric studies in the derivation of designs or design rules, and to validate or calibrate mathematical models. He outlines some general principles of model testing, illustrated by some intriguing examples in geotechnics. These principles include: difficulties frequently occur when different physical properties obey different modelling laws and require ingenuity and flair in reconciling and in interpreting the results to extrapolate to full-scale; physical models frequently reveal new mechanisms that would not have been detected from field tests or conceived in the formulation of mathematical models; feed back from field tests is essential for the continuing validation of model-test techniques and of mathematical models; dynamic modelling is usually more difficult than static.

The two papers on the use of scaled models in hydrodynamics provide good illustrations of some of these principles. Thus the paper by Robinson, Murray and McCann demonstrates how ingenious and sophisticated experimental techniques can be combined with a thorough understanding of the phenomena involved to extract the maximum information. I was also impressed by the manner in which an apparent difficulty had cleverly been exploited to advantage. Thus, the fact the damping term had two elements which depended on different powers of one variable had been used to extract the separate influences. Eatock Taylor and Knoop show just how effectively a highly idealised experiment can be used to illustrate a phenomenan that had been deduced theoretically and thus to underwrite the use of the theory to study design implications. But they also show how some unexpected results reveal the presence of other phenomena that had not been incorporated in the theory - I refer to their results in irregular waves which may be associated with the slowly varying drift forces observed by others.

Although the applications are different, the papers by Jeary & Ellis, Boswell & Taylor, and Lerput and Narzul form an interesting group. They all deal with the use of mathematical models for the dynamic analysis of structures and the prediction of full-scale behaviour. They all illustrate that in seeking the objective of predicting the behaviour of prototype structures, one is never fully satisfied with the model and never relies on one single aspect of modelling. The model can nearly always be improved by the feed-back from field tests and by the use of data derived from well-conceived tests on scaled models. The papers demonstrate that it is the combination of all three (mathematical model, field tests and scaled models) that gives the engineer such a powerful tool. The paper by Jeary and Evans show how the mathematical model is constructed from a knowledge of the dynamics of the structure and how the model is used both in application to specific designs and for sensitivity analysis of the influence of imperfections in the structural data. This analysis in turn provides a guide to where research is needed and where integrity monitoring is most important. Lerput and Narzul deal with an application (jacket type offshore structure) for which the full-scale behaviour cannot be predicted either from model tests or from a mathematical model. Their object was to make field measurements to improve the mathematical model. Boswell and Taylor still have to make a comparison between their mathematical model and physical measurements.

Finally, the papers by Evans and Lee and by Dr. Reed illustrated the complementarity of small scale models and field tests for the wind loads on structures. Evans and Lee showed how necessary it is to calibrate and monitor models by reference to field measurement. Dr. Reed presents some field data and, in particular, new methods for analysing them.

# Small-scale modelling of reinforced concrete structures to resist earthquake motions

## 11

Daniel P Abrams
*University of Colorado at Boulder, USA*

SUMMARY

Modelling techniques used in a series of experimental studies are
evaluated. Tests of numerous ten-story small-scale structures are
described from a perspective of experiment design. Results of the
dynamic tests are presented to demonstrate the applicability of
experimental procedures, and to introduce the worth of the type of
testing for developing seismic design procedures. Moreover, results
of a separate series of tests of large- and small-scale components
(beam-column assemblies, walls, and columns) subjected to the same
distortion histories are presented to discern similarities in response
between models and actual structures.

## INTRODUCTION

Research on the behavior of multistory reinforced concrete buildings
is not new. Many investigators during the past two decades have
examined response of either physical or numerical models of structures
in an attempt to improve methods of analysis and design for dynamic
loads. Attempts which have proven fruitful have been encumbered by
limitations inherent in the type of model. Numerical models have
resulted in numerous solutions available for study, but have not
revealed forms of response other than those perceived initially by
the analyst. Physical models have generated new thoughts of behavior,
but have been limited because of doubts with regard to scaling.

This paper describes a series of studies using small-scale building
models (1,2,3,4), and another experimental study which evaluates the
usefulness of the models for extrapolation to actual structures.
The small-scale models were approximately one-twelfth scale and were
subjected to simulated earthquake motions. Components of the models
were tested as well as large-scale replicas of the components to
compare hysteresis characteristics and to assess differences
attributable to scale.

## EXPERIMENTAL STUDIES USING SMALL-SCALE STRUCTURES

### Design of Test Structures

Test structures were designed so that features of response might
suggest improvements for the design of actual structures. The
configuration (Fig. 1) of the structures and the base motions were
established so that response could be interpretted clearly with a
numerical model. Response was intended to be governed primarily by
flexure within a single plane. Masses at each of the ten levels

(not shown in figure) were made extremely stiff so that transfer of lateral load between the two three-bay frames and the central wall would be related solely to the relative stiffness of each component. Amounts of mass at each level were established to be consistent with intensities of base motions so that yield of reinforcement would occur. Simulation of gravity loads was not a consideration. Connections were designed so that story weights would be attached to frames and wall at discrete locations to simulate an ideal lumped mass condition.

Strength of members was established according to the amount of nonlinear deformation intended. Beams were reinforced lighter than would be according to a linear analysis using cracked-section stiffnesses and were expected to yield during the design earthquake simulation. Columns were designed to remain linear. Walls of different structures were designed to behave either linearly or nonlinearly and were reinforced either heavily or lightly. The design mehtod used was the substitute-structure method (5).

Materials were chosen to replicate those used in actual construction. Model concrete was a mortar consisting of cement, fine sand, and coarse sand in the proportions of 1.00: 0.96: 3.83. The water-to-cement ratio was 0.80. Nominal compressive strength was 32 MPa. Annealed plain wire with a nominal yield stress of 350 MPa was used for longitudinal reinforcement in beams, columns, and wall. Redrawing of the wire to attain a diameter of 2.3 mm resulted in a well defined yield point at a relatively high strain rate (0.005 strain per second). Amounts of reinforcement in beams and columns ranged from 0.5% to 1.1% of the cross-sectional area.

Experimental Program

The series of tests consisted of twelve ten-story specimens which were each subjected to three progressively increasing base motions. The simulated earthquake motions were tailored versions of ground motions measured at El Centro, California in the 1940 Imperial Valley Earthquake (NS component), and at Taft, California in the 1952 Tehachapi Earthquake (N21E component). Smoothed spectral-response curves representing these motions were used to establish scaling parameters of the simulated motions. Durations of the motions were scaled by a factor of 2.5 so that fundamental frequencies would dominate response, and cause structures to accelerate less with increasing amounts of damage. Amplitudes of the motions were scaled according to the design analysis to result in intended levels of nonlinear behavior. After each earthquake motion, a low-amplitude sinusoidal motion was input at the base which varied in frequency across the range of the fundamental frequency of the test structure. Additionally, before any forced-vibration test, each test structure was set into free vibration at a very small amplitude. Measurements consisted of accelerations, displacements, and forces resisted by the wall at each of the ten levels. A total of 48 waveforms were recorded for each dynamic test.

OBSERVED RESPONSE OF SMALL-SCALE STRUCTURES

Selected samples of measured response (Fig. 2) are presented to demonstrate the applicability of the experimental procedures, and to introduce the attractiveness of using a small-scale physical model.

Procedures used for design of the experiment can be verified by examining response of the test structures. Displacement and acceleration response (Fig. 2a, 2b, and 2e) indicated that the structures were excited within a reasonable range for purposes of the study. Displacements were on the order of one percent of the height of the structure for the design motion which was representative of the drift in an actual building. Maximum accelerations reflected an amplification of base acceleration as large as 1.6. Forces resisted by the wall (Fig. 2c and 2d) indicated that the wall was sized appropriately so that interaction with the frames would be perceptible. Amplitudes of force resisted by the wall were similar in magnitude to the total lateral force resisted by the overall structure. A reversal of force was observed at the tenth level indicating a relatively flexible wall at top and a stiff wall at base.

Frequency contents of measured response indicated that first and second modes were excited and that scaling of time was done appropriately. Examination of the force measurement at the sixth level (Fig. 2c) shows a much larger participation of the second mode than other measurements indicating that the test structures were sensitive to higher modes.

Observed response of the small-scale models revealed occurrences of nonlinear behavior as selected in the design process. Widths of cracks at ends of beams and at base of wall suggested that yielding of reinforcement had occurred at these locations. The size and pattern throughout the height of residual components of wall response (Fig. 2c and 2d) corroborated this observation. According to an influence study, force applied to the wall at the first level was found to be sensitive to rotation at the base of the wall, and forces applied to the wall at intermediate levels were sensitive to rotations at the beam ends at the same levels.

The relevance of using small-scale models for analysis of a particular structural form or loading may be seen with a few sample conclusions deduced from the measured response.
(a) Consideration of nonlinear behavior in the design process was economical with no loss of serviceability. Different structures designed with walls to respond linearly (Fig. 2a) or nonlinearly (Fig. 2b) deflected nearly the same amount despite a difference of four times the amount of reinforcement placed at the base of the wall.
(b) Displaced shapes of the structures were essentially the same for all amplitudes of motion suggesting that a single-degree-of-freedom approximation to response may be made with a nonlinearly behaving structure.
(c) Response could be characterized with modal properties despite behavior within the nonlinear range. Structures responded at frequencies which could be related to average slopes of member hysteresis relations. Frequency contents of measured response suggested modal participation as would be calculated with a linear analysis.
(d) Relations of base moment and displacement maxima could be approximated using a linear model suggesting a simplified design approach for nonlinearly behaving structures.
(e) Distributions of lateral forces resisted by individual frames or wall were not as simple as an engineer might think. Measured wall forces (Fig. 2c and 2d) contained components at many different frequencies and sizeable residual components which would make predictions of force distributions nearly impossible.

## COMPARISON OF LARGE- AND SMALL-SCALE RESPONSE

Although conclusions deduced from measured response may prove to be important to the development of analysis and design methods, engineers may be reluctant to revise established procedures on the basis of tests of such a small scale. A separate investigation has been initiated to study the usefulness of small-scale building models. The study has examined correlations between structures constructed at a small scale (6,7) and at a scale nine times larger. Beam-column assemblies have been constructed and subjected to slowly applied loading reversals to examine differences in resistance mechanisms that may be attributable to scale. Similarities in response of small-scale test structures and actual buildings have been examined using a numerical model with hysteresis characteristics established from either large- or small-scale experiments.

Differences attributable to scale have been evaluated in terms of the conclusions deduced from small-scale test structures. Salient results of the study are noted.

(a) Stiffness and strength characteristics of small-scale specimens were similar to those of large-scale specimens. Comparison of measured behavior (Fig. 3 and 4) which has been normalized revealed the same tendencies with respect to cracking, yielding of reinforcement, and slip of reinforcement in loading, unloading and load-reversal regions.

(b) The most significant difference because of scale was attributable to bond resistances of reinforcement to concrete. Steeper slopes in unloading regions and sharper reductions in load-reversal regions for small-scale specimens suggested that slip of reinforcement was more prevalent.

(c) Because of the problem of scaling bond resistance, it may be concluded that correlations in behavior of different size specimens may be more dependent on the particular type of configuration than on the scale. For example, interior beam-column assemblies (Fig. 4) did not model as well as exterior beam-column assemblies (Fig. 5). Demand for more bond resistance was inherent for the interior specimen because of the simultaneous push and pull on beam reinforcement across the width of the column.

(d) Comparison of the influences of large- and small-scale behavior on overall response can be extrapolated from the component tests using results of a nonlinear dynamic analysis (Fig. 5). The numerical model consisted of properties of the interior-joint specimen to obtain simple results which represented an upper bound comparison. A direct simulation was not observed, however conclusions deduced from the small-scale test structures appear to be valid for actual structures. Large-scale structures would behave more linear than small-scale structures, particularly in ranges of response with smaller amplitudes. Conclusions regarding the use of simple linear analysis methods to represent nonlinearly behaving structures could be extrapolated to actual structures.

(e) Because of differences in scaling different types of components (interior joints or exterior joints or even walls) overall response that is sensitive to relative stiffnesses of these components may be modeled incorrectly. Lateral forces resisted by the wall of a wall-frame structure would be suspect of mis-representation with a small-scale model. Conclusions made from such measurements should be regarded with caution.

## CONCLUDING REMARKS

Merits and limitations of using small-scale models to analyse the response of multistory reinforced concrete buildings to strong earthquakes have been discussed. It has been demonstrated using experimental results that a model structure may be designed which will be sensitive enough to examine the complex behavior of a reinforced conrete structure subjected to a transient motion at the base. Conclusions may be made from the test data of such a model to improve current analysis and design procedures and to stimulate future research. Substantiation of these conclusions for structures of a large scale may, in some cases be made, by comparing response of large- and small-scale components and extending the comparison to an overall structure using a numerical model.

## ACKNOWLEDGEMENTS

Earthquake-simulation tests using small-scale structures were done at the University of Illinois as part of a research project under the direction of Professor Mete Sozen. Funding was provided by the National Science Foundation. Tests of the usefulness of small-scale building models were done at the University of Colorado and were funded by the National Science Foundation, grant number PFR-80-07094. Credit is due to John Stewart, former graduate student, for testing and evaluating results of the large-scale components.

## REFERENCES

(1)  ABRAMS, D P, and M A SOZEN, 'Experimental Study of Frame-Wall Interaction in Reinforced Concrete Structures Subjected to Strong Earthquake Motions', Civil Engineering Studies, SRS No. 460, University of Illinois, Urbana, (May 1979).

(2)  HEALEY, T J, and M A SOZEN, 'Experimental Study of the Dynamic Response of a Ten-Story Reinforced Concrete Frame with a Tall First Story', Civil Engineering Studies, SRS No. 450, University of Illinois, Urbana, (August, 1978).

(3)  MOEHLE, J P, and M A SOZEN, 'Earthquake-Simulation Tests of a Ten-Story Reinforced Concrete Frame with a Discontinued First-Level Beam', Civil Engineering Studies, SRS No. 451, University of Illinois, Urbana, (August 1978).

(4)  MOEHLE, J P, and M A SOZEN, 'Experiments to Study Earthquake Response of R/C Structures with Stiffness Interruptions', Civil Engineering Studies, SRS No. 482, University of Illinois, Urbana, (August 1980).

(5)  SHIBATA, A, and M A SOZEN, 'The Substitute-Structure Method for Seismic Design in R/C', Journal of the Structural Division, ASCE, Vol. 102, No. ST1, (January, 1976).

(6)  KREGER, M E, and D P ABRAMS, 'Measured Hysteresis Relationships for Small-Scale Beam-Column Joints', Civil Engineering Studies, SRS No. 453, Univeristy of Illinois, Urbana, (August 1978).

(7)  GILBERTSEN, N D, and J P MOEHLE, 'Experimental Study of Small-Scale R/C Columns Subjected to Axial and Shear Force Reversals', Civil Engineering Studies, SRS No. 481, Univeristy of Illinois, Urbana, (July, 1980).

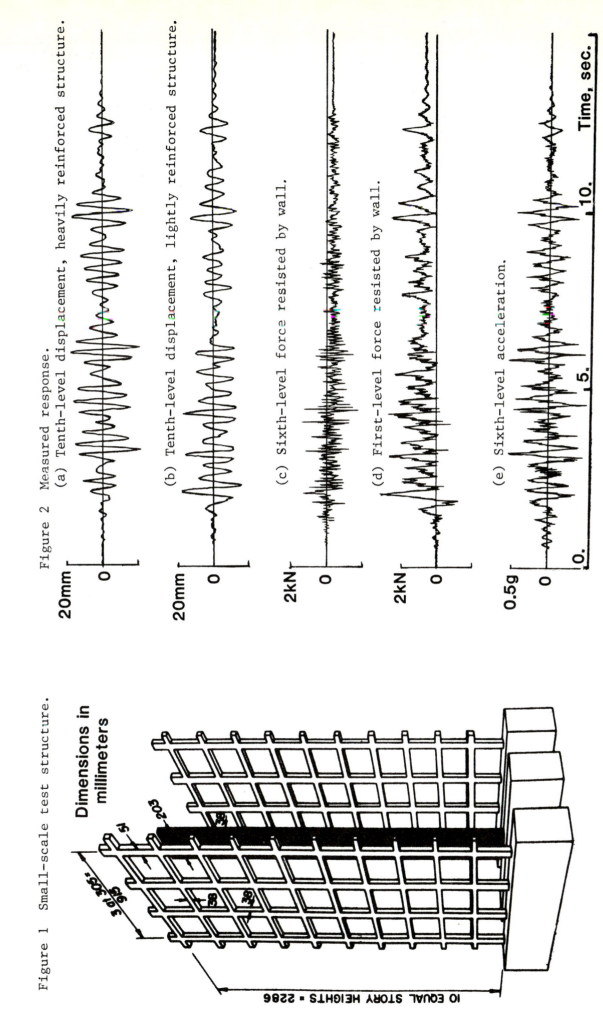

Figure 2  Measured response.
(a) Tenth-level displacement, heavily reinforced structure.
(b) Tenth-level displacement, lightly reinforced structure.
(c) Sixth-level force resisted by wall.
(d) First-level force resisted by wall.
(e) Sixth-level acceleration.

20mm    0
20mm    0
2kN    0
2kN    0
0.5g    0

0.    5.    10.    Time, sec.

Figure 1  Small-scale test structure.

Dimensions in millimeters

203
54
305
916
38
38
38

10 EQUAL STORY HEIGHTS = 2286

91

Figure 3   Comparison of behavior for exterior beam-column specimens.

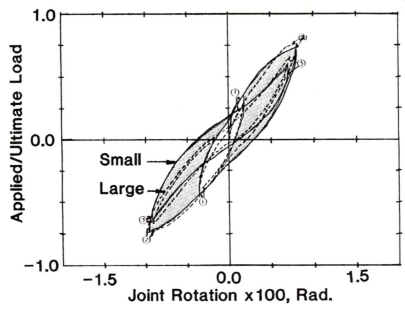

Figure 4   Comparison of behavior for interior beam-column specimens.

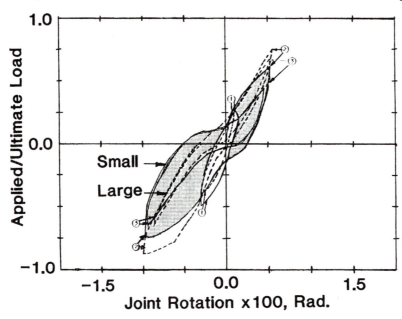

Figure 5   Comparison of response for large- and small-scale structures.

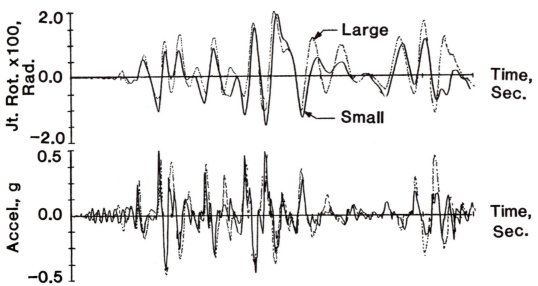

# Analytical and physical modelling of reinforced concrete frame-wall/coupled wall structures

## 12

A E Aktan and V V Bertero
*University of California, Berkeley, California, USA*

## SUMMARY

The common idealizations in the analytical modelling of R/C frame-wall structures subjected to earthquake excitation is discussed.  A 15-story frame-
coupled wall structure and a 7-story frame-wall structure, which are the
subjects of continuing integrated analytical and experimental research on the
seismic response of frame-wall systems at Berkeley, are used to exemplify
these idealizations.  Results of tests on a 4-1/2 story, 1/3 scale sub-
assemblage of a coupled wall system of the 15-story structure are used to
complement the assessment of the state of the art in analytical modelling.

## INTRODUCTION

Integrated analytical and experimental research on the seismic response of
R/C frame and wall/coupled wall structures has been in progress at the
University of California, Berkeley(4).  Difficulties encountered in correlating analytical predictions with experimental results indicated the need
to investigate the reasons of poor correlation.  This has been done by
reviewing the state of the art in predicting such response.

Objective of the Paper:  To present an assessment of the state of the art in
predicting the seismic response of R/C frame-wall structures.  Emphasis is
given to the importance of modelling in such prediction.

Scope:  To achieve this objective observations and results obtained in the
analytical and experimental studies concerning the seismic response of the
buildings shown in Figs. 1 and 2 are reviewed.  Emphasis is given to the
discussion of importance of realistic analytical modelling of the structure.
Results obtained from experiments conducted on a 1/3 scale 4-1/2 story
sub-assemblage of one of the coupled wall systems of the 15-story building
are also used to reinforce and/or complement some of the conclusions arrived
at from the assessment of the role of analytical models.

## ANALYTICAL MODELS

The analytical models of the two structures of Figs. 1 and 2 are shown in
Figs. 3 and 4.  These models were constructed in conjunction with the computer
code Drain-2D(7).  Analytical modelling is not an exact science and different
models of the same structures may be analyzed by the same computer code.  No
matter which code is used the engineer should recognize that what he is
analyzing is a model and not the real structure.  Therefore, the engineer
should be aware of the limitations and/or idealizations that were introduced
to model the real structure.  A brief discussion of some of the most
important idealizations follows.

### Interactions

(a)  Three-dimensional interactions.  These arise from the simultaneous
excitation of the ground in three orthogonal directions and/or three-
dimensional response of structures.  Studies have indicated that interactions
between bi-axial flexural effects may result in significant differences in
the planar response characteristics of R/C members, particularly when large
inelastic distortions are expected in columns and walls (2, 9).  Even in the
"elastic range" of R/C structures the interactions between the three translational components of the ground motions and/or the three-dimensional response of the structure lead to considerably different axial force time-
histories in the vertical members as compared with that obtained from planar

model response.  This different axial force history can significantly affect the stiffness strength, and deformational capacity of these members.  The vertical component of the ground motion can also affect the dynamic response of long span horizontal members.

Torsional Mode of Response.  In the actual building, this type of response can occur due to one or combinations of different sources:  torsional (rotational component) ground motion; eccentricity between centers of mass and resistance; and accidental torsion.  Computer codes which incorporate torsion in an idealized manner exist(6).  These codes neglect the interactions between torsional and flexural as well as axial distortions on the inelastic response of R/C which are complex phenomena requiring further research before realistic analytical modelling may be carried out.  In practice, the contribution of the torsional modes of response even for structures which are symmetric with respect to both principal axes is recognized through a specified accidental torsion(14).

(b)  Planar Interactions.  Even if the three-dimensional effects discussed above are neglected, the modelling of the planar behavior require several assumptions regarding actual planar interaction.  For example, in the planar modelling of the structural systems of Figs. 1 and 2, shown in Figs. 3 and 4, the following assumptions regarding planar behavior were made:  The floor systems at each level were assumed to be axially infinitely rigid diaphragms.  The floor slab and the transverse beams framing into the planar frame were assumed to have zero flexural stiffness.  These assumptions lead to equal horizontal displacements at each floor level as well as neglecting the axial effects imposed on columns by the transverse beams due to different vertical distortion characteristics of walls and frames at the same level.

The two analytical models shown in Fig. 3 were constructed to study the effects of frame-wall interaction on response.  In the coupled wall-frame model, all the frames were lumped in one, cut along the middle (central axis) and folded into two halves to reduce the degrees of freedom.  As the columns remained elastic, the altering of the axial force histories by folding the frame into two was justified.  Analyses of models in Fig. 3 using the El-Centro and Derived Pacoima Dam ground motions have indicated that the wall base shears were increased and the moment to shear ratios were decreased when the frames were included in the analysis(1).  Present Code(14) does not consider these effects of interaction.  The increased redundancy due to the frames, however, was beneficial in retaining the lateral stability of the structure once sufficient plastic hinges have developed in the coupled walls to convert them into a mechanism.

The interaction between the exterior and the interior frames as well as between the wall and the outriggering side columns of the interior frame in Fig. 4 is being investigated.  The outriggering side columns form a different lateral load resisting mechanism than the exterior frames, due to their geometry and relation with the wall.  Folding them or lumping them together with the other frames can lead to significant errors.

(c)  Structural and Nonstructural Component Interaction.  The interacting effect of the required external wall and partitions, as well as stairways have been neglected in the analytical model.  This usual assumption has been shown to lead to significant errors(3).

Deformation Characteristics of the Soil and the Foundation

By assuming infinitely rigid foundations, these effects were neglected in the models.  Preliminary studies including foundation rocking has shown that the inelastic deformation demands on the coupling beams can increase significantly due to this effect.  The uncertainties involved in modelling soil, foundation, and their interaction are even larger than those involved in the modelling of R/C structures.

Mass Characteristics

The total mass of each floor was lumped at that floor.  No rotational inertia was considered.  Sensitivity of response to different mass modelling schemes requires further investigation.

## Topological Characteristics of the Elements

One-dimensional geometry was assumed for the wall elements as well as the beam and column elements. Representing the wall elements by a number of two-dimensional finite elements may be considered desirable, however, the increases in computational cost are usually prohibitive. On the other hand, one-dimensional idealization of the wall members lead to limitations in simulating observed deformation patterns and failure modes of wall members. Furthermore, the fluctuation of the wall neutral axis cannot be represented and the different demands between the exterior and interior edge members of coupled walls cannot be differentiated between when one-dimensional idealization is used. Construction of models of more representative topology for at least the lower stories of the walls are required for a realistic simulation of response, as discussed in subsequent sections.

## Connections and Joints

Although experimental results indicate that R/C joints can undergo significant deformations(5), these joints were idealized in the analytical models by rigid connections and joints, i.e., the regions of the wall, beam and column elements at the joints were considered as axially and flexurally infinitely rigid zones. Deformable connection elements exist in the element library of Drain-2D, but these were not used. A softening effect may also be artificially simulated by decreasing the length of the rigid end eccentricities of the elements in Figs. 3 and 4. This is usually unjustified as the lack of experimental information renders any assumption a speculative one.

## Nonlinear Response

(a) _Geometric nonlinearities_. Preliminary estimation of the effects of these nonlinearities (beam-column and P-$\Delta$) show that they could be neglected for the two structures under consideration, particularly in the case of the 7-story structure.

(b) _Material nonlinearities_. These were introduced at the element level, by assuming that both the beam elements, which represent the beams of the analytical models, and the beam-column elements, which represent the column and wall elements, develop concentrated plastic hinges of zero length at each end. Such lumped plasticity models enable economically feasible time-history analysis as compared to distributed inelasticity models. The actual spread of yielding is an important response characteristic of R/C that should be considered in detailed studies of local behavior(11). A brief discussion of how the lumped plasticity model accounts for the two main nonlinear responses, axial-flexure and shear, follows.

Axial-Flexural Response: The beam-column elements in Drain 2D are two-component elements, incorporating the effect of axial force on flexural yield through a pre-determined axial-flexural yield interaction curve. Hysteresis is elastic-plastic with deformation hardening. The beam elements are single component. While degrading stiffness hysteresis characteristics are incorporated for the beam plastic hinges, a constant flexural yield level has to be specified. Studies that included the effects of changing axial force on the flexural yield level of these elements exist(10). However, none of these studies have incorporated the effects of the axial force on the flexural and shear stiffnesses, which was observed to be significant, particularly for accurate prediction of coupled-wall responses.

The correct estimation of the required input quantities for both types of elements demands a thorough understanding of section moment-curvature responses and axial-flexural interactions. In generating beam inputs, assumptions on effective slab width and a representative axial force level are required (this last can be particularly important for coupling beams). In general, the flexural rigidity and the deformation hardening slope to be used for modelling the behavior of the critical regions of the elements may not be directly transferred from the moment-curvature responses. As the moment-rotation response at these regions should include concentrated end rotations resulting from bond slippage of the main reinforcing bars which the moment-curvature computations do not include, a careful synthesis of section response is necessary to prepare realistic inputs for member response. Some computer codes recognize the contribution of concentrated end rotations by introducing a plastic hinge particularly for this phenomenon(8).

For the study of the models of Fig. 3 and 4, the primary moment-rotation and force-deformation relations of both types of elements were assumed to be bi-linear. The effects of incorporating a tri-linear primary curve for member response may be significant. Another idealization was regarding the axial force-axial distortion relations which were assumed to remain linear. Studies have indicated a strong coupling between axial and flexural responses(12) so this may be an erroneous idealization for coupled walls. Finally, the effects of bond slippage, shear and fluctuating axial force on flexural hysteresis were completely neglected. It is known that these effects can be of importance in the study of the flexural resistance and particularly stiffness degradations that have been observed in seismic response of real structures.

Shear Response: The shear response of both the beam and the beam-column members were incorporated by adding the elastic shear deformation terms to the appropriate terms in the stiffness matrix. It is usually assumed that the shear deformation continues to be represented by the elastic term even after flexural yielding of the wall (plastic hinge formation in the analytical model). This is not an adequate representation of the observed behavior. Test results have shown that as soon as flexural yielding occur, there is a considerable deterioration of shear stiffness and strength. The post-yield flexural deformation capacity depends upon the intensity of the shear stress at first flexural yielding. The larger the shear stress the smaller is the amount of this deformation capacity that can take place before a sliding shear failure mechanism starts to develop. This mechanism can limit the re-distribution of shear and moments as well as the strength and the deformation capacity to values considerably smaller than those that may be computed through the beam-column model. There were attempts to overcome this deficiency by introducing additional plastic hinges at the ends of the beam elements and defining a yield shear value for these hinges(13). In this manner, the shear force resulting from the moment gradient along a wall member is checked and limited to a certain value at which the shear distortions of these hinges become plastic. Although this model checks the level of shear in the wall members, it cannot represent the predominantly sliding deformation mode during inelastic response. Representation of this deformation pattern was observed to be a significant requirement for analytical modelling when reliable post-yield response information is sought. This is especially the case when the shear stress is high and the post-yield wall response governs the structural response.

As an illustration of the effects that the actual shear response can have, as well as the effect of planar interaction, some of the results obtained on the structure of Fig. 2 are presented. The distribution of shears and moments along the wall at 2.70 and 2.85 seconds of response to the Derived Pacoima Dam Motion are shown in Fig. 5. The plastic hinging pattern corresponding to these times are also indicated. The wall-frame interaction is observed to result in a significant lowering of the inflection point along the wall at 2.85 seconds, and correspondingly increase the shear force at the lower two floors, as compared to the response at 2.70 seconds. The wall-frame interaction is based on the different deformation mechanisms of the wall and the frame. If the shear stress at first flexural yielding would have been high, the incorporation of the predominantly sliding shear mode of response at the base of the wall would be significantly consequential on the wall-frame interaction, as the wall, in this case, would have deformation characteristics closer to that of the frame. Consequently, the redistribution of moments and shears after yielding of the wall would not have been as it is indicated by the results obtained.

Numerical Aspects of the Models

The integration of the equations of motion were carried out assuming constant acceleration in each time step(7) without iteration on element stiffnesses. Correct element states were determined after integration and any equilibrium unbalances were corrected during the subsequent time step. The error arising from these unbalances may be significant, depending not only on the time step but also on the nonlinear characteristics of the structural response. The time step should therefore be selected not only on the basis of the periods of structure, as it is usually done, but also considering the possible interaction between the numerical idealizations and the nonlinearities in response.

## PHYSICAL MODEL

A 4-1/2 story, 1/3 scale subassemblage of a coupled wall system of the 15-story building was tested under a load history synthesized from the generated Derived Pacoima Dam response of the structure(1). Equal lateral forces were applied to each wall of the test specimen as shown in Fig. 6. The four vertical actuators were coupled to the lateral actuators to apply overturning moments and coupling axial forces in addition to the gravity loads at the top of the test specimen, in order to satisfy the force boundary conditions imposed by the upper part of the structure on the subassemblage. During the test the internal forces at the midpoint of the girders were monitored by special force transducers, which enabled the analysis of the test specimen as a statically determinate system at any load stage(1).

The development of the crack patterns in the two walls were observed to be significantly different depending on the magnitude of axial forces acting in each wall. As a consequence of the different axial forces, the shear forces of the walls were also significantly different, and, all these led to differences in the deformation profiles of the two walls. During cyclic loading exterior edge member of the wall under compression had consistantly larger lateral displacement than the exterior edge member of the wall under tension. It was measured that the tension wall was only resisting 10% of the total base shear even at service load response.

## CONCLUSIONS

The testing of the physical model led to a number of observations that did not agree with the predictions obtained from the analytical model. The measured distribution of internal force and distortion in the test specimen were not represented by the analytical model. The most consequential idealizations in the analytical model that led to the discrepancies between the analytically generated and observed responses were: (1) Modelling the walls with one-dimensional elements which did not permit to incorporate the actually observed planar behavior of the edge members and of the panel of the wall as affected by axial, shear and flexure. (2) Neglecting the effect of axial force on flexural stiffness which led to equal amount of shear in the walls, and (3) Axially infinitely rigid diaphragm which led to equal lateral displacements.

Considerable advances in the state of the art regarding analytical modelling is required before representative magnitudes, distributions and time-histories of force and distortion throughout frame-wall structures can be analytically predicted during seismic response. Only comprehensively integrated experimental and analytical research can lead to improvements in the analytical modelling and therefore prediction.

## ACKNOWLEDGEMENTS

The research reported in this paper has been conducted under grants by the National Science Foundation. The authors are grateful to a large number of graduate students and technical staff that were involved in the conduct of this research.

## REFERENCES

(1)  AKTAN, A E, and BERTERO, V V, 'The Seismic Resistant Design of R/C Coupled Structural Walls', Report No. UCB/EERC-81/07, Earthquake Engineering Research Center, University of California, Berkeley, California, (1981).

(2)  AOYAMA, H, and YOSHIMURA, M, 'Tests of RC Shear Walls Subjected to Bi-axial Loading', Proceedings, 7WCEE, Istanbul, Turkey, (September 1980).

(3)  AXLEY, J W, and BERTERO, V V, 'Infill Panels: Their Influence on Seismic Response of Buildings', Report No. UCB/EERC-79/28, Earthquake Engineering Research Center, University of California, Berkeley, California, (1979).

(4)  BERTERO, V V, 'Seismic Behavior of R/C Wall Structural Systems', Proceedings, 7WCEE, Istanbul, Turkey, (September 1980).

(5)  BERTERO, V V, 'Seismic Behavior of Structural Concrete Linear Elements (Beams, Columns) and their Connections', State-of-the-Art Report, AICAP-CEB Symposium on Structural Concrete Under Seismic Actions, Bulletin D'information No. 131, CEB, (April 1979).

(6)  GUENDELMAN-ISRAEL, R, and POWELL, G H, 'Drain-Tabs, A Computer Program for Inelastic Earthquake Response of Three-Dimensional Buildings', Report No. UCB/EERC-77/08, Earthquake Engineering Research Center, University of California, Berkeley, California, (1977).

(7)  KANAAN, A E, and POWELL, G H, 'Drain 2-D, A General Purpose Computer Program for Dynamic Analysis of Inelastic Plane Structures', Report No. EERC 73-6 and EERC 73-22, College of Engineering, University of California, Berkeley, California, (final revision August 1975).

(8)  OTANI, S, 'SAKE- A Computer Program for Inelastic Response of R/C Frames to Earthquakes,' Civil Engineering Studies, SRS Report No. 413, University of Illinois at Urbana-Champaign, Urbana, Illinois, (1974).

(9)  PECKNOLD, D and SUHARWARDY, M, 'Effects of Two-Dimensional Earthquake Motions on Response of R/C Columns', Workshop on Earthquake-Resistant Reinforced Concrete Building Construction (ERCBC), University of California, Berkeley, (July 11-15, 1977).

(10) SAATCIOGLU, M, DERECHO, A T, and CORLEY, W G, 'Coupled Walls in Earthquake Resistant Buildings - Modelling Techniques and Dynamic Analysis', Report to the National Science Foundation, Portland Cement Association, (1980).

(11) SUCUOGLU, H and AKTAN, A E, 'Hysteretic Response of Reinforced Concrete Frames', ACI Special Publication SP-63, Reinforced Concrete Structures Subjected to Wind and Earthquake Forces, (1980).

(12) TAKAYANAGI, T and SCHNOBRICH, W C, 'Computed Behavior of Reinforced Concrete Coupled Shear Walls', Civil Engineering Studies, SRS Report No. 434, University of Illinois at Urbana-Champaign, Urbana, Illinois, (1976).

(13) TAKAYANAGI, T, DERECHO, A T, and CORLEY, W G, 'Analysis of Inelastic Shear Deformation Effects in Reinforced Concrete Structural Wall Systems', Nonlinear Design of Concrete Structures, Study No. 14, Solid Mechanics Division, University of Waterloo, (1980).

(14) Uniform Building Code, 1973, 1976 and 1979 Editions, International Conference of Building Officials, Whittier, California.

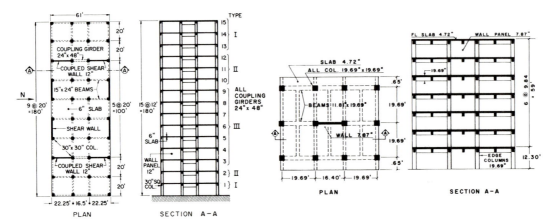

Figure 1  The 15-Story Prototype          Figure 2  The 7-Story Prototype

Figure 3 Analytical Models of the 15-Story Prototype

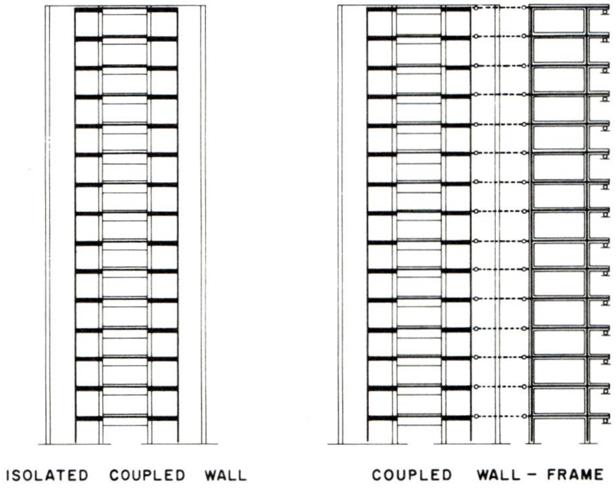
ISOLATED COUPLED WALL
COUPLED WALL - FRAME

Figure 4 Analytical Model of the 7-Story Prototype

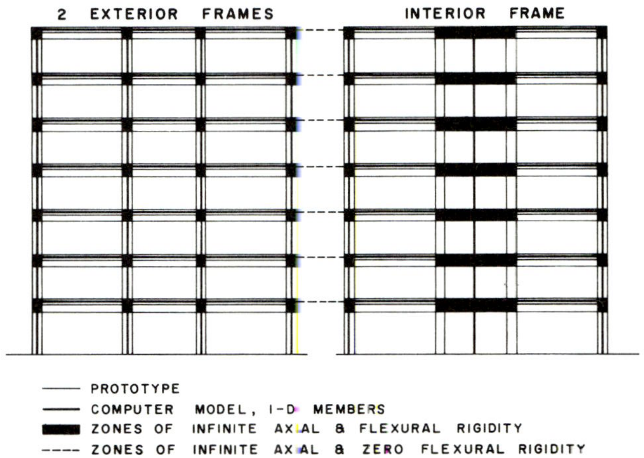
2 EXTERIOR FRAMES
INTERIOR FRAME
PROTOTYPE
COMPUTER MODEL, I-D MEMBERS
ZONES OF INFINITE AXIAL & FLEXURAL RIGIDITY
ZONES OF INFINITE AXIAL & ZERO FLEXURAL RIGIDITY

Figure 5 Shears and Moments of the 7-Story Wall at 2.70 and 2.85 Seconds of Derived Pacoima Dam Response

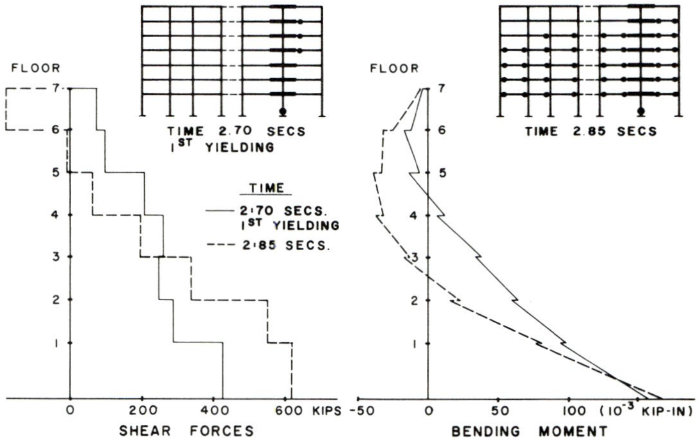
FLOOR
TIME 2.70 SECS
1ST YIELDING
TIME
2.70 SECS.
1ST YIELDING
2.85 SECS.
0
200
400
600 KIPS
SHEAR FORCES
FLOOR
TIME 2.85 SECS
-50
0
50
100 ($10^{-3}$ KIP-IN)
BENDING MOMENT

Figure 6 4-1/2 Story, 1/3 Scale Subassemblage-Physical Model

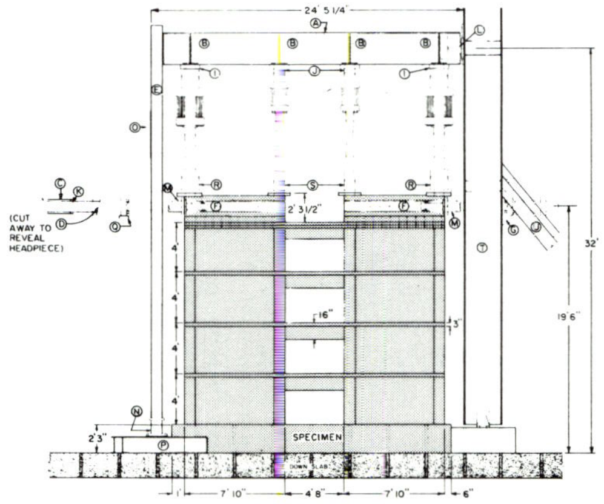
24' 5 1/4"
(CUT AWAY TO REVEAL HEADPIECE)
2' 3 1/2"
16"
32'
19' 6"
SPECIMEN
2' 3"
7' 10"
4' 8"
7' 10"
6"

# Contribution to discussion of Paper No. 12

A E Aktan and V V Bertero
*University of California, Berkeley, USA*

The authors would like to use the additional space for discussions to elaborate further on the results of the experimental investigation conducted on the physical model presented in Fig. 6 of the paper. One objective of the experiments was to <u>measure</u> the actual distribution of force and distortion in the test structure and compare this with analytically predicted distributions.

The total lateral force first-floor displacement envelopes of the two walls are presented in Fig. 7. The test structure was subjected to a lateral force history consisting of full cycles with full reversal of forces in the elastic range followed by predominantly half cycles of increasing peak forces, applied equally to both walls at the top of the 4-1/2 stories. The lateral displacements of specimens were recorded by measuring the horizontal displacement of the two external (extreme) edge columns at each floor level. It is the envelopes of total lateral force vs. the measured edge displacements at the first floor level which are illustrated in Fig. 7. As is clearly depicted by this figure, the lateral displacements of the wall in compression were consistently larger, indicating a substantial growth of the diaphragm system. This growth increased with each consecutive limit state: (1) flexural cracking of tension wall, (2) diagonal cracking of tension wall, (3) diagonal cracking of compression wall, (4) yielding of the coupling girders, (5) yielding of the tension wall, (6) yielding of compression wall and spalling of the concrete cover of its exterior edge column, (7) crushing of the panel of the compression wall due to the combination of high axial and shear stresses near the exterior edge column, and (8) failure of the exterior edge column of the compression wall.

It was measured that most of the growth of the diaphragm occurred along the wall in compression (approximately 60%). The growth along the beam was approximately 30% and that of the wall in tension approximately 10%.

Another significant observation from Fig. 7 regards the overstrength of the structure. The building shown in Fig. 1 from which the test sub-assemblage was obtained was designed with the provisions of the 1973 Uniform Building Code (UBC) for the factored base shear of 6.3% of the weight of the building. The actual lateral force capacity of the coupled wall system was obtained as 20.3% of the building weight.

The typical shear force redistributions at the base of the coupled wall system during a half cycle of loading is illustrated in Fig. 8. It is observed that the wall in compression attracts 85% of the total base shear even at the serviceability load level defined by 1973 UBC. The contribution of the wall in tension to the maximum base shear is only 10%. The average nominal shear stress at failure was $8.9\sqrt{f'c}$, based on the assumption that walls contribute equally, while in reality, the walls under compression and tension were subjected to $16.2\sqrt{f'c}$ and $1.6\sqrt{f'c}$ nominal shear stress, respectively.

I imagine the authors were motivated by a desire to persuade others to use the method and so kept their mathematics to a minimum, in such a circumstance I apologise for complicating the argument, my only justification being their comment that, "the implications (of the method) are not fully appreciated".

I believe the real merit of the technique lies in the analysis of highly repetitive substructures since Jennings (1) has clearly shown that the simultaneous or subspace iteration technique used in SAPIV (and other programs) to extract a small number of the important natural frequencies effectively performs the same steps as a Guyan reduction in its first iteration and more importantly improves its accuracy in each succeeding iteration.

Reference

1. Jennings, A. "Mass Condensation and Simultaneous Iteration for Vibration Problems", Int. J. Num. Meth. Eng., 3 pp, 13-24.

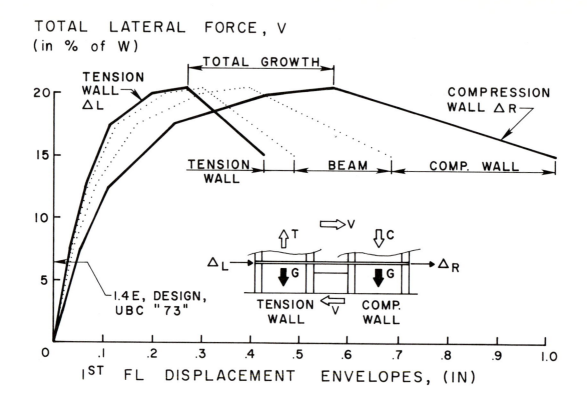

Figure 7  Diaphragm Growth

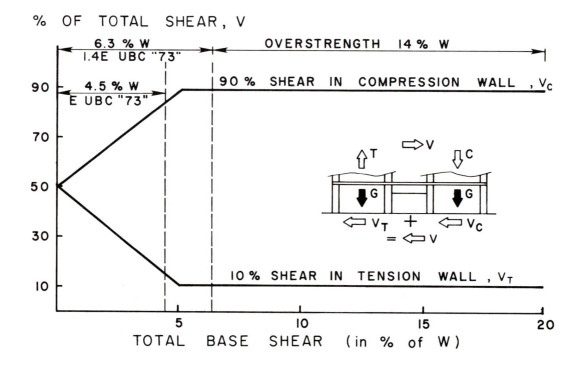

Figure 8  Redistribution of Total Shear

It is hoped that the above two figures and their discussion clarified and emphasized the significance of the conclusions formulated in the paper.

102

# Earthquake resistance of horizontal joints and simple shear walls in precast concrete large panel buildings  13

H G Harris, B E Abboud and G J J Wang
*Department of Civil Engineering, Drexel University, Philadelphia, PA 19104, USA*

## SUMMARY

This paper presents the results of an experimental program dealing with the seismic resistance of precast concrete large panel (LP) buildings. The development of small scale modeling techniques to study the cyclic shear behavior of horizontal joints and isolated shear walls of typical LP building construction is described. Results of sixteen complete horizontal joints and six five-story shear walls tested under monotonic and reversed shear loading is discussed. Exploratory vibration test data on three-story and five-story precast shear walls using a small shake table facility are also presented. Information on the character of the shear versus slip, lateral displacement and panel separation was obtained as well as the effect of cyclic degradation as a function of the amount of vertical tie steel. As a result of the experimental study, it was found that the model interior horizontal joint showed low energy absorption capability before failure and an increase in energy absorbing ability at lower shear capacity after large slip had occurred. However, this increased energy absorbing capability was not a stable one. Results of the five-story shear walls, loaded under a quasi-static earthquake load distribution showed large changes in cyclic shear resistance and ductility with increasing tie steel area. The vibration testing of small scale model shear walls indicated that these are very stiff structural systems when vibrating in their elastic range.

## BACKGROUND

Large panel industrialized concrete buildings have become steadily more viable forms of construction in the United States in the last few years (1), (2). Their increasing use and extension into the more active seismic regions of the country requires an understanding of their behavior under dynamic loading. Experimental data of this type, although meager, are available for some types of large panel systems used in other parts of the world. The construction details of American type large panel (LP) buildings however, differ in many ways from these. The need therefore, exists for studying in depth the dynamic characteristics of these relatively new forms of residential construction which gives promise of lowering the cost of housing through the industrialization process.

Large panel reinforced and prestressed concrete buildings are basically bearing wall structures arranged in a box type layout. They consist of precast vertical wall panels with precast floors and roofs of panels or planks. Large panel buildings for residential applications are predominantly of the cross-wall construction, where the load bearing elements are placed perpendicular to the longitudinal axis of the structure.

In most LP systems, the bearing walls transfer their loads directly to the sub-structure without an intermediate frame. This form of construction is well suited for multistory housing applications where walls of substance are needed between dwelling units to ensure fire resistance and noise suppression. Under seismic loading, LP systems remain mainly elastic except for the joint regions which undergo yielding and slip. The joint regions become critical therefore, in their ability to absorb the input energy.

HORIZONTAL JOINTS

Typical layouts of precast concrete LP buildings of the cross-wall type are shown in Fig. 1. A variety of simple and coupled shear walls can be mobilized in such structures for the purpose of resisting the horizontal components of an earthquake. In this study focus is made on the simple shear wall (Fig. 1(b)), and its horizontal wall-floor cast-in-place joints.

The principal function of a horizontal wall-floor joint, is to transfer the applied vertical and horizontal loads from the wall panel above and the floor slabs at the joint to the lower panel or foundation, and to prevent the differential displacement between adjacent bearing wall panels in the horizontal direction. In general, the horizontal joint (Fig. 2) will be transmitting the vertical and horizontal forces or a combination of these forces, which will be acting on the joint, to the adjacent wall panels. To be viable, therefore, as an energy dissipative mechanism during an earthquake, the horizontal joint must be able to show stable hysteretic behavior without significant pinching under cyclic shear loading.

The direct small scale modeling technique was used in this experimental study to investigate the general behavior of the interior horizontal joints designed for vertical loading and to satisfy progressive collapse requirements while subjected to simulated seismic loading. The validity of the small scale direct modeling techniques was demonstrated in earlier work (3), (4). Based on past experience, a number of factors which could influence the model joint type, size and configuration were considered in the model preliminary design stages. First, the use of interior "American" type horizontal joint in this experimental study was due to the fact that this type of joint is the most predominant in large panel structures in the United States, and also there is no available test data on the behavior of complete interior horizontal joints subject to repeated cyclic loading.

A total of sixteen 3/32 (1/10.67) scale models of the interior American type horizontal joint (Fig. 2) were fabricated and tested (5). The main variables investigated toward analyzing the general trends of this type of joint were the loading history which was either monotonically increasing load or cyclically reversing load, the magnitude and distribution of normal forces across the shear plane and the amount of reinforcement across the shear plane (vertical ties). The load set up used in testing the horizontal joints is shown in Fig. 3. Typical load vs slip behavior in the pre-slip and post-slip regions is shown in Fig. 4. Note the relative instability of the hysteretic behavior when the joint goes plastic (Fig. 4(b)). Significant loss of shear strength and stiffness was demonstrated by the cyclically loaded specimens as compared to their monotonically loaded companion specimens (Fig. 5).

## SIMPLE SHEAR WALLS

An LP simple shear wall (Fig. 1(b)) can be characterized as a cantilever beam and is expected to behave essentially in the same manner
depending on its slenderness ratio. The shear wall (Fig. 6) as a
single stack cantilever, will be subjected to bending moments and
shear forces originating largely from lateral loads and to axial compression caused by gravity loads. The reinforcement and construction
details for the individual walls are shown in Fig. 7. The individual
walls were assembled in a vertical stack fashion similar to prototype
construction with the vertical tie reinforcement attached to both
panels by means of a 1" square tube (Fig. 7). A triangular quasi-
static horizontal loading, approximating the first mode inertial
loads, was applied to six five-story 3/32 scale models through a
"whiffle" tree arrangement as shown in Fig. 8 (6).

Mechanical jacks were chosen to perform the push-pull type loading.
In order to restrain the shear wall model to move in a single vertical
plane during testing, a set of rollers were placed on opposite faces
at each story level and attached to a rigid vertical frame as shown
in Fig. 9. The restraint provided by this guide system was minimal.
The instrumentation for measuring the horizontal, sidesway and lift-
up displacements consisted of sixteen dial guages having an accuracy
of 0.01 mm. Three different sizes of continuously threaded rod
vertical ties having the properties shown in Table 1 were used for
the 6 specimens tested in companion sets, one monotonic and one
cyclic.

The load-deflection curves of the three monotonically loaded specimens are shown in Fig. 10. A gradually curving nonlinear behavior is
indicated, reflecting the stress-strain characteristics of the
threaded rod vertical ties used in order to facilitate construction
of the models. Cyclic shear behavior is typified by the hysteretic
curves obtained for the intermediate size vertical ties shown in
Fig. 11. Failure for this specimen occurred during the beginning of
the 18th cycle. A summary of the shear wall model tests is given in
Table 2.

A considerable reduction in strength due to cycling was indicated by
the two larger vertical tie sizes. The smallest tie size showed no
appreciable reduction in strength due to load reversal. The mode of
failure of the two specimens with intermediate size vertical ties is
shown in Fig. 12. In both of these tests failure occurred through
rupture of the vertical tie, however, note that considerably more
cracking occurred in the companion cyclically loaded specimen (Fig.
12 (b)). Quasi-static tests such as the above are very useful prior
to earthquake loading on the shaking table because they may point to
potential problems.

A series of vibration tests on uncracked three-story and five-story
shear wall models were conducted on a small 4' x 6' electro-magnetic
shaking table facility. Dynamic similitude was preserved in the 3/32
scale models by the addition of steel ballast plates as shown in
Fig. 13. The walls were stabilized on the shaking table by means of
two "whiffle" trees with an adjustable tension arrangement (Fig. 13).
The dynamic test setup is shown in Fig. 14.

Frequency response curves were obtained for each model tested in
order to determine the resonant frequencies more accurately and to
calculate the damping coefficients. First mode shapes were easily
measured, but higher modes were harder to identify. A comparison of
the first mode shapes of the five-story models with two different

sizes of vertical ties is shown in Fig. 15. The static mode shapes shown in Fig. 15 were obtained from the inverted triangular load tests at three different load levels. The solid lines representing a load distribution in the elastic range at .04 P and .05 P appear to correlate best with the dynamic test results.

## CONCLUSIONS

Based on the present experimental investigation the following general conclusions can be made:

(1) The ultimate shear capacity of the horizontal joint subjected to cyclically reversing shear loading is lower than that of an identical joint subjected to monotonically increasing shear. The reduction to the shear strength due to cyclic loading was found to vary in general between 5 and 33 percent depending on the parameters investigated.

(2) Before failure, the joint subjected to cyclically reversing shear indicated a low energy absorption due to the nature of the pinched shape of the hysteresis curves for a complete cycle of loading. After failure (very large slip during which applied shear cannot be increased), the joint had a rectangular shape of hysteresis loops indicating a substantial increase in energy absorbing ability. However, this elasto-plastic behavior was not stable in all the cyclically loaded specimens. It is anticipated that another type of vertical tie steel reinforcement, with different yield characteristics than the threaded rods used for convenience in this program, will alter the above conclusion.

(3) Vibration testing of 3/32 scale model shear walls indicates that these are very stiff structural systems when vibrating in the elastic range.

(4) Quasi-static tests simulating an earthquake type loading applied to six five-story shear wall 3/32 scale models show large changes in cyclic shear resistance as the area of the tie steel is increased up to 3.6 times the value needed to meet the design requirements for progressive collapse.

(5) As the area of vertical ties is increased, the shear resistance of the models became more ductile and the hysteresis loops became more pinched.

(6) The ratio of cyclic to monotonic shear resistance decreased as the vertical tie steel area was increased, however, the total shear load (both monotonic and cyclic) was increased.

## REFERENCES

(1) BERTERO, V V, Organizer, "Proceedings of a Workshop on Earth-quake-Resistant Reinforced Concrete Building Construction," (July 11-15), 1977, Vol. III.

(2) APPLIED TECHNOLOGY COUNCIL, "Proceedings of ATC-8 Seminar on Design of Prefabricated Concrete Buildings for Earthquake Loads," (April 27-28, 1981), Los Angeles, Applied Technology Council, Berkeley, California.

(3) HARRIS, H G and MUSKIVITCH, J C, "Report 1: Study of Joints and Sub-Assemblies-Validation of the Small Scale Direct Modeling Techniques," Nature and Mechanism of Progressive Collapse in Industrialized Buildings, Office of Policy Development and Research, Department of Housing and Urban Development, Washington D.C, (October 1977), also Department of Civil Engineering, Drexel University, Philadelphia, PA 19104.

(4) HARRIS, H G and MUSKIVITCH, J C, "Models of Precast Concrete Large Panel Buildings," Journal of the Structural Division, ASCE, Vol. 106, No.ST 2, Proc. Paper 15218, (February 1980), pp 545-565.

(5) HARRIS, H G, and ABBOUD, B E, "Cyclic Shear Behavior of Horizontal Joints in Precast Concrete Large Panel Buildings," Design of Prefabricated Concrete Buildings for Earthquake Loads, (April 27-28, 1981), Los Angeles, Applied Technology Council, Berkeley, California.

(6) HARRIS, H G, and WANG, J J G, "Static and Dynamic Testing of Model Precast Concrete Shear Walls of Large Panel Buildings," ACI Special Volume Dynamic Modeling of Concrete Structures, to be published by American Concrete Institute, Detroit, Michigan.

ACKNOWLEDGEMENT

The results of this study were supported by the National Science Foundation under Grant No. NSF-79-24723 and the internal research and development effort of the College of Engineering, Drexel University.

TABLES

Table 1  Vertical Tie Properties

| Model No. | Vert. Tie Design | Area at Thread Root, in$^2$ | Tie Steel Per Cent | Yield Str. ksi | Ult.Str. ksi |
|---|---|---|---|---|---|
| V, VI | 6-32 | 0.00891 | 0.074 | 77 | 79 |
| II,III | 10-24 | 0.01755 | 0.146 | 74 | 81.5 |
| I, IV | 1/4"-20 | 0.03173 | 0.266 | 79 | 83.2 |

Table 2  Summary of Test Results

| Shear Wall | Vert. Tie | Total Horiz. Load Mono. (lb) | Cyclic (lb) | O.T. Moment Mono. (in.-lb) | Cyclic (in.-lb) | Total No. of Cycles | Mode of Fail. |
|---|---|---|---|---|---|---|---|
| 1 | 2 | 3 | 4 | 5 | 6 | 7 | 8 |
| I | 1/4"-20 | 2500 | -- | 89,650 | -- | -- | TUBE |
| IV | 1/4"-20 | -- | 1900 | -- | 68,140 | 18 1/2 | TUBE |
| III | 10-24 | 2100 | -- | 75,301 | -- | -- | TIE |
| II | 10-24 | -- | 1800 | -- | 64,550 | 17 1/4 | TIE |
| V | 6-32 | 1200 | -- | 43,030 | -- | -- | TIE |
| VI | 6-32 | -- | 1200 | -- | 43,030 | 11 1/4 | TIE |

Fig. 1 Typical LP plans.

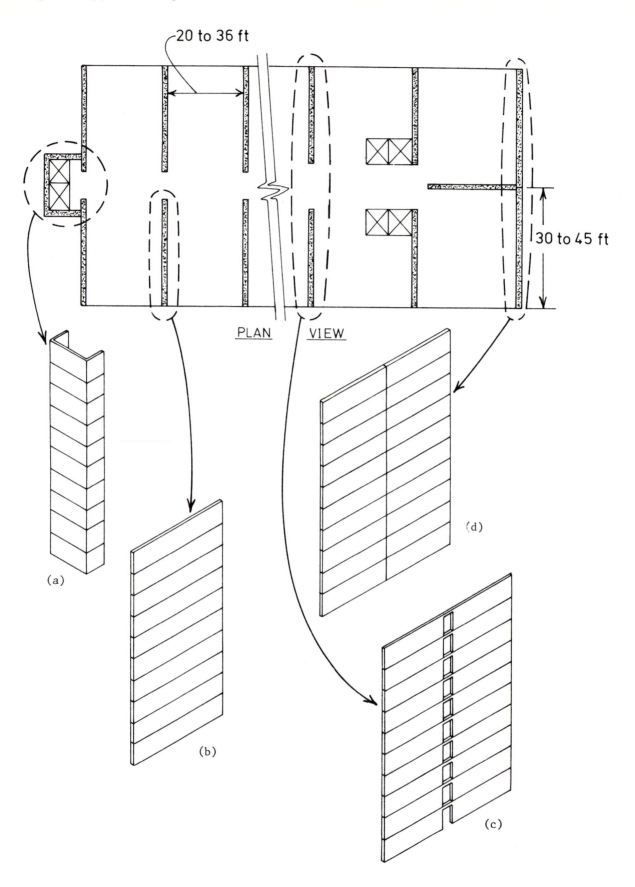

Fig. 2  Horiz.wall-floor joint.

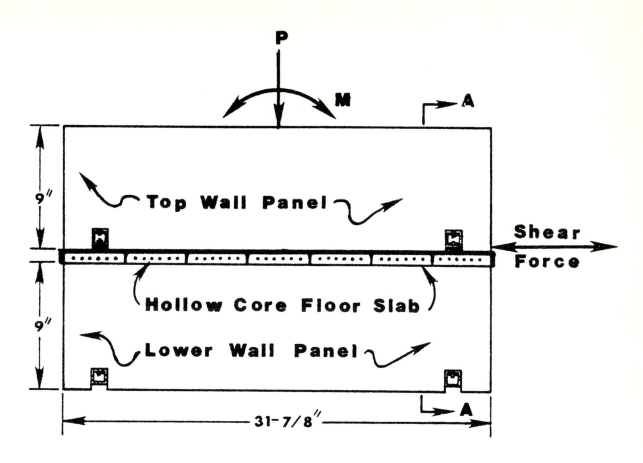

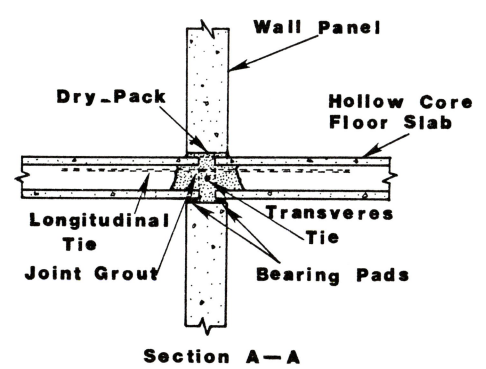

Section A—A

109

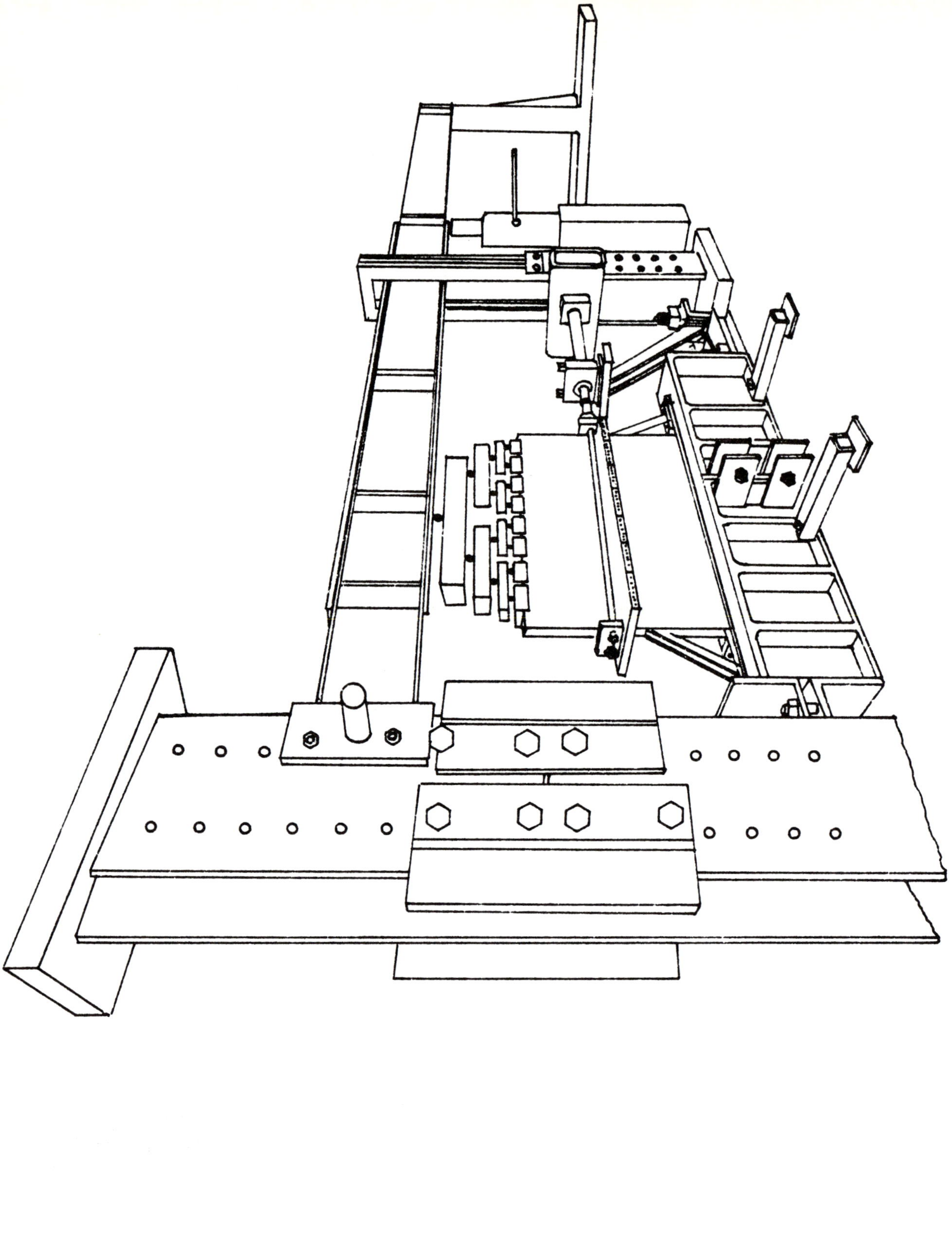

Fig. 3 Test Setup.

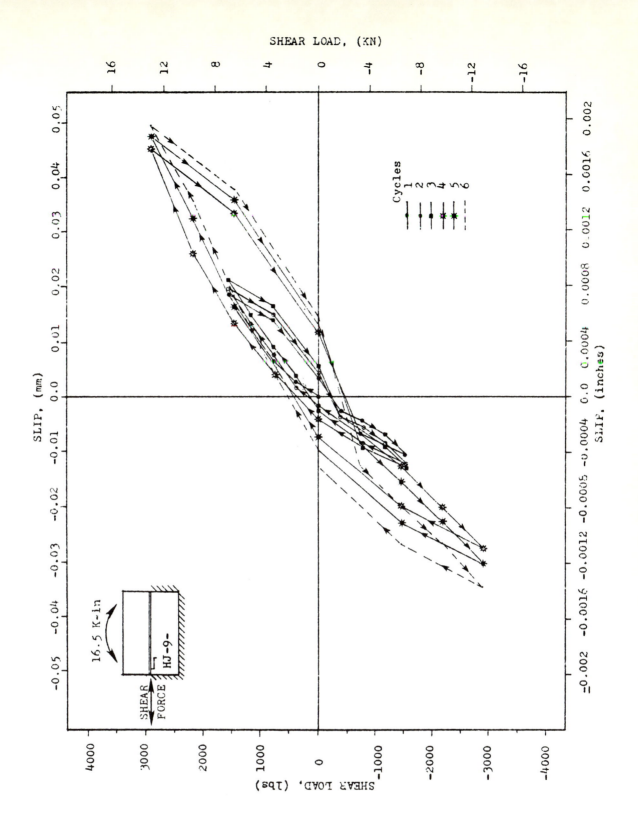

Fig. 4 Shear-slip curves for Joint HJ-9.

(a) Cycles 1-6.

111

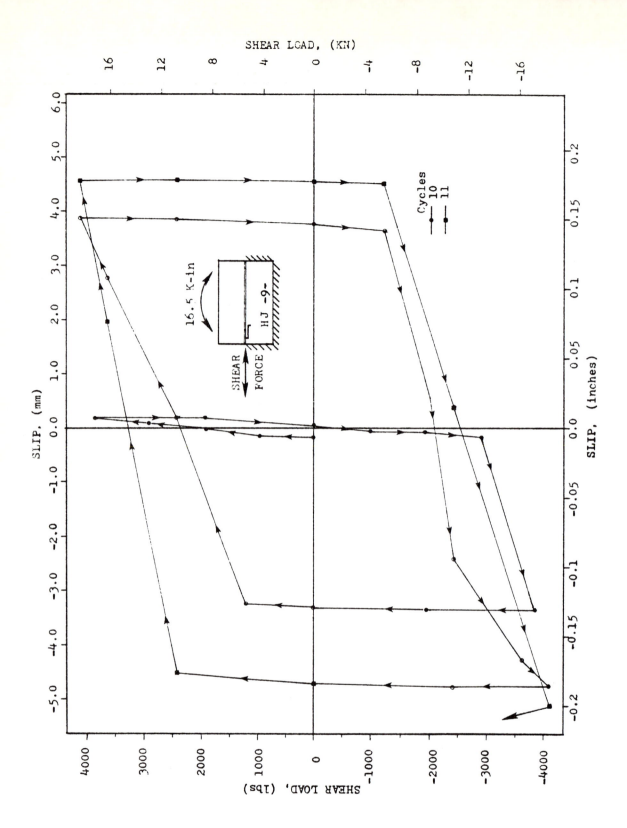

Fig. 4 Shear-slip curves for Joint HJ-9.

(b) Cycles 10-11.

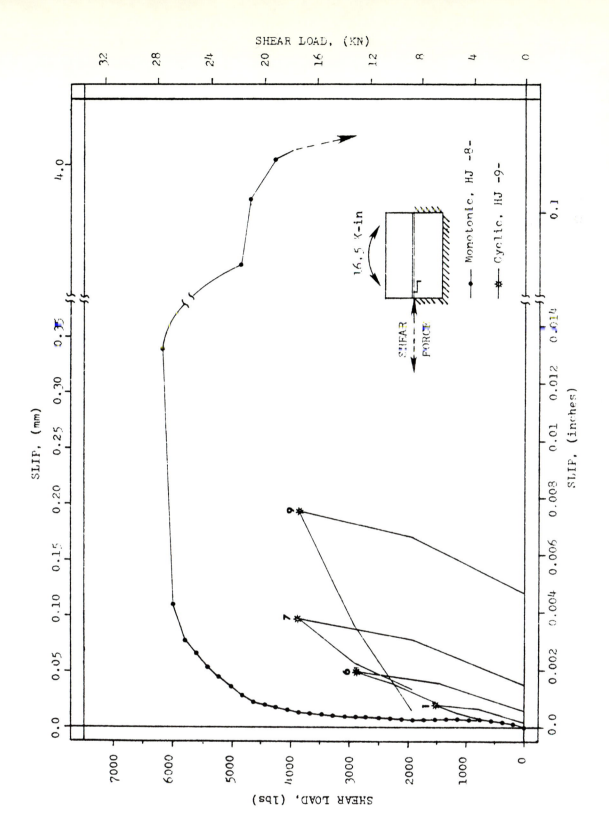

Fig. 5  Monotonic vs. cyclic behavior.

113

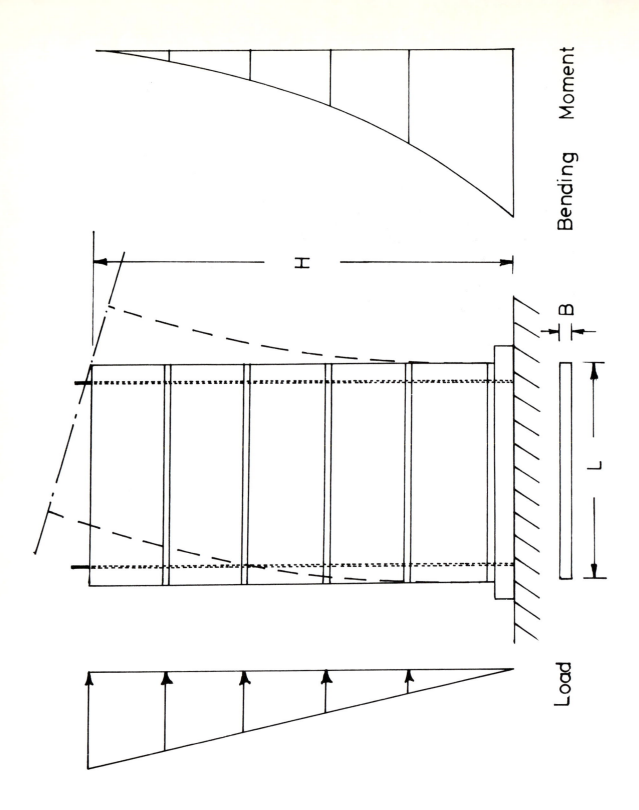

Fig. 6  Simple Shear Wall.

114

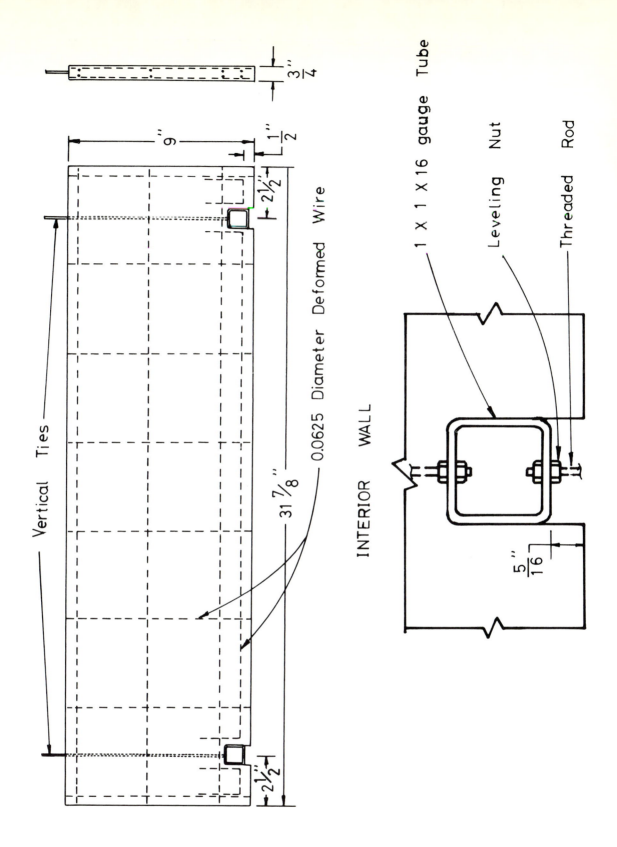

Fig. 7  Wall Panel Details.

115

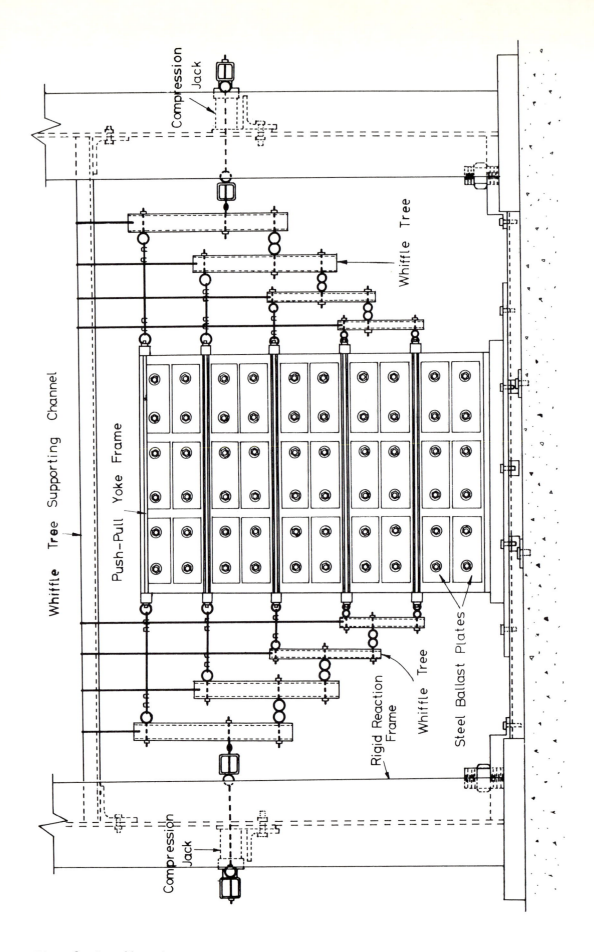

Fig. 8 Loading Arrangement.

116

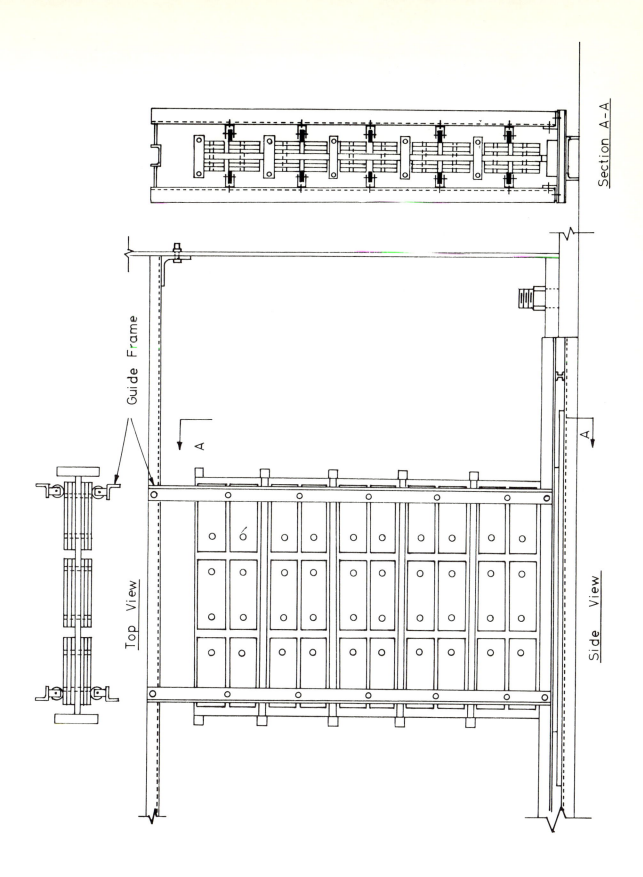

Fig. 9  Stabilizing Frame.

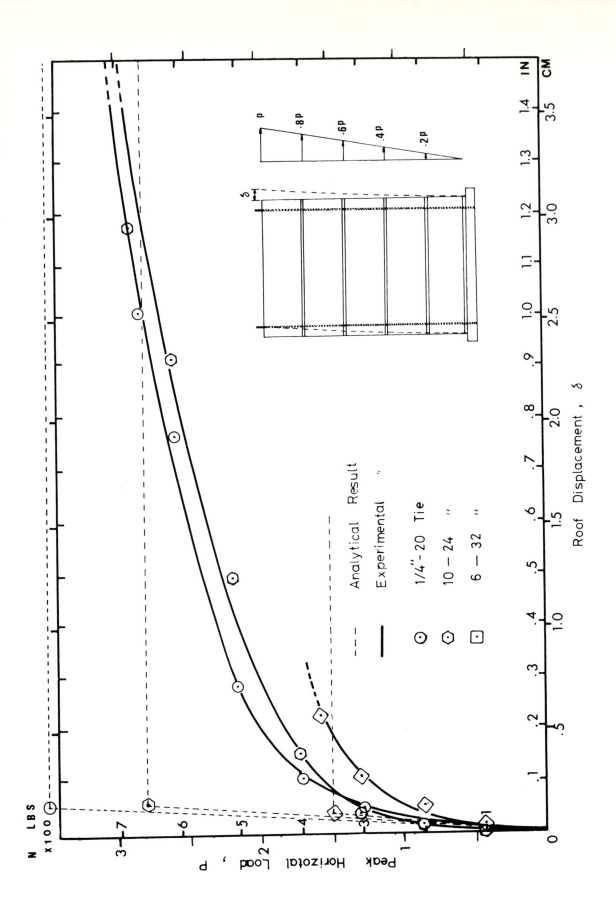

Fig. 10  Monotonic Tests.

118

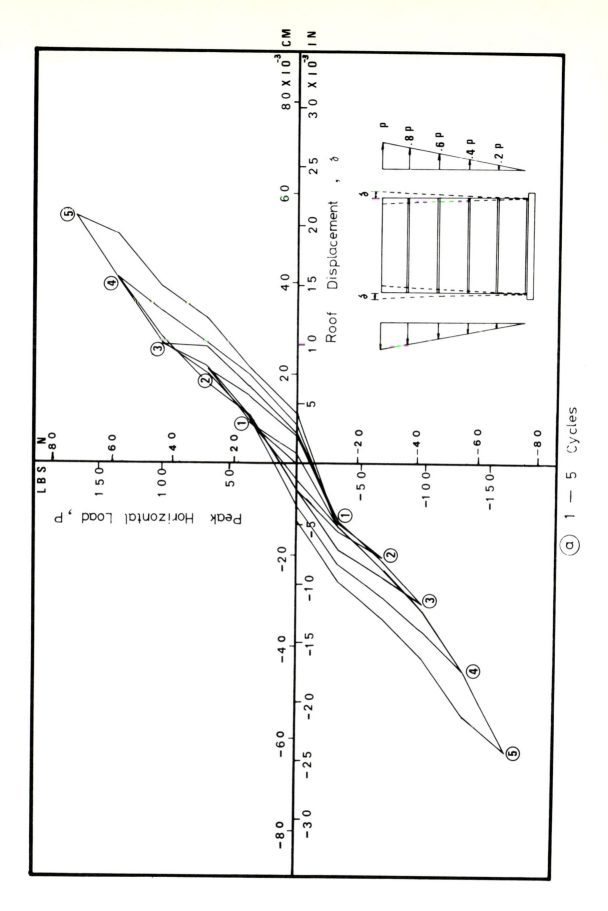

Fig. 11  Cyclic Load-deflection.

(a) Cycles 1-5.

119

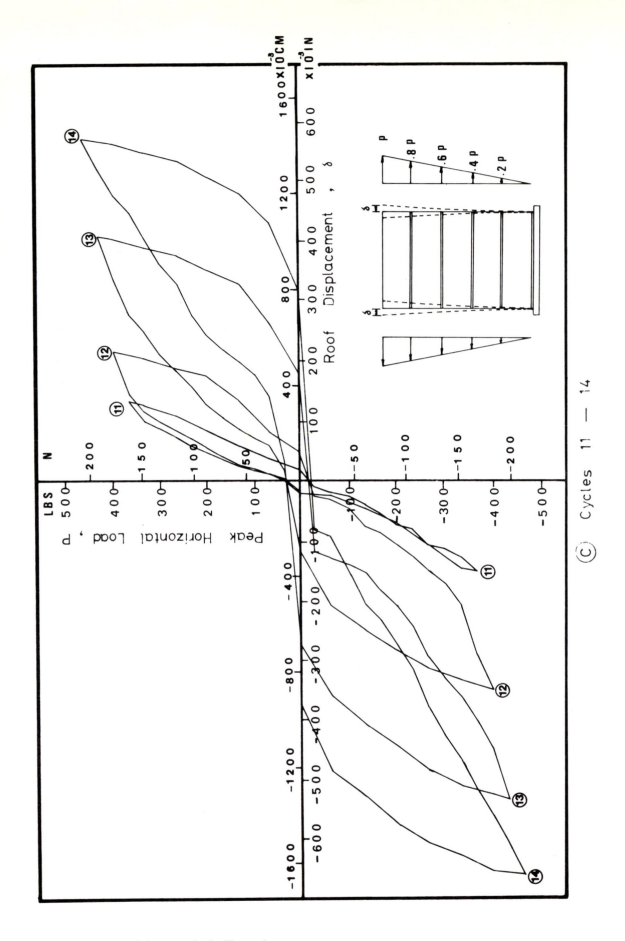

Fig. 11  Cyclic Load-deflection.

(b)  Cycles 11-14.

120

Fig. 12  Mode of Failure.

   (a)  Monotonic.

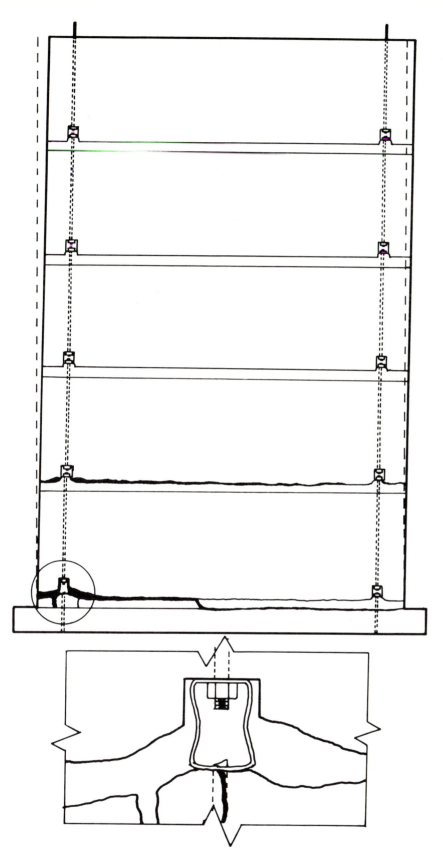

Fig. 12  Mode of Failure.

(b) Cyclic.

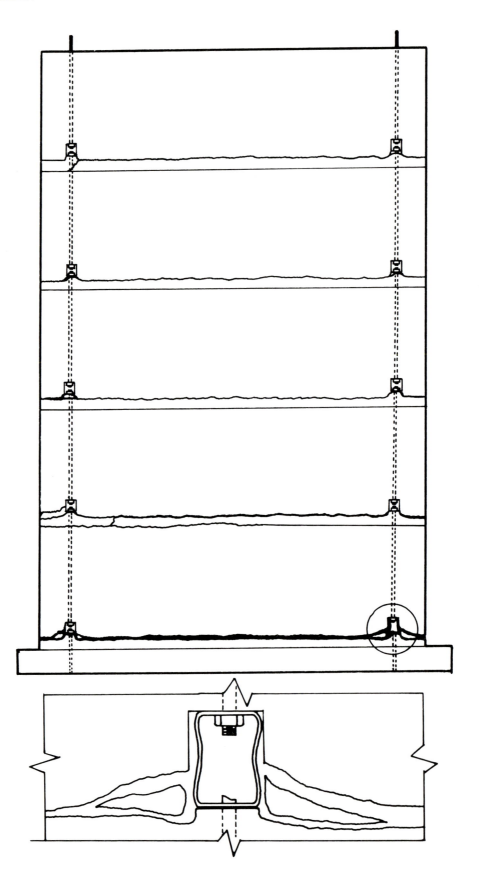

Fig. 13 Precast Shear Wall Mdl.

Steel Cover

Cables

Wall

Whiffle Tree

Steel Plates

Horizontal Joints

Load Cell

Footing

1/2" Aluminum Plate

End Elevation

Steel Angle

Shake Table

Side Elevation

123

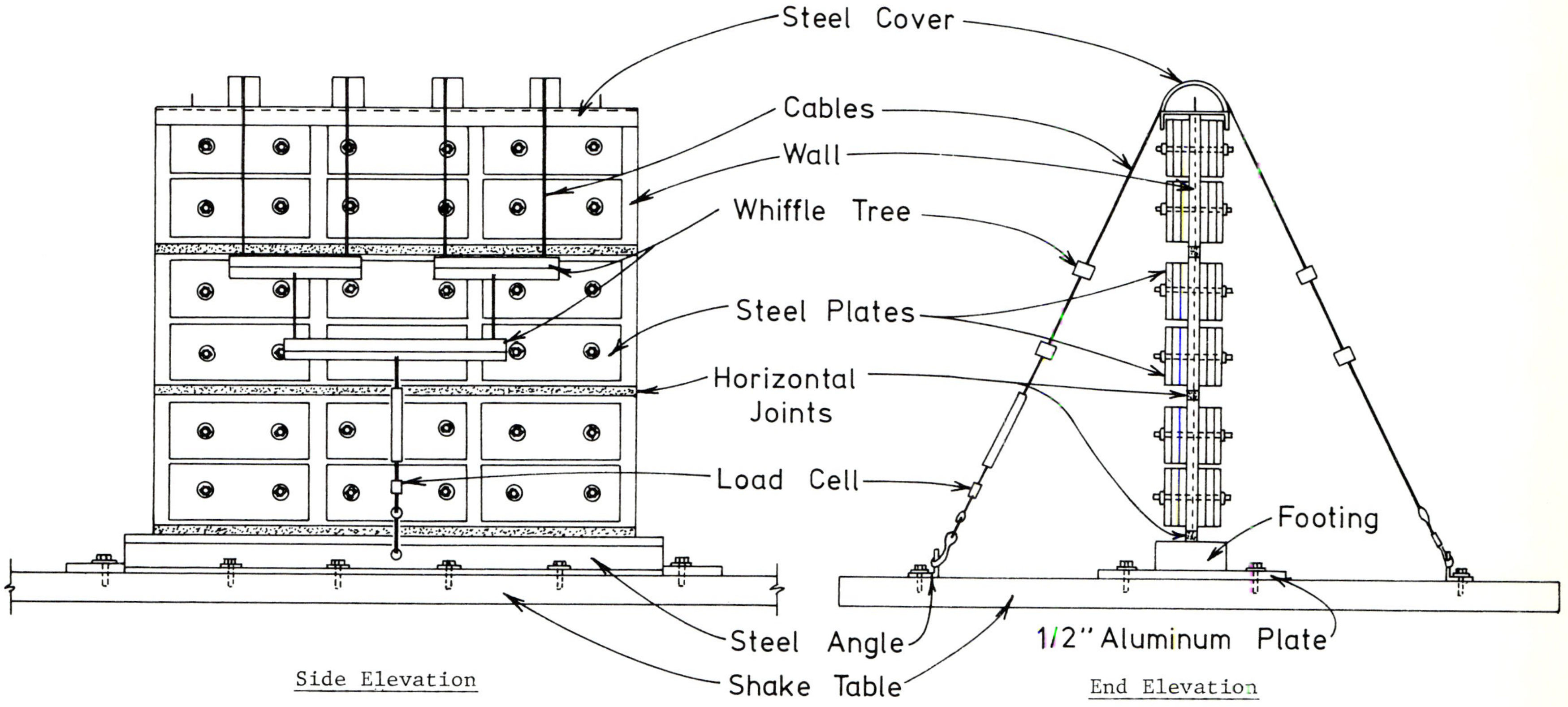

Fig. 14 Instrument & Test Setup.

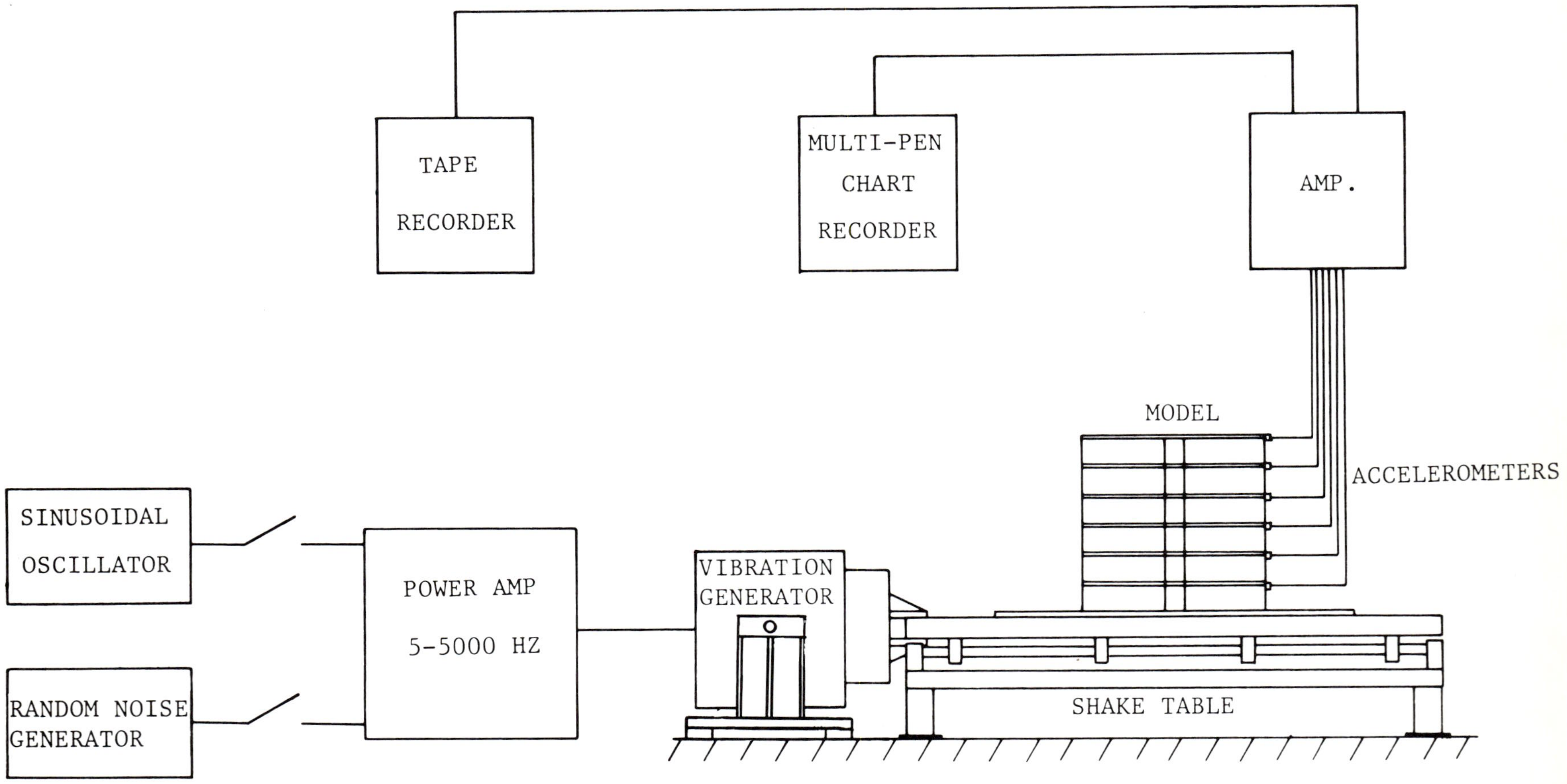

124

Fig. 15  Static vs dynamic
          comparisons.

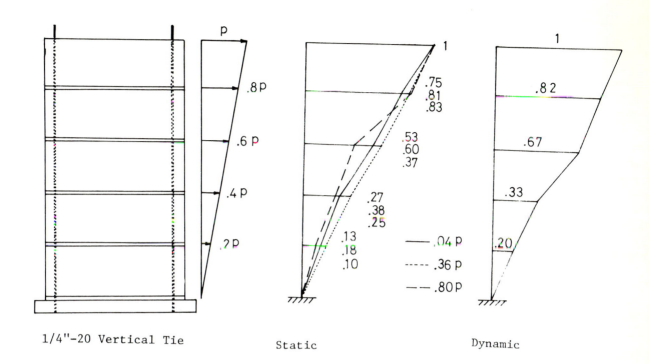

1/4"-20 Vertical Tie                Static                    Dynamic

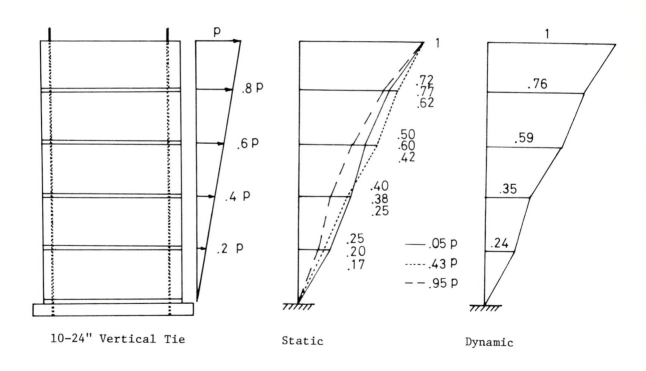

10-24" Vertical Tie                 Static                    Dynamic

125

# One-fifth scale model of a seven-storey reinforced concrete frame-wall building under earthquake loading

H G Harris, V V Bertero, R W Clough
*Department of Civil Engineering, University of California, Berkeley, CA 94720, USA*

## SUMMARY

A one-fifth scale "replica" model of the full-scale prototype seven-story reinforced concrete frame-wall building tested at the Large-Size Structures Laboratory, Building Research Institute, Tsukuba, Japan has been designed for testing on the 20' x 20' U.C. Berkeley Earthquake Simulator under realistic conditions. An ultimate strength model was required for this purpose to enable the study of large deformations and mechanisms of failure. In order to achieve this, a reinforced concrete model has been selected with careful duplication of the prototype material characteristics. Additional non-structural weights, of a significant magnitude, are added to every floor and roof level in order to preserve dynamic similitude. Critical deformations, accelerations, end rotations and internal and surface strains will be measured in the planned model tests for direct correlation with the prototype and with analytical results. A unique feature of the model tests will be an attempt to determine the participation of the wall in resisting the seismic loads by direct measurement of the time history of the internal forces in all of the free standing first-story columns.

## BACKGROUND

The research effort described in this paper was initiated as the first of a series of several tasks by the planning group of the U.S.-Japan Co-operative Program on Large Scale Testing. The master program, as envisioned by the planning group, consists of full-scale tests, small-scale tests, component tests and analytical studies (1). The program's overall objective is to improve seismic safety practices through studies to determine the correlation of all of the various experimental and analytical techniques for predicting seismic response.

As part of the first joint research effort, a seven-story reinforced concrete frame-wall earthquake resistant building has been designed, constructed and tested at the Large-Size Structures Laboratory, Building Research Institute, Tsukuba, Japan. The full-scale seven-story building has three spans in the direction of loading and two spans in its transverse direction. A full height shear wall is located at the mid-span of the center line frame.

In buildings of this type, it is generally recognized that the properties of the shear wall will greatly influence the aseismic performance of the building. Preliminary analysis of this structure in Japan (2) has indicated that the shear wall would resist 85% of total base shear in the linear elastic range and 60% at its post structural yielding. Somewhat larger values were obtained in a Berkeley analysis (3).

Testing of the full scale structure utilized a procedure intended to simulate dymanic response to prescribed seismic excitations. A computer actuator on-line (pseudo-dynamic) test procedure was used with computer control of the roof displacement and story forces linearly related to their heights. Essentially the seven-story structure was tested as an equivalent single-degree-of-freedom system.

## ONE-FIFTH SCALE MODEL

The layout of the "replica" 1/5-scale model, which follows all the geometric similitude considerations of the prototype building, is shown in Fig. 1. Two important features are evident: the stabilizing walls added in sections 1 and 4 to both model and prototype structures to increase the torsional resistance of the structure during testing, and the addition of transducers in all the free standing first-story columns at mid-height in the model structure to enable the experimental determination of the first story shear.

The particular scale chosen for the model test specimen was dictated by the limitations of the UC Berkeley shaking table (4, 5). A simplified drawing of the shaking table and its dynamic performance limitations is shown in Fig. 2. The maximum capacity of the table is 110,000 lbs. Based on this limitation and the desire to study the building failure mechanism, a 1/5-scale model of the prototype structure was chosen. The design of the model footing is different from that of the prototype in order to enable proper attachment to the 20' x 20' shaking table slab. Although the footing weight is important in that it adds to the limiting table weight, it does not enter into the scaling of the moving masses of the structure. A total weight of model and ballast of 113.8 kips was calculated and this is believed to be within the capabilities of the shaking table in achieving the stated test objectives.

The design of the model was chosen to comply with similitude requirements (6), in addition to shaking table limitations. For a direct model of the type required to simulate the existing prototype structure, a summary of scale factors that must be satisfied for earthquake loading is shown in Table 1. In this table, the scale factor, "$S_i$" is defined as the ratio of the "$i$th" prototype quantity to the "$i$th" model quantity. Three types of models are suggested in Table 1 : a true replica model (Col. 4), a model which uses materials with the same properties as the prototype materials but needs additional masses of non-structural nature (Col. 5), and lastly a model identical to the above but with gravity forces neglected (Col. 6). The one-fifth scale model for the US-Japan Program satisfies similitude with regard to geometric and loading parameters, also complies to all material requirements except for the mass density. The latter is augmented by means of lead ballast weights attached to the roof and floor slabs in such a manner as to cause little or no influence to the structural stiffnesses.

## MODEL MATERIALS

The selection of model materials that will satisfy the similitude requirements given in Table 1 is perhaps the single most difficult step in a successful model investigation. For simulation of the complete elastic and inelastic behavior up to physical failure (collapse) the mechanical properties of the model materials must be identical to those of the prototype. Model and prototype steel reinforcement must

TABLE 1  Summary Of Scale Factors For Earthquake Response of Structures (6).

| 1 | 2 | 3 | SCALE FACTORS | | |
| --- | --- | --- | --- | --- | --- |
| | QUANTITY | DIMEN. | TRUE REPLICA MODEL | ARTIF. MASS SIMULATION | GRAV.FORCES NEG.PROTO.MAT |
| | 2 | 3 | 4 | 5 | 6 |
| LOADING | Force, P | F | $S_E S_\ell^2$ | $S_E S_\ell^2$ | $S_\ell^2$ |
| | Pressure, q | $FL^{-2}$ | $S_E$ | $S_E$ | 1 |
| | Acceleration, a | $LT^{-2}$ | 1 | 1 | $S_L^{-1}$ |
| | Gravitational Acceleration, g | $LT^{-2}$ | 1 | 1 | neglected |
| | Velocity, v | $LT^{-1}$ | $S_\ell^{1/2}$ | $S_\ell^{1/2}$ | 1 |
| | Time, t | T | $S_\ell^{1/2}$ | $S_\ell^{1/2}$ | $S_\ell$ |
| GEOM. | Linear Dimension, $\ell$ | L | $S_\ell$ | $S_\ell$ | $S_\ell$ |
| | Displacement, $\delta$ | L | $S_\ell$ | $S_\ell$ | $S_\ell$ |
| | Frequency, $\omega$ | $T^{-1}$ | $S_\ell^{-1/2}$ | $S_\ell^{-1/2}$ | $S_\ell^{-1}$ |
| MAT. PROP. | Modulus, E | $FL^{-2}$ | $S_E$ | $S_E$ | 1 |
| | Stress, $\sigma$ | $FL^{-2}$ | $S_E$ | $S_E$ | 1 |
| | Strain, $\epsilon$ | | 1 | 1 | 1 |
| | Poisson's Ratio, $\nu$ | | 1 | 1 | 1 |
| | Mass Density, $\rho$ | $FL^{-4}T^2$ | $S_E/S_\ell$ | * | 1 |
| | Energy, EN | FL | $S_E S_\ell^3$ | $S_E S_\ell^3$ | $S_\ell^3$ |

\* $\left(\frac{g\rho\ell}{E}\right)_m = \left(\frac{g\rho\ell}{E}\right)_p$

have similar yield, plastic plateau, strain hardening and elongation characteristics, i.e., similar stress-strain relations up to failure under monotonically increasing as well as cyclic deformation (hysteretic behavior).  In addition, local bond  characteristics must also be simulated by the model bars by means of external deformations similar to those used in prototype steel.

The model concrete must have the same stress-strain characteristics as the prototype concrete in compression and tension.  Experience with model concrete applications has shown that although the compressive strength and elastic modulus in compression can be closely matched on model size standard specimens, the tensile strength of model concretes tends to be higher and the discrepancy increases with decreasing size of maximum aggregate.  In order to minimize this deficiency, the maximum size of aggregate that can physically fit the model without causing placement problems should be used.

Typical reinforcing details are shown in Fig. 3.  The properties of the reinforcement are summarized in Table 2.

All columns in the model are reinforced with 8 D4.4 (4.4 mm) bars which are continuous without splices.  All model beam reinforcement is of D3.8 (3.8 mm) annealed bars and all wall, slab, hoop and stirrup reinforcement is of D2 (2 mm) annealed "bright basic" wire.  Model reinforcement has been annealed to achieve stress-strain characteristics as close as possible to those of the prototype.  A comparison of the model and prototype stress-strain curves for the column reinforcement is shown in Fig. 4.  The D22 prototype reinforcement, which satisfies the Japanese grade SD35 (equivalent to US Grade 50 steel), was tested at two different dates, as indicated.

The model reinforcement D4.4 was obtained from the Portland Cement Association and was subsequently annealed at 1075°F for 2 hours and then air cooled in the furnace.  As can be seen in Fig. 4, the average yield strength and the yield plateau range of the model reinforcement is very close to that of the prototype steel.  However,

Table 2  Reinforcement Properties.

(1)Mill Sheet 80-05-10;  (2)Tested 2-4-81;  (3) Tested 3-23-81.

| DESIGNATION | | PROTOTYPE | | AVAIL. 1/5 SCALE MODEL | | | PROTOTYPE PROPERTIES | | |
|---|---|---|---|---|---|---|---|---|---|
| Prot. | Model | Diam. | Area | Source & Model | Diam. | Area | Yield Str. | Ult. Str. | Elong. |
| | | in. | in.$^2$ | | in. | in.$^2$ | ksi | ksi | % |
| D22 | D4.4 | .866 | .5890 | PCA/D2.5 | .178 | .0249 | 61.3(1) | 87.33 | 24.25 |
| | | | | | | | 57.8(2) | 88.91 | 24.21 |
| | | | | | | | 50.2(3) | 81.78 | 19.1 |
| | | | | | | | 56.4 | 86.0 | 22.52 |
| D19 | D3.8 | .748 | .4394 | PCA/D2 | .159 | .0199 | 58.1(1) | 83.05 | 24.51 |
| | | | | | | | 52.2(2) | 62.24 | 22.9 |
| | | | | | | | 51.9(3) | 81.5 | 21.4 |
| | | | | | | | 54.1 | 75.6 | 22.9 |
| D10 | D2 | .394 | .1219 | D-W Co. Basic | .079 | .00488 | 55.4(1) | 79.1 | 28.7 |
| | | | | | | | 52.6(2) | 77.3 | 20.4 |
| | | | | | | | 55.0(3) | 81.2 | 21.0 |
| | | | | | | | 54.3 | 79.20 | 23.37 |

Table 3  Concrete Properties Of Field Cured Specimens (3/23/81).

| Floor Level | $f'_c$ psi | $\varepsilon_{f'_c}$ % | $E_o$ x10$^6$ psi | $E_{1/3}$ x10$^6$ psi |
|---|---|---|---|---|
| 1 | 2 | 3 | 4 | 5 |
| 1F | 4116 | .218 | 3.74 | 3.39 |
| 2F | 4148 | .240 | 3.69 | 3.36 |
| 3F | 3897 | .228 | 3.49 | 3.14 |
| 4F | 4130 | .225 | 3.50 | 3.01 |
| 5F | 4189 | .210 | 3.50 | 3.33 |
| 6F | 2047 | .185 | 2.53 | 1.97 |
| 7F | 2686 | .192 | 2.85 | 2.47 |

Note:  All quantities are the average of 3 specimens.

the strain hardening slopes are somewhat lower for the model rein-forcement.  This lower strain hardening stiffness can be accepted because except for the regions of the columns adjacent to the ground floor at no other column region is the steel expected to undergo inelastic deformations.  Furthermore, the significant inelastic regions at the column bases are expected to be the last to develop.

All model reinforcement has been deformed by passing the smooth bar through sets of knurls.  The column and beam reinforcement has been provided in the knurled condition and the wall and slab reinforcement was knurled in the laboratory at UC, Berkeley using a knurling device originally developed at Cornell University (7).

The prototype concrete used in the full scale model has a specified compressive strength of 225 kg/cm$^2$ (3630 psi) for the first 4 stories and 270 kg/cm$^2$ (3840 psi) for the rest  of the height and uses a smooth rounded aggregate with a maximum size of 1 inch.  Both standard cured and field cured 6" x 12" specimens were cast with the full scale model.  Strength and modulus variation with height of the building using the field cured specimens is given in Table 3.  The initial modulus (Col. 4) for the whole building has an average value of 3.33 x 10$^6$ psi.  The secant modulus at 1/3 $f'_c$ (Col. 5) is 2.95 x 10$^6$ psi.  The average compression strength is given in Col. 2.  Very weak concrete strengths are indicated at floor levels 6 and 7.  Although standard cured specimens do not reflect this discrepancy in strength, the field cured specimens are thought to reflect better the

the actual strength of the concrete in the structure. No attempt is made to simulate these strength variations in the 1/5-scale model. Rather, average strengths of the whole structure are modeled.

Model concrete has been cast using local top sand having the gradation curve shown in Fig. 5 and a local pea gravel with a size of less than 1/4" in the proportion of gravel to sand of 0.2 by weight. The model concrete mix uses Type III "high early" strength cement and Pozzolith 300 R admixture to increase workability. The mix proportions are indicated in Fig. 6 which shows the strength-age curves of field and standard cured 3" x 6" cylinders.

A comparison of the stress-strain behavior of the prototype and model concrete is shown in Fig. 7. Note the variation of strength and stiffness of the prototype concrete with floor height.

## INSTRUMENTATION AND DATA ACQUISITION

The one-fifth scale model will be tested in a constrained manner with only horizontal motion in the direction of the main wall. Figure 8 indicates the position of various measuring devices on the exterior of the model. Fig. 9 shows the location of the various strain gauges to measure the steel strains in the critical regions.

The basic data acquisition system, a permanent facility at the Earthquake Simulator Laboratory, consists of multiple signal conditioners preceding a 128 channel signal scanner and analog to digital data conversion unit coupled to a mini-computer and disc storage unit on line with a magnetic tape deck. The balanced, nulled or conditioned signals from all of the individual instruments are read in bursts at 0.01952 second intervals with a phase lag of fifty microseconds between sequential channels and the signals are converted to digital data and sent to disc storage. Thus, on this system the data read rate is 51.23 times per channel per second (51.23 Hz).

A second data acquisition unit is to be used in testing the 1/5-scale model in parallel with the system described above to provide an expanded data collection capability of 80 channels. The equipment configuration of the system will be as nearly identical to that above as possible.

## CONCLUSIONS

From this study the following general conclusions can be made:

1. Development of reliable analytical methods for predicting a building's response to severe earthquake ground motions is very difficult. An integrated analytical and experimental procedure is therefore required in which testing of realistic structures under simulated earthquake motions and measurement of the nonlinear response can be made.

2. Tests such as the above can be best performed on specially built earthquake simulators (shaking tables). Even for the largest available earthquake simulator in the US, i.e. U.C. Berkeley, the study of large segments of realistic structures requires the use of scaled models.

3. In the present study, a one-fifth scale model of the seven-story reinforced concrete frame-wall structure was the largest possible model that could be used in order not to exceed the capacity of the shaking table.

4. The simulation of inelastic behavior up to collapse of "ductile" structures require a true replica of the stress strain characteristics of the materials (steel and concrete) as well as their bond-slippage laws. This simulation offers serious difficulties when small scale models are used.

ACKNOWLEDGEMENT

The work reported herein has been supported by the National Science Foundation under Grant No. PFR-80-09478.

REFERENCES

(1) U.S-JAPAN PLANNING GROUP, "Recommendations for a US-Japan Co-operative Research Program Utilizing Large Scale Testing Facilities" Report No. UCB/EERC-79/26, Earthquake Engineering Research Center, College of Engineering, University of California Berkeley, CA.

(2) CHAVEZ Z, J W, "Study of the Seismic Behavior of Two-Dimensional Frame Buildings, A Computer Program for the Dynamic Analysis: INDRA," Bulletin of the International Institute of Seismology and Earthquake Engineering, Vol. 18 (1980), pp.41-62, Building Research Institute, Ministry of Construction, Japan.

(3) BERTERO, V V, "US/Japan JTCC Meeting January 19-22, 1981," Memo to R.D. Hanson (University of Michigan), Department of Civil Engineering, University of California, Berkeley, CA.

(4) STEPHEN, R M, BOUWKAMP J G, CLOUGH R W, and PENZIEN J, "Structural Dynamic Testing Facilities at the University of California, Berkeley" Report No. EERC 69-8, Earthquake Engineering Research Center, College of Engineering, University of California, Berkeley, CA, August 1969.

(5) REA, D, ABEDI-HAYATI, S, and TAKAHASHI, Y, "Dynamic Analysis of Electrohydraulic Shaking Tables," Report No. UCB/EERC-77/29. Earthquake Engineering Research Center, College of Engineering, University of California, Berkeley, CA, (December 1977).

(6) KRAWINKLER, H, MILLS, R S, and MONCARZ P D, "Scale Modeling and Testing of Structures for Reproducing Response to Earthquake Excitation," John A. Blume Earthquake Engineering Center, Department of Civil Engineering, Stanford University, (May 1978).

(7) HARRIS, H G, SABNIS G M, and WHITE R N, "Small Scale Direct Models of Reinforced and Prestressed Concrete Structures," Report No. 326, Dept. of Structural Eng., School of Civil Eng., Cornell University, Ithaca, New York (September 1966).

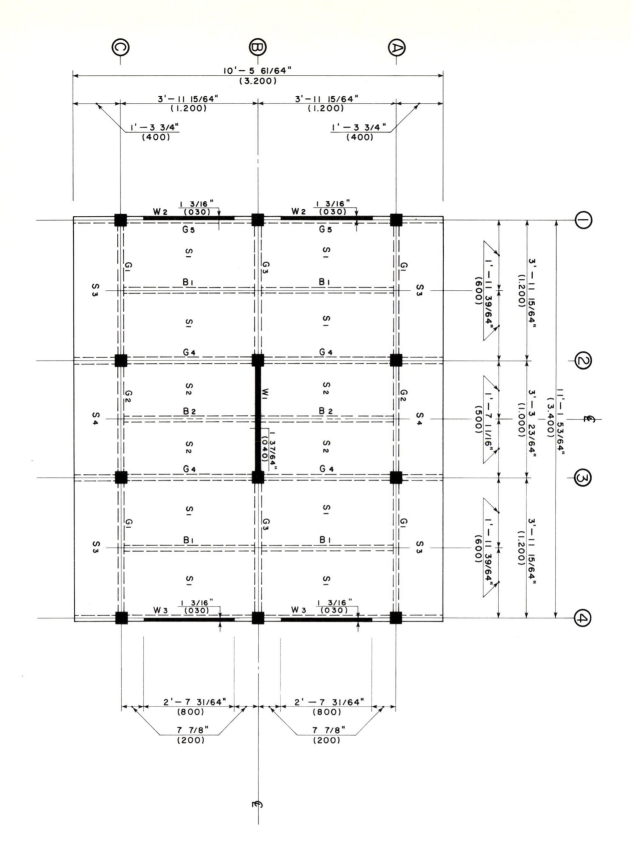

Fig. 1  One-fifth scale model.

(a) Plan View.

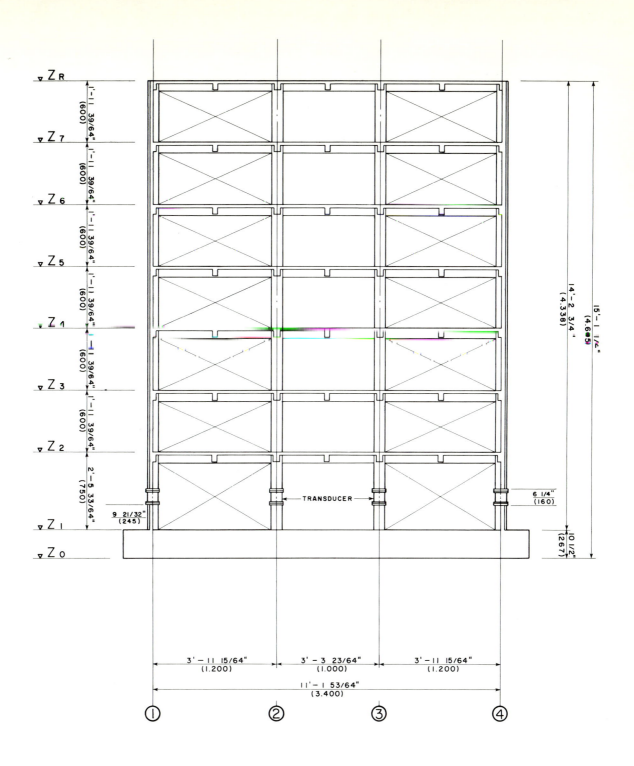

Fig. 1  One-fifth scale model.

(b) Longitudinal Section A or C.

133

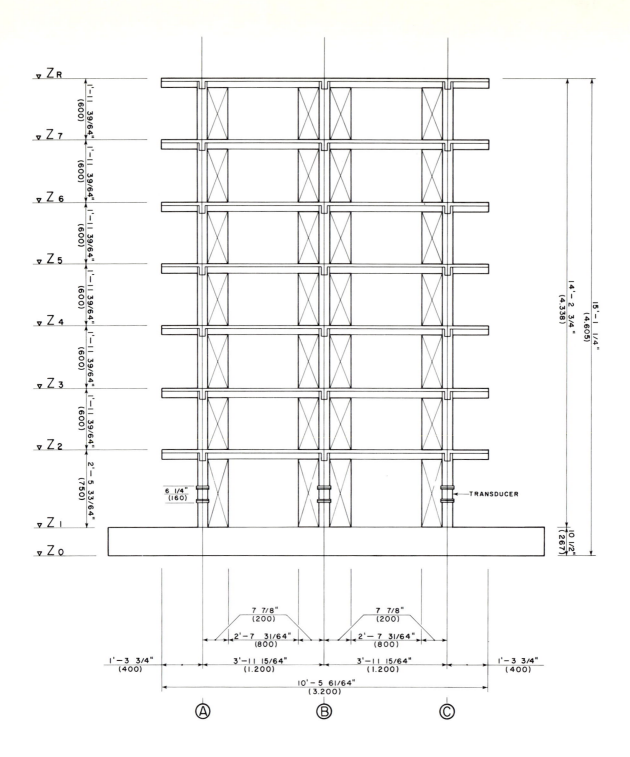

Fig. 1 One-fifth scale model.

(c) Transverse Section 1 or 4.

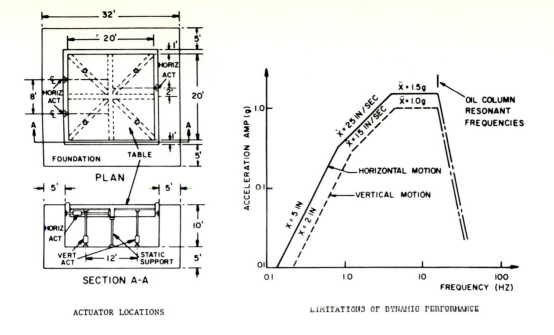

Fig. 2 Shaking Table Motion Limits.

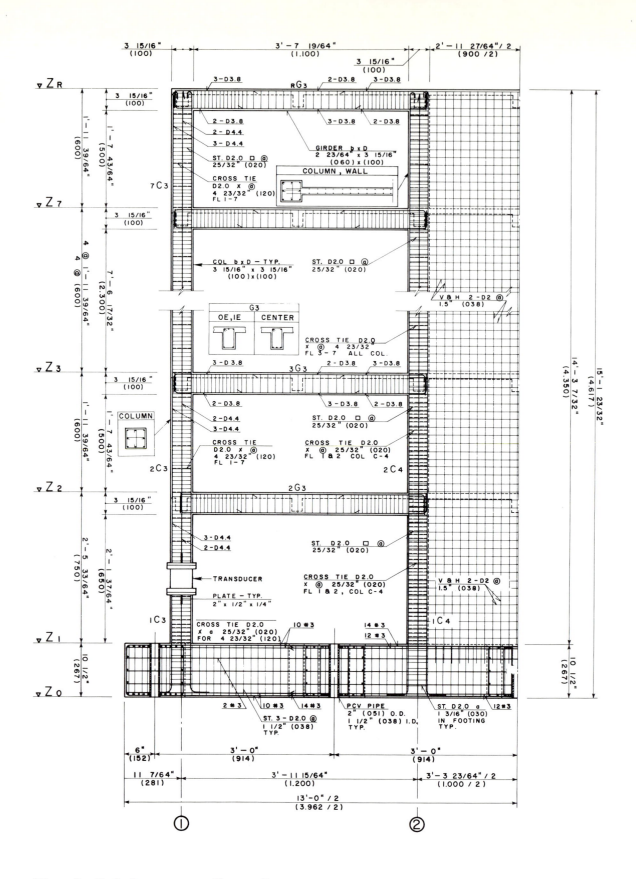

Fig. 3 Reinforcement: Frame B.

136

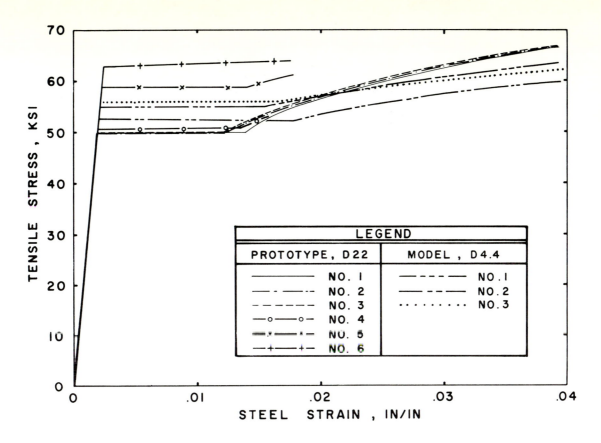

Fig. 4  Stress-strain curves for Col.
         Reinf.

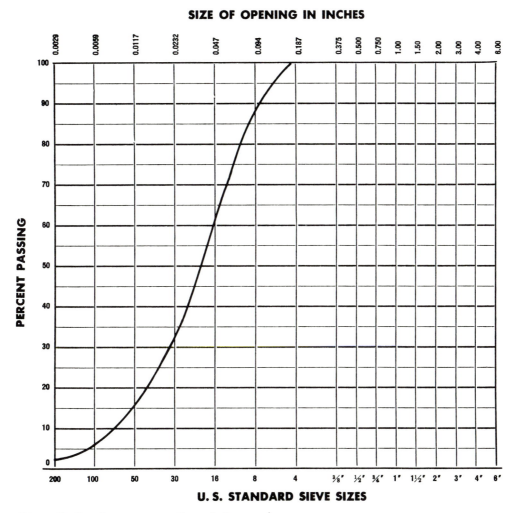

Fig. 5 Grad. curve of model sand.

137

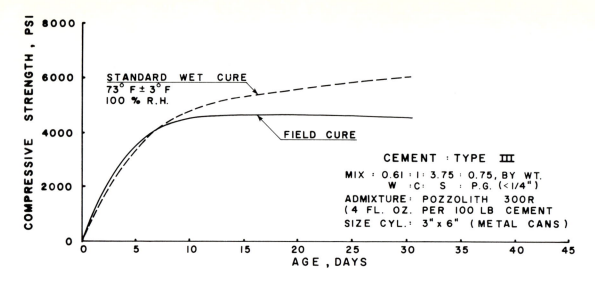

Fig. 6 Strength–age curves of
model concrete.

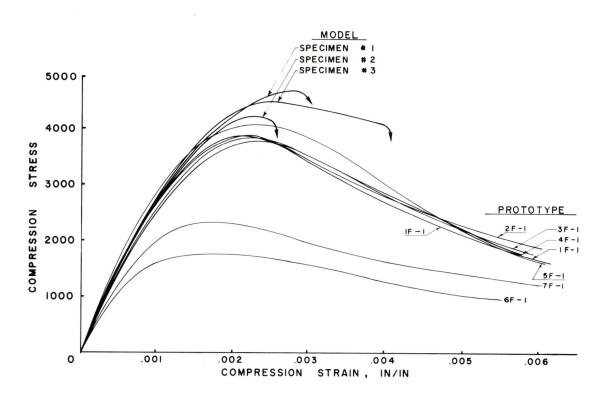

Fig. 7 Stress-strain curves for concrete.

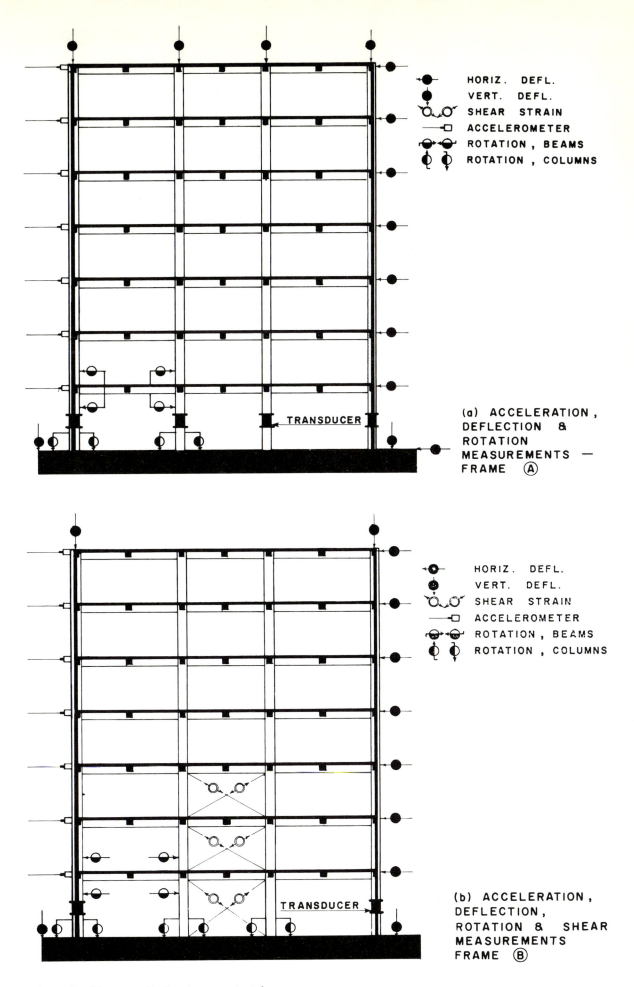

**HORIZ. DEFL.**
**VERT. DEFL.**
**SHEAR STRAIN**
**ACCELEROMETER**
**ROTATION, BEAMS**
**ROTATION, COLUMNS**

TRANSDUCER

**(a) ACCELERATION, DEFLECTION & ROTATION MEASUREMENTS — FRAME Ⓐ**

**HORIZ. DEFL.**
**VERT. DEFL.**
**SHEAR STRAIN**
**ACCELEROMETER**
**ROTATION, BEAMS**
**ROTATION, COLUMNS**

TRANSDUCER

**(b) ACCELERATION, DEFLECTION, ROTATION & SHEAR MEASUREMENTS FRAME Ⓑ**

Fig. 8 External instrumentation.

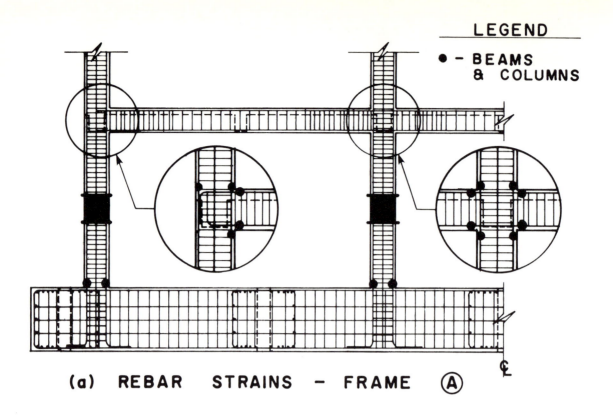

LEGEND

• – BEAMS
     & COLUMNS

**(a) REBAR STRAINS – FRAME Ⓐ**

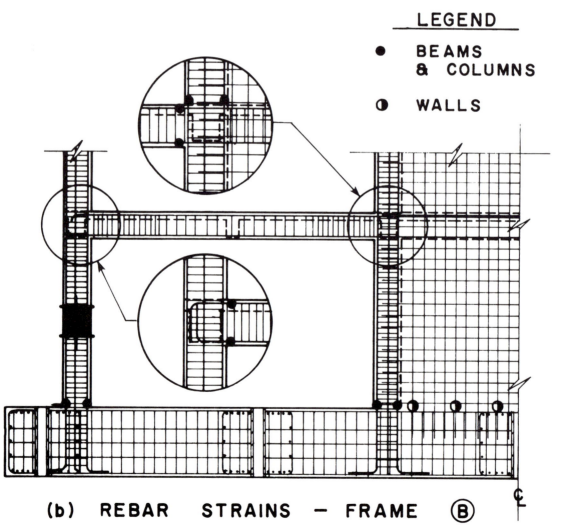

LEGEND

● BEAMS
     & COLUMNS

◐ WALLS

**(b) REBAR STRAINS – FRAME Ⓑ**

Fig. 9  Internal instrumentation.

# Simulation of seismic loadings for reduced scale models

## 15

A Zelikson, J Bergues, P Habib, M P Luong
*Laboratoire de Mécanique des Solides, Ecole Polytechnique, 91128 Palaiseau Cedex, France*

SYNOPSIS

The hydraulic gradient similarity method was used to represent gravity forces by seepage ones, and shock tubes discharged to simulate earthquakes by producing running waves in a model of saturated sand representing at a scale of 1 : 75 a body of soil 135m x 60m and 30m deep. Tests seem to indicate this a suitable method for studying earthquakes in the laboratory, especially for very large models in cases where boudary echoes must be prevented.

## INTRODUCTION

When soil structure interaction during earthquakes is studied by scale modelling the problem of echoes from the boundary of the soil's cell becomes a major one. In fact radiation of elastic energy to infinity is by far the main cause for damping of struture's vibrations. Inside the model's cell waves should be dispersed by the walls and conditions of black body created in order to prevent standing waves non existing in situ. Accordingly the structure's model is several times smaller than the soil's one. Times of dissipation of excessive pore pressures in the saturated soil are as the square of lengths and distort similitude unless corrections are made which are only possible at moderate scales. Thus structures like nuclear power plants or offshore platforms require huge models which surpass the capacities of centrifuges. The main problem in applying the hydraulic gradient similarity method to large models is in providing for the large flows of water in and out of the model needed for maintaining the high gradient, which become prohibitive for steady state experiments. However for short period transient cases like earthquakes this is no problem at all. Thus it seems that at present the hydraulic gradient method might be the only solution for very large sand models.

## SIMILITUDE

(The scale "A (model) by A" is $A^*$)

The basic demands besides conservation of material are conservation of deformation scale $\varepsilon^*$ and stress scale $\sigma^*$. Volume energy is composed of $\sigma\varepsilon$ and $\rho v^2$ ($\rho$ density, $v$ velocity) $\rho* = 1$ so $v* = 1$, thus $t^* = \ell^* = (a^*)^{-1}$ (a, acceleration) In considering the action of gravity on the saturated sand grains the submerged unit weight should be taken the value of which relative to water $\gamma_{sr}$ is about 1. As $a^* = (\ell^*)^{-1}$ we must have $\gamma_{sr}^* = (\ell^*)^{-1}$. This is provided by producing inside the model of a constant vertical field of seepage. The head gradient $\nabla h$ represents $\gamma_{sr}$ giving $\gamma_{sr}^* = (\nabla h + \gamma_{sr}) : \gamma_{sr} \cong \nabla h$ for

$\gamma_{sr} \cong 1$ and $h \gg 1$. Dissipation of dynamical pore pressure $P_d$ is according to the law $q = - k \, grad \, P_d$. As $P_d^* = 1$, we must have $k^* = \ell^*$ in order to satisfy $q^* = v^* = 1$. This additional condition is achieved by sand's reconstitution or change of fluid and is hardly possible for $\ell^*$ smaller than 1 : 100, whence the need for huge models. Similitude of ground motion during earthquakes is according to $v^* = 1$ $a^* = (\ell^*)^{-1}$ and $T^* = \ell^*$ for the periods T. Also a state of running waves should be created. This is done by discharging of shock tubes at given intervals the air pressures thus created being modified by filters and impinging on a rubber membrane in contact with the sand.

## EXPERIMENTAL ARRANGEMENTS (FIGURE 1)

The soil was composed of fine-grained Fontainebleau sand and 20% additional finely ground sand reducing permeability by 50. The sand was placed above at 0,07m filter inside the 1.8m x 0.6 x 0.5 rigid steel cell. An echo box is fixed at a taperred end of the long cel. It has a rubber membrane on the soil's side, 3 shock tubes fixed on the opposite side, and filters fixed inside. The shock tubes are sealed by thin calibrated steel foils punctured by needles at desired moments after the tubes had been filled with air under high pressure. All the above arrangements contribute to the signal's form. After saturating the soil, small flow is maintained by gravity, the watar level being several centimeters above the soil's surface. Just before firing the tubes compressed air is blown into the space between the water surface and the steel cover pushing water under high gradient through the soil. When the water level reaches the soil's level the earthquake is generated uder the condition of free soil's surface. The structure chosen shows typical problems of the hydraulic gradient method (fig. 2) it is a buried vertical cylinder. In order to keep flow vertical near the structure its top and bottom are covered with porous stones connected between them by flexible tubes calibrated to have the same hydraulic resistance as the soil replaced by the structure. The inside of the structure was under air pressure compensating the water pressure outside.

## RESULTS

We present as an example accelerogrammes for a single tube discharge at a gradient of 75 giving $\ell^* = 1 : 75$ (the model representing a structure of 12m in situ). Fig. 3, shows accelerations "in-situ" N°4,5 show the horizontal component in the soil at mid hight and at 0.1m on either side of the structure. The 7%g earthquake was produced by 4.5 MPa in the shock tube. N° 1 (vertical) and N° 2 (horizontal) at the base of the structure in front of the shock show vibrations of the structure as a whole at amplitudes half those of the soil. N° 3 (horizontal) fixed on the lining at mid hight shows interference patern of two lining modes of vibration around 10 Hz in situ and amplitudes twice those of the soil. Movement dies out rapidly showing good damping and little echoes. The tubes were designed to be fired at up to 20 MPa whereby earthquakes of 0.3g in situ are expected to be studied at $\ell^* = 1 : 75$ (or 0.6g at $\ell^* = 1 : 38$). On the basis of the experience gathered from the numerous test carried out so far it can be estimated that the basic technical problems have been solved and laboratory earthquakes can be simulated by the hydraulic gradient method.

REFERENCES

MANDEL, J, Essais sur modèles réduits en Mécanique des terrains, Etude des conditions de similitude, Revue de l'Industrie Minérale, n° 44, (1962).

ZELIKSON, A, Sur un procédé de similitude nouveau, applicable notamment en Mécanique des Sols, C.R. Acad. Sci., Paris, tome 256 (1963).

ZELIKSON, A, Geotechnical models using the hydraulic gradient similarity method, Géotechnique, 19, n° 4, (1969).

ZELIKSON, A, LEGUAY P, BADEL D, Représentation d'un séisme par des séquences de tirs explosifs en centrifugeuse, Structural Mechanics in Reactor Technology (SMIRT 6), Paris (Août 1981).

ZELIKSON, A, Artificial two dimensional sands, shock tube centrifuge, as experimental tools in tests of cavities subject to explosions, Military Applications of Blast Simulation, (MABS 6), Cahors, (June 1979).

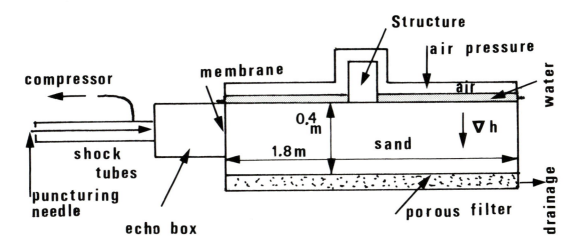

**Experimental Cell longitudinal section – schematic**
**Fig.1**

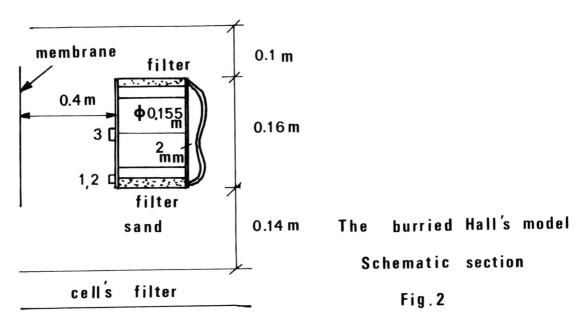

**The burried Hall's model**

**Schematic section**

**Fig.2**

143

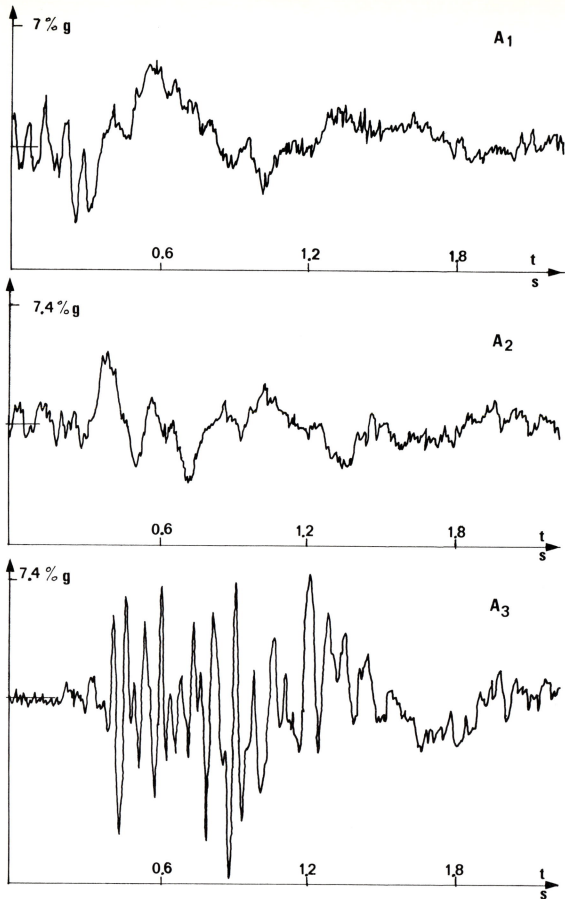

**Fig.3 Acceleration vs time**

144

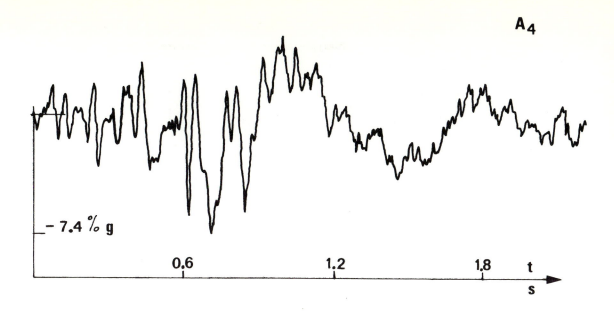

$A_4$

$-7.4\% \text{ g}$

0.6   1.2   1.8   t
s

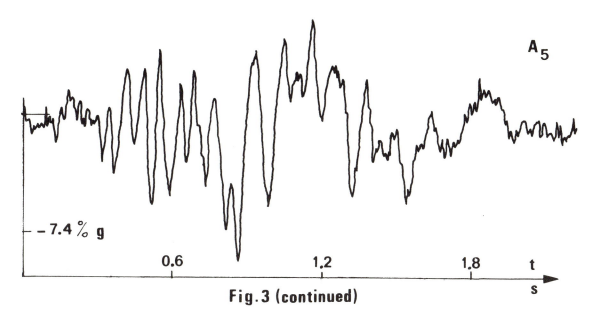

$A_5$

$-7.4\% \text{ g}$

0.6   1.2   1.8   t
s

**Fig. 3 (continued)**

Fig. 4  General view

145

Fig. 5    Shock tubes. Installation.

Fig. 6  Tube's sealing arrangement.

Fig. 7  Echo box.

Fig. 8   Rubber membrane.

# Modelling considerations in seismic and hydrodynamic experimental investigations

16

David Williams
*URS/John A Blume & Associates, Engineers, San Francisco, USA*

## SUMMARY

The restrictive nature of the similitude requirements for small-scale model studies are described for situations in which inelastic dynamic response is being investigated, as is the case for most earthquake-related studies. Development of feasible models to satisfy the major requirements are discussed. For offshore structures engineering, similitude conflicts arise due to the basic incompatibility of satisfying both Reynolds number criteria (important for modelling fluid resistance) and Froude number criteria (essential for inertial similitude). This conflict and a possible approach to circumvent the problem are described.

## INTRODUCTION

Current development in design technology demands a better knowledge of the reliability of structures (particularly new design concepts) in severe seismic environments and extreme sea states. The sought knowledge can be provided in several ways: by the analysis of failures; by "borrowing" relevant experience from other industries; by performing theoretical studies; and by undertaking experimental investigations. The experimental studies can take the form of either full-scale experiments (monitoring instrumented structures during severe events) or experiments with models of either complete structures or components subjected to simulated environmental loads. The above techniques complement each other and must be used in combination to attain a good understanding of "new" problems. Model experiments are particularly suited to identifying and examining important parameters affecting response, to observing and recording trends of behaviour, and to generating experimental data for the purpose of correlating it with theoretical response predictions. Concepts that are difficult to model analytically can often be investigated most fruitfully by experiment.

The advent of large shaking tables capable of simulating intense seismic excitations and large wave tanks capable of simulating severe sea states, has created renewed interest in studying the behaviour of various types of structures by use of physical models. Especially when used in conjunction with theoretical dynamic analyses the approach can provide valuable information. A review of the use of model analysis in structural design and a discussion of scale effects and material properties is provided by Godden (1). The special requirements imposed for seismic and hydrodynamic model studies are briefly outlined in this paper.

Generally, for small-scale models of complex structures (where a number of parameters are being studied simultaneously) attention must be paid to similitude requirements. A structure in which 3-dimensional nonlinear response is of interest implies a certain degree of complexity. Of course, for large-scale models or for the case where separate effects can be uncoupled, similitude requirements are much less restrictive to the extent that the model may often be thought of as a small prototype. However, model studies are usually restricted to situations where complex structural interactions are taking place or the interactions are unknown. An initial assumption of uncoupled behaviour between the various effects could defeat the purpose of the model study.

For offshore structures, experimental model studies with full structural systems are desirable for verifying more refined analytical predictive procedures as they become available. Full-scale measurement of environment effects on offshore prototype structures is rarely feasible, mainly because of the infrequent occurrence of significant events at a particular site but also because of economic factors. The oil industry cooperative Ocean Test Structure project (2), which consisted of a 1/3- to 1/6-scale model of a conventional jacket-type platform installed in the Gulf of Mexico to study actual ocean waves and the forces they produce, was an attempt to circumvent this problem as it relates to wave effects. A comparable approach for seismic effects is not possible.

Simulation, on the other hand, allows economic, controlled, and repeatable generation of extensive significant data, provided its practicality can be established. Instrumentation is very much simpler, and on-site data-acquisition capability is generally available. Correlation between measured and predicted responses, in order to verify and compare analytical procedures and design practice, can best be satisfied by relevant model studies. Correlation between model test results and full-scale measurements can always be undertaken subsequently when and if the latter becomes available.

SIMILITUDE

Seismic Modelling

The restrictive nature of the conflicting criteria to be satisfied for seismic studies are illustrated with reference to a reported experimental model study on the seismic response of long curved reinforced concrete overcrossing structures (3,4). In this example, a representational model was designed so that, although not an exact scaled version of a particular bridge, it incorporated the significant structural characteristics that contributed to the partial collapse of an interchange structure during the San Fernando, California earthquake of 9 February, 1971. In the study a 20 x 20 ft (6.1 x 6.1 m) shaking table was used, capable of applying prescribed base motions in one horizontal and in the vertical direction. Details of this facility have been reported elsewhere (5).

It was not possible within the constraints of the available facilities to meet the design requirements for a model of the complete structure. This would have required a geometric scale of 1/60, a scale that would have presented constructional problems and put some of the significant natural frequencies of the model outside the performance limits of the table. A feasibility study suggested a minimum scale of 1/30, which meant that only half of the prototype could be modelled. A simplified model that included the significant dynamic characteristics of the prototype in the region of the collapsed spans, was developed.

The development of a model to satisfy necessary similitude requirements of a complex structure such as a long curved bridge within the space and performance limitations of the shaking table can prove difficult. The major problem in designing such a model, apart from construction, is that of satisfying the similitude requirements for gravitational and dynamic force effects. An inelastic study requires the use of a true-scale model in which all length quantities are to the same geometric scale $L_r$. (Note: The subscript 'r' is used throughout to denote the ratio of any two homologous quantities in model and prototype.) To provide equal strains in the model and prototype ($\varepsilon_r = 1$), generally necessary for studying nonlinear geometric and material effects, all force quantities must be brought to a scale factor given by

$$F_r = \varepsilon_r E_r L_r^2 = E_r L_r^2$$

and when the same stress-strain relationships are used in model and prototype this reduces to $F_r = L_r^2$.

Gravitational forces are given by

$$F_g = \rho L^3 g$$

Obviously, such forces were important in this structure; they affect the frictional forces at the expansion joints, the torsional lifting of the deck, and the yielding of the columns. The inertial forces are given by

$$F_i = \rho L^3 a$$

Equating these three relations leads to the requirements

$$a_r = g_r = 1$$

hence the time ratio, $T_r = \sqrt{L_r}$

and the effective mass density ratio, $\rho_r = 1/L_r$

These requirements represent a true-scale model in which materials of the same properties are used in model and prototype, the applied ground accelerations are equal in model and prototype, the time scale is $\sqrt{L_r}$, and the effective mass density of the model material must be artificially increased to bring the mass density to a value of $1/L_r$.

Clearly, in such a study where yielding, ductility, impacting and fracture will occur, it is necessary to maintain equal stresses and equal material properties in model and prototype where possible; that is, the true-scale 'realistic' model is the only viable solution. Other options that occur in the field of structural modelling, which include distorting section or material properties, are generally only used where the linearity of structural deflections or material properties are preserved, or in particular cases where it is possible to maintain $\varepsilon_r = 1$ with materials of similar $\sigma/\varepsilon$ characteristics but of different strengths.

Nondimensional material properties such as Poisson's Ratio ($\nu$), and the damping factor ($\xi$) also have to be considered, and their values kept equal in model and prototype if they have a significant effect on the response. Clearly damping is influential in any dynamic problem, requiring that $\xi_r = 1$. This again calls for the use of the same basic materials in model and prototype.

Adopted Model

The model as conceived and duly constructed was a 1/30 true-scale weight-distorted model of a symmetrical simplified version of the prototype. Microconcrete was chosen as the material most appropriate for simulating reinforced concrete despite scale effects. For a seismic investigation, design philosophy implies that highly nonlinear response should prevail before collapse; that is a large reserve of strength and energy dissipating capacity exists beyond the limit of elastic behaviour. In such cases similarity between model and prototype material of the load-deflection characteristics for the complete range of displacement amplitudes up to failure (including cyclic behaviour) is essential. For the reinforced concrete bridge study the development of a similar microconcrete proved most satisfactory.

The effective mass density ratio of $1/L_r$ required in this solution was accomplished by adding external weights to the model in such a way that neither the section stiffnesses nor the system damping were appreciably affected (the mass of the model was increased to 30 times its self-weight - based on the prototype box girder section - by attaching lead blocks). The resulting model ratios of this structure when compared with the hypothetical prototype bridge were

$$L_r = 1/30, \quad T_r = 1/5.5, \quad a_r = 1 \text{ and } F_r = 1/900$$

Hydrodynamic Modelling

The dynamic response of offshore structures differs from that of nonsubmerged structures in that "drag" constitutes a significant portion of the resistance to response. Generally, correct simulation for complete structural systems requires satisfying several conflicting similitude criteria simultaneously. In the past, model studies have been undertaken for some structural types or individual components (6,7) on the assumption that certain parameters can be ignored (for example, gravity platforms where drag forces have been neglected so that the Froude model law is applicable) (8,9). However, for many offshore structures (in particular, the commonly used tubular steel template structures and some recent innovative design concepts) drag and inertial forces are of comparable significance. Accordingly, both force effects should be included in experimental studies. It should also be noted that, previously, many model studies have been undertaken for logistics purposes, e.g. to verify the adequacy of transportation and installation procedures. As such, they have not been concerned with similitude problems, and no attempt to quantify environmental design forces has been made. A more detailed review of the similitude problem as it pertains to model studies for offshore structures follows.

In such situations, inertial, gravitational, and drag forces may be significant. In dynamic problems, inertia forces are always present and these become the reference for scaling other forces. Inertia forces, $F_i$, due to acceleration, $a$, of the structural mass and/or the fluid mass, $m$, are given by

$$F_i = ma = mv^2/L$$

151

where $v$ is velocity and $L$ is a length dimension.

For fluid surface problems, such as wave excitation or stability problems where deadload effects are significant, gravitational forces, $F_g = mg$, must be taken into consideration. To represent both of these force types satisfactorily, they must be to the same scale in the model, unless the effects can be uncoupled. Consequently, it is necessary to satisfy the requirement that the Froude number, $\overline{F}$, which is their ratio

$$\overline{F} = F_i/F_g = v^2/Lg$$

must be maintained the same in both model and prototype; that is

$$\overline{F}_r = 1 \qquad\qquad (1)$$

Drag (or viscous) forces, $F_d = \mu L^2 dv/dL$, occur at structure-fluid interfaces and are significant when wake formation is important. Reynolds number, $\overline{R}$, is the ratio of the inertial force to viscous force

$$\overline{R} = F_i/F_d = \rho Lv/\mu$$

where $\rho$ is the mass density and $\mu$ is the viscosity of the fluid.

When drag forces are significant, dynamic similitude requires that the Reynolds number be maintained at corresponding values for model and prototype in addition to observing geometric similitude. Thus, in a model where both $F_i$ and $F_d$ are important, it may be necessary to satisfy the condition

$$\overline{R}_r = 1 \qquad\qquad (2)$$

In cases where all three force types $F_i$, $F_g$ and $F_d$ are important, it may be necessary to meet the requirement that $\overline{F}_r = \overline{R}_r = 1$, which may be difficult or even impossible to achieve. For example, assuming $g_r = 1$, equation (1) yields

$$v_r = \sqrt{L_r}$$

and, assuming $\rho_r = \mu_r = 1$, equation (2) yields

$$v_r = 1/L_r$$

Clearly, to satisfy this conflict, $L_r = 1$, indicating that model and prototype must be the same size.

Where it is not possible to satisfy both requirements, some judgement is required in modelling, and there may be justified doubt regarding the validity of the study. The usually large dissimilarity between the Reynolds number in the laboratory and in nature prompts many engineers to feel that the realistic forces cannot be simulated.

Possible Problem Solution

A potential solution to the conflict of the Froude number and Reynolds number requirements when both are important consists of restricting experiments to ranges of Reynolds numbers such that a constant drag coefficient prevails. This was the method employed by Kim and Hibbard (10), who appreciated the scaling problem for a jacket-type platform.  From their measurements, they concluded that a 1/3- to 1/6-scale structure in an appropriately scaled sea state allowed simulation of typical design wave forces.  Of course, this represents a very large-scale model that could not be tested with existing wave tank or shaking table facilities.

When such scales are impractical, artificial roughening of fluid/ structural surfaces to alter the Reynolds number ranges of flow might be considered.  Here the object is to ensure that a super-critical (or transcritical) flow condition, typified by a constant drag coefficient, occurs over the entire structure.  This is a procedure that has been adopted in model studies to investigate the wind response of large scale cooling tower shells as reported by Farrell and Patel (11).  Experience from this field could be drawn upon for future model studies.

CONCLUSION

Detailed engineering investigations can often be facilitated by use of physical model studies in which case similitude must be considered.  Despite the restrictive nature of the conflicting requirements for situations concerned with inelastic dynamic response, feasible models can be developed.  For seismic studies using shaking tables true-scale weight distorted models made of material similar to the prototype are often the best compromise.  For hydrodynamic studies special attention should also be paid to maintaining correct flow conditions and to this end the use of artificial roughening might be considered.

ACKNOWLEDGEMENTS

The material presented in this summary paper results from previous model studies performed at the Earthquake Engineering Research Center, University of California, Berkeley, and ongoing model work concerned with seismic and hydrodynamic effects at URS/Blume.  The author is grateful for the discussions and collaboration with colleagues from both organizations.  He particularly wishes to acknowledge the contribution of Professor William G. Godden, University of California, Berkeley, who was faculty investigator for the highway bridge model studies.

REFERENCES

(1)    GODDEN, W G, 'Model analysis and the design of prestressed concrete nuclear reactor structures', Nuclear Engineering and Design, North-Holland Publishing Co., Amsterdam, Vol. 9 (1969).

(2)    HARING, R E, HULETT, J M, PEARCE, B K, and POMONIK, G M, 'Improving platform design', Ocean Industry, Gulf Publishing Co., Houston  (May 1977).

(3)    WILLIAMS, D, and GODDEN, W G, 'Seismic response of long curved bridge structures: Experimental model studies', Earthquake Engineering and Structural Dynamics, Vol. 7 (1979).

(4)    WILLIAMS, D, and GODDEN, W G, Experimental Model Studies on the Seismic Response of High Curved Overcrossings, Report No. EERC 76-18, Earthquake Engineering Research Center, University of California, Berkeley (1976).

(5)    REA, D, and PENZIEN, J, 'Structural research using an earth-

153

quake simulator', Proceedings of the Structural Engineers Association of California, Monterey, California (1972).

(6)    SARPKAYA, T, 'Lift, drag, and added-mass coefficients for a circular cylinder immersed in a time dependent flow', Journal of Applied Mechanics, Transactions of the ASME (March 1963).

(7)    SARPKAYA, T, and GARRISON, C J, 'Vortex formation and resistance in unsteady flow', Journal of Applied Mechanics, Transactions of the ASME (March 1963).

(8)    HAFSKJOLD, P S, TØRUM, A, and EIE, J, 'Submerged offshore concrete tanks', Proceedings, XIX Congress, Permanent International Association of Navigation Congresses, Ottawa, Canada (1972).

(9)    GARRISON, C J, TØRUM, A, IVERSON, C, LEIVSETH, S, and EBBES-MEYER, C C, 'Wave forces on large volume structures - A comparison between theory and model tests', Proceedings, Offshore Technology Conference, Houston (1974).

(10)   KIM, Y Y, and HIBBARD, H C, 'Analysis of simultaneous wave force and water particle velocity measurements', Proceedings Offshore Technology Conference, Houston (1975).

(11)   FARRELL, C, and PATEL, V C, 'Flow around rounded rough-walled structures: Experimental and analytical studies', American Society of Civil Engineers Annual Convention, San Francisco (1977).

W Gene Corley
*Portland Cement Association, USA*
Andrew Scanlon
*Portland Association, USA*

## SUMMARY

A combined experimental and analytical investigation of behavior of structural walls under seismic loading has been done at the Portland Cement Association. One-third scale models of structural walls were tested under cyclic lateral loading to provide information on strength and ductility of walls over a wide range of structural parameters. Test results were used in development of force-deformation hysteretic loops for use in computer modelling of dynamic response. Significant findings of the combined investigation are presented.

## INTRODUCTION

Experience gained from recent earthquakes indicates that during seismic motion, well designed buildings stiffened by reinforced concrete structural walls perform well from the standpoint of both safety and damage control. However, because of concern for possible lack of ductility, this type of construction is penalized in current codes[1] by requirements for higher seismic design loads than are required for ductile moment resistant frames.

A combined analytical and experimental research program to investigate seismic performance of structural walls was done at the Construction Technology Laboratories of the Portland Cement Association. The overall objective of the program is to develop design criteria for reinforced concrete structural walls in earthquake-resistant buildings.

## EXPERIMENTAL INVESTIGATION

In the experimental portion of the work, reversing lateral loads were applied at the top of 1/3-scale models of structural walls. Tests were made on isolated walls[2], walls connected by beams[3], and walls pierced with openings[4]. Wall cross-section shapes considered in the test program were rectangular, barbell, and flanged as shown in Fig. 1. Also shown in Fig. 1 is a view of the test apparatus. The test program was undertaken to provide information on strength and deformation characteristics of structural walls under cyclic loading.

Selected results from the isolated wall tests are shown in Figs. 2 to 5. These results were obtained for a rectangular wall with 1.47% vertical reinforcement placed close to each end. First yield occurred after 10 cycles. The test was stopped after 30-1/2 cycles.

A continuous load-deflection plot shown in Fig. 2 indicates increase in deflection with increasing cycles at loads above yield level.

Decreasing stiffness on reloading as cyclic loading progresses is evident.

Moment-rotation loops shown in Fig. 3 indicate pinching in the reloading cycles. Shear-shear distortion loops shown in Fig. 4 exhibit more significant pinching. Pinching occurs when reloading commences on a cracked section with all load being resisted by the reinforcing bars. When the cracks close the section stiffness increases. Moment-rotation envelopes for a series of wall tests are shown in Fig. 5.

Free vibration tests[5] were carried out to compare frequencies and damping at initial load with frequencies and damping after several lateral load cycles near flexural yield. These tests indicated that:

1.  Prior to application of lateral load, measured frequencies ranged from 64 to 82% of the frequency calculated based on uncracked section properties.
2.  Measured frequencies decreased by an average factor of 2.2 from initial loading to load cycles near flexural yield. This reduction results from the decrease in stiffness due to extensive cracking at high load levels.
3.  For the same conditions as indicated in Item 2, the average damping coefficient changed from 3.4% to 8.5% of critical.

The following general observations were made from tests on isolated walls.

1.  Structural walls designed according to the 1971 ACI Building Code without special details for seismic load will attain the ACI flexure and shear design strengths. In addition they possess considerable inelastic deformation capacity (ductility) when subjected to reversing loads.
2.  For walls subjected to low shear stresses ($v_{max} < 0.25 \sqrt{f'_c}$ N/mm$^2$), flexural bar buckling and loss of compression concrete are limiting factors in inelastic performance. A significant number of inelastic cycles can be sustained prior to bar buckling.
3.  For walls subjected to high shear stresses ($v_{max} > 0.6 \sqrt{f'_c}$ N/mm$^2$), web shear distress is the limiting factor in inelastic performance. However even in a wall stressed to $0.7 \sqrt{f'_c}$ N/mm$_2$, a significant number of inelastic cycles can be sustained prior to loss of strength.
4.  Stiff boundary elements increase inelastic performance of walls.
5.  Confinement reinforcement placed in boundary elements within the hinging region of a structural wall significantly increases inelastic performance.
6.  Displacement due to shear distortions are a significant portion of total lateral inelastic displacements in walls subjected to reversing loads.

ANALYTICAL INVESTIGATION

Concurrent with the experimental program to determine strength and ductility of walls subjected to cyclic loading, an analytical investigation was carried out to determine force and ductility demands on walls subjected to seismic excitation. Computer program DRAIN-2D[6] was used to perform dynamic inelastic response analysis of structural walls.

In the computer analysis, walls and beams are modelled as line elements. Inelastic action is modelled by formation of plastic hinges at ends of line elements. Dynamic inelastic response to seismic loading consisting of an acceleration-time history is determined using step-by-step integration. Constant acceleration is assumed during each time step.

An important part of the analytical investigation consisted of selecting paramaters defining force-deformation hysteresis loops. Moment-rotation and shear force-shear distortion hysteresis loops were based on a model proposed by Takeda[7] as shown in Fig. 6. This model was modified to incorporate pinching and strength decay as shown in Figs. 7 and 8. For analysis of coupled wall systems, the basic model was modified to include effects of changing levels of axial load as shown in Fig. 9.

The initial elastic portion of the bilinear primary force-deformation relationship up to yield level can be determined from cross-sectional properties of the member with appropriate modifications to incorporate effects of cracking. Tests show that effective flexural stiffness prior to yielding can be as low as 30 to 50% of uncracked stiffness. Effective shear stiffness can be reduced to 10 to 50% of uncracked stiffness.

Post-yield stiffness, as well as unloading and reloading stiffness are more difficult to define. Results of the experimental test program were used to determine appropriate ranges for these parameters. Tests show that post-yield stiffness can be expected to be in the range of 3 to 10% of effective initial stiffness.

To determine effects of variations in force-deformation parameters as well as structural and input motion parameters, a large number of dynamic inelastic analyses were performed on isolated walls[8,9] and coupled walls[10,11]. Significant conclusions from these analyses are as follows.

1.  Stiffness properties of members before and after yielding should be specified as accurately as possible. However, moderate variations in post-yield loading, unloading, and reloading branches do not significantly affect dynamic response.
2.  Axial forces induced in coupled walls can significantly affect dynamic response. Wall stiffness and strength depend on axial force level. This aspect of hysteresis loop modelling should be incorporated in dynamic inelastic response analysis of coupled wall systems.
3.  Strength decay in hysteresis loops appears to be significant only in the case of coupling beams loaded well into the inelastic range.
4.  Dynamic response does not appear to be significantly affected by pinching or shear yielding.

CONCLUDING REMARKS

The combined experimental and analytical investigation has provided much useful information on behavior of structural walls subjected to seismic loading. This information is being used to develop design information for reinforced concrete structural wall systems. The investigation confirms field observations that properly designed and detailed structural walls possess adequate strength and ductility to resist significant seismic loading.

ACKOWLEDGEMENT

Support provided by the National Science Foundation through Grant No. ENV77-15333 is gratefully acknowledged. Any opinions, findings, and conclusions are those of the authors and do not necessarily reflect the views of the National Science Foundation.

REFERENCES

(1) Uniform Building Code, 1979 Edition, International Conference of Building Officials, 5360 South Workman Mill Road, Whittier, California 90601.

(2) OESTERLE, R G, FIORATO, A E, JOHAL, L S, CARPENTER, J E, RUSSELL, H G, AND CORLEY, W G, 'Earthquake-Resistant Structural Walls-Tests of Isolated Walls', Report to National Science Foundation, Portland Cement Association, November 1976.

(3) SHIU, K N, ARISTIZABAL-OCHOA, J D, BARNEY, G B, FIORATO, A E, AND CORLEY, W G, 'Earthquake Resistant Structural Walls - Coupled Wall Tests', Report to National Science Foundation, Portland Cement Association, July 1981.

(4) SHIU, K N, DANIEL, J I, ARISTIZABAL-OCHOA, J D, FIORATO, A E, CORLEY, W G, 'Earthquake Resistant Structural Walls - Tests of Walls with and without Openings', Report to National Science Foundation, Portland Cement Association, 1981.

(5) ARISTIZABAL-OCHOA, J D, 'Analysis of Free Vibration Tests of Structural Walls', Report to National Science Foundation, Portland Cement Association, February 1980.

(6) KANAAN, A E, AND POWELL, G H, 'General Purpose Computer Program for Inelastic Dynamic Response of Plane Structures', Report No. EERC 73-6, Earthquake Engineering Research Center, University of California, Berkeley, April 1973.

(7) TAKEDA, T, SOZEN, M A, AND NIELSEN, N N, 'Reinforced Concrete Response to Simulated Earthquakes', Journal of the Structural Division, ASCE, Vol. 96, No. ST-12, December 1970, pp 2557-2573.

(8) DERECHO, A T, FUGELSO, L E, AND FINTEL, M, 'Structural Walls in Earthquake-Resistant Buildings, Dynamic Analysis of Isolated Structural Walls, INPUT MOTIONS', Report to National Science Foundation, Portland Cement Association, December 1977.

(9) DERECHO, A T, GHOSH, S K, IQBAL, M, FRESKAKIS, G N, AND FINTEL, M, 'Structural Walls in Earthquake Resistant Buildings, Dynamic Analysis of Isolated Structural Walls, PARAMETRIC STUDIES', Report to National Science Foundation, Portland Cement Association, March 1978.

(10) SAATCIOGLU, M, DERECHO, A T, AND CORLEY, W G, 'Coupled Walls in Earthquake Resistant Buildings, Modeling Techniques and Dynamic Analysis', Report to National Science Foundation, Portland Cement Association, June 1980.

(11) SAATCIOGLU, M, DERECHO, A T, CORLEY, W G, PARMELEE, R A, AND SCANLON, A, 'Coupled Walls in Earthquake Resistant Buildings, Parametric Investigation and Design Procedure', Report to National Science Foundation, Portland Cement Association, July 1981.

## Fig. 1    Test Apparatus and Specimen Dimensions

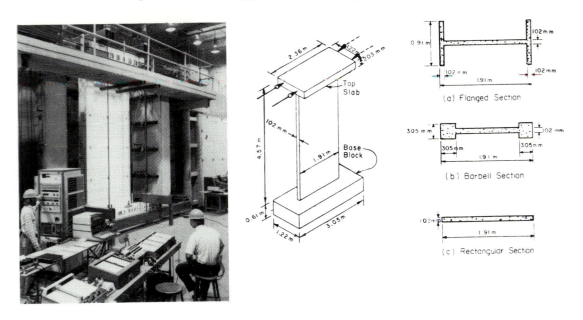

(a) Flanged Section

(b) Barbell Section

(c) Rectangular Section

## Fig. 2    Load–Deflection Plot

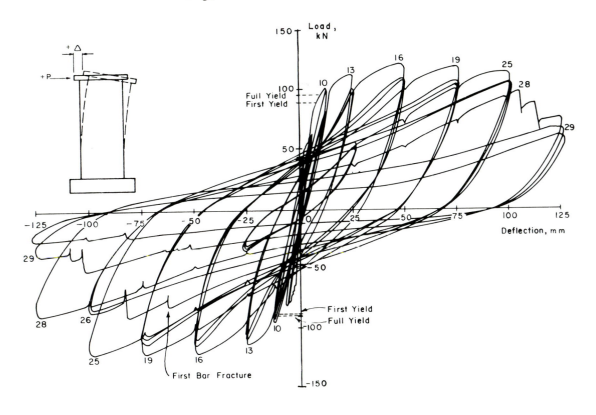

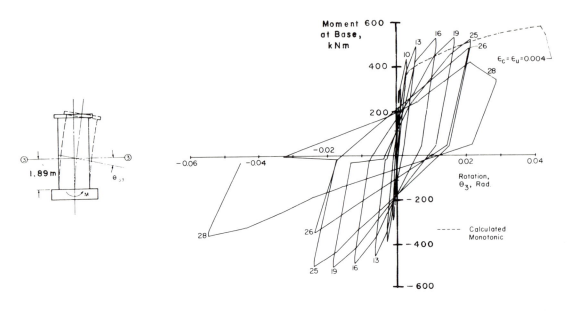

Fig. 3    Moment-Rotation Plot

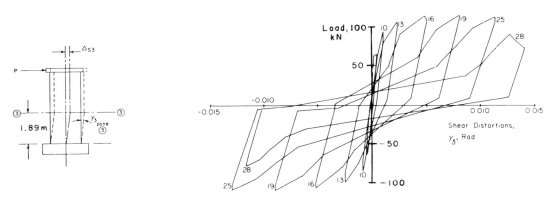

Fig. 4    Shear-Shear Distortion Plot

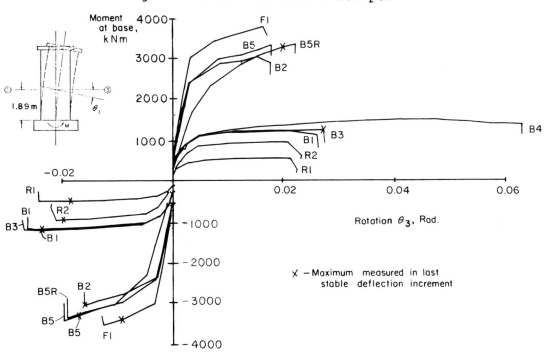

Fig. 5    Moment-Rotation Envelopes

160

**Fig. 6   Takeda's Hysteresis Loop**

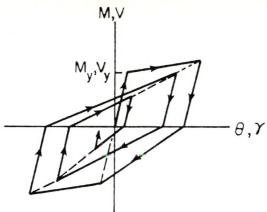

**Fig. 7   Pinching**

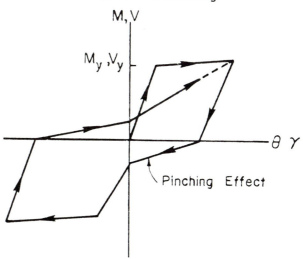

Pinching Effect

**Fig. 8   Strength Decay**

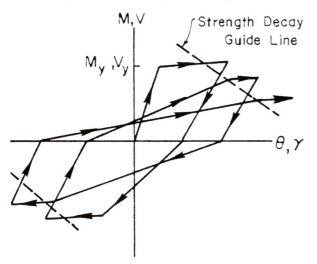

Strength Decay Guide Line

**Fig. 9   Flexure-Axial Load Interaction**

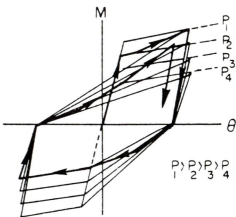

# Modelling of steel and reinforced concrete structures for seismic response simulation

## 18

Piotr D Moncarz
*Failure Analysis Associates, Palo Alto, California, USA*
Helmut Krawinkler
*Stanford University, Stanford, California, USA*

## SUMMARY

This paper summarizes the results of a four-year study on the feasibility and limitations of experimental model analysis in earthquake engineering. The relevant similitude laws for different types of dynamic models are summarized in a condensed tabular form. A more detailed discussion is presented on material simulation and the results of model studies on elements and simple structures.

## INTRODUCTION

In earthquake studies, where post-elastic response characteristics and safety against failure are of primary interest, analysis of many structures requires the application of experimental methods. The utilization of medium and small size earthquake simulators often necessitates studies with reduced scale physical models. Once the size of the model is established, modelling laws will define the desired characteristics of model materials, geometry and load history.

With the greatest degree of simplification, a typical stress $\sigma$ in the structure can be expressed through a functional relationship of the form

$$\sigma = F(\vec{r}, t; \rho, E, a, g, \ell, \sigma_0, \vec{r}_0) \qquad (1)$$

using the notation $\vec{r}$ = position vector, $t$ = time, $\rho$ = density, $E$ = material stiffness, $a$ = acceleration, $g$ = acceleration of gravity and $\ell$ = length. The subscript o refers to initial conditions. The functional relationship presented in Eq. 1 can be converted to a dimensionless form by application of dimensional analysis

$$\frac{\sigma}{E} = F\left(\frac{\vec{r}}{\ell}, \frac{t}{\ell}\sqrt{\frac{E}{\rho}}; \frac{a}{g}, \frac{g\ell\rho}{E}, \frac{\sigma_0}{E}, \frac{\vec{r}_0}{\ell}\right) \qquad (2)$$

Using this relationship a set of scaling laws for models suitable for seismic studies has been derived [1] and is summarized in Table 1.

Results of case studies involving steel and reinforced concrete prototype structures are presented. This study is concerned mainly with post-elastic response of the materials and the structures.

## MODELLING OF STEEL STRUCTURES

### Materials

Materials suitable for modelling of steel structures are either prototype structural steel or other metals of suitable characteristics. In this study, copper alloys with a specific stiffness of about half that of steel are considered. Monotonic stress-strain diagrams of annealed phosphor bronze (alloy 510) and cartridge brass exhibit a satisfactory similarity to the diagram of structural steel (Fig. 1). Weldability requirements eliminate cartridge brass as a suitable model material since the welding heat causes a considerable decrease in the strength of the material.

Since time is scaled in dynamic model tests (Table 1), the effects of strain rates on strength and stiffness properties need to be evaluated. Figure 2 presents results obtained for the effect of strain rates on the yield strength of structural steel and of annealed phosphor bronze. The effect of strain rates on the material stiffness was found to be negligible.

Simulation of the monotonic stress-strain diagram does not assure similitude of response to cyclic load. Figure 3 presents a comparison of hysteresis loops for structural steel and phosphor bronze. The loops are of similar shape but phosphor bronze exhibits a higher work hardening. This is even more evident from the skeleton curves of Fig. 4. The influence of those differences in work hardening on the behavior of the model will depend strongly on the number and amplitudes of inelastic excursions during the load history.

## Component and Structures

Certain phenomena can be studied only through component tests. An example of results from tests carried out on a structural steel cantilever and its 1:1.9 scale phosphor bronze model is presented in Fig. 5. The scale ratio was dictated by the scaling laws for true replica models, i.e., $\ell_r = (E/\rho)_r$ (Table 1). Except for the first few post yielding cycles (lower strength in the heat affected zone at weldments in the phosphor bronze model), the peak forces at the different displacement amplitudes and the hysteresis loops for the two specimens matched rather closely. Nevertheless, the higher amount of hysteretic energy dissipated per cycle by the phosphor bronze model indicates that phosphor bronze is not an ideal material for models of welded steel structures. Cycling at various frequencies has shown that the responses are not sensitive to cycling frequency.

Two single-degree-of-freedom structures, a steel pseudo-prototype and its phosphor bronze model, were subjected to the appropriately scaled N-S component of the 1940 El Centro acceleration record. Figure 6 provides the comparison of the column shear - story displacement responses of both structures to 177% of the earthquake record. As was expected from the material and component tests, the yield strength of the phosphor bronze model is smaller and its hysteresis loops fatter than in the steel structure. Nevertheless, the comparison is quite satisfactory as for the number of major inelastic excursions and their force and displacement amplitudes, thus demonstrating that phosphor bronze models can provide a reliable prediction of the peak response characteristics of steel prototype structures.

## MODELING OF REINFORCED CONCRETE STRUCTURE

## Model Materials

In modeling of R/C structures, microconcrete and steel wire gained the widest acceptance as model materials. A study on the compressive properties of microconcrete was performed using cylindrical specimens instrumented with four extensometers oriented at 90 degrees (Fig. 7). Tensile tests of the mixes showed a similar tensile/compressive strength ratio as in prototype concrete. Figure 8 presents compression stress-strain curves obtained for different strain rates. Figure 9 represents the summary of results obtained from over 100 test specimens tested under different strain rates. The same effect of strain rate on the compressive strength can be seen regardless of the strength of the mixes. Mild steel wire deformed by knurling offers bond properties similar to prototype reinforcement. To provide for the desired yield strength and to simulate the strain hardening present in prototype reinforcement, heat treatment following the deforming process is desirable.

## Component Testing

The simulation of inelastic cyclic response and cycling rate effects for beam elements were studied on eighteen 1:14.4 scale models (Fig. 10) of previously tested cantilever beam. Two beams were tested to study the simulation of the prototype response using its scaled load history. The load-displacement responses initially showed close resemblance in stiffness, strength and hysteretic behavior between the prototype and the model (Fig. 11a). However, the model beams showed considerably larger strength deterioration under large deformation reversals (Fig. 11b). This is due to the lack of strain hardening in the model reinforcements which did not permit a redistribution of stresses once a major flexural crack had formed.

The feasibility of small-scale model tests of beams failing in a mode controlled primarily by shear was investigated in the second test series (4 beams). Figures 12 and 13 show test results for one of the beams with L/d = 2.0. Both the pinched hysteresis loops and the crack pattern are representative of prototype behavior.

Dynamic rate effects were addressed in the third test series in which 12 beams were subjected to an identical displacement history (3 cycles per amplitude) with different frequencies. A comparison of the first hysteresis loops at two displacement amplitudes is presented in Fig. 14. Figure 15 presents the quantification of the influence of cycling frequency on the average peak loads at the two displacement amplitudes. There is no significant difference between the strength deterioration rate for cycles of the same amplitude carried out at different frequency levels. However, considerable deterioration can be observed in the low frequency test (0.025 Hz) at the first large inelastic excursion ($P_{21}$).

## CONCLUSIONS

Adequate simulation of material properties presents the most difficult and important aspect of small-scale modeling. Monotonic and cyclic stress-strain properties and properties significant for the particular model (e.g., weldability) have to be considered. Phosphor bronze was found to be suitable for dynamic model studies of steel structures despite the discrepancies in the cyclic stress-strain behavior. Carefully designed models can provide reliable information on overall dynamic inelastic response characteristics and may provide in many cases a rather accurate prediction of localized failure mechanisms.

## REFERENCES

1. Moncarz, PD, and Krawinkler, H, "Theory and application of Experimental Model Analysis in Earthquake Engineering," Report No. 50, The John A. Blume Earthquake Engineering Center, Department of Civil Engineering, Stanford University, Stanford, California, June 1981.

Table 1 Basic similitude requirements.

| Model Type / Scaling Parameters | | True Replica (1) | Artificial Mass Simulation (2) | Gravity Forces Neglected* any material (3) | Gravity Forces Neglected* prototype mat. (4) | Strain Distortion (5) |
|---|---|---|---|---|---|---|
| length | $\ell_r$ | $\ell_r$ | $\ell_r$ | $\ell_r$ | $\ell_r$ | $\ell_r$ |
| time | $t_r$ | $\ell_r^{1/2}$ | $\ell_r^{1/2}$ | $\ell_r(E/\rho)_r^{-1/2}$ | $\ell_r$ | $(\epsilon_r\ell_r)^{1/2}$ |
| frequency | $\omega_r$ | $\ell_r^{-1/2}$ | $\ell_r^{-1/2}$ | $\ell_r^{-1}(E/\rho)_r^{1/2}$ | $\ell_r^{-1}$ | $(\epsilon_r\ell_r)^{-1/2}$ |
| velocity | $v_r$ | $\ell_r^{1/2}$ | $\ell_r^{1/2}$ | $(E/\rho)^{1/2}$ | $1$ | $(\epsilon_r\ell_r)^{1/2}$ |
| gravitational acceleration | $g_r$ | $1$ | $1$ | neglected | neglected | $1$ |
| acceleration | $a_r$ | $1$ | $1$ | $\ell_r^{-1}(E/\rho)_r$ | $\ell_r^{-1}$ | $1$ |
| mass density | $\rho_r$ | $E_r/\ell_r$ | $-$ | $\rho_r$ | $1$ | $\epsilon_r E_r \ell_r^{-1}$ |
| strain | $\epsilon_r$ | $1$ | $1$ | $1$ | $1$ | $\epsilon_r$ |
| stress | $\sigma_r$ | $E_r$ | $E_r$ | $E_r$ | $1$ | $E_r\epsilon_r$ |
| modulus of elasticity | $E_r$ | $E_r$ | $E_r$ | $E_r$ | $1$ | $E_r$ |
| specific stiffness | $(E/\rho)_r$ | $\ell_r$ | $-$ | $(E/\rho)_r$ | $1$ | $\ell_r\epsilon_r^{-1}$ |
| displacement | $\delta_r$ | $\ell_r$ | $\ell_r$ | $\ell_r$ | $\ell_r$ | $\ell_r\epsilon_r$ |
| force | $F_r$ | $E_r\ell_r^2$ | $E_r\ell_r^2$ | $E_r\ell_r^2$ | $\ell_r^2$ | $E_r\ell_r^2\epsilon_r$ |
| energy | $(EN)_r$ | $E_r\ell_r^3$ | $E_r\ell_r^3$ | $E_r\ell_r^3$ | $\ell_r^3$ | $E_r\ell_r^3\epsilon_r^2$ |

*Can always be used for linear elastic models.

The scale ratios underlined are chosen by the investigator.

Figure 1 Normalized stress-strain diagrams.

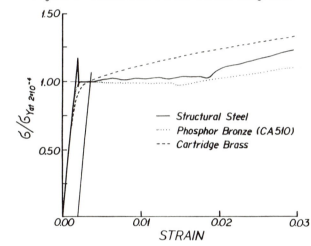

— Structural Steel  
⋯ Phosphor Bronze (CA510)  
--- Cartridge Brass

Figure 2 Effect of strain rates on yield strength of steel and annealed phosphor bronze.

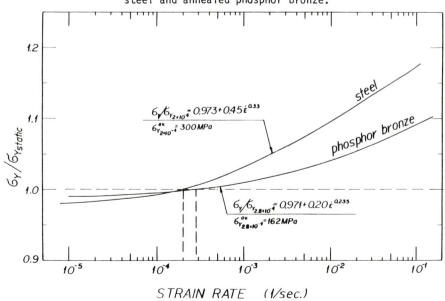

$$\frac{\sigma_Y}{\sigma_{Y_{2\times10^{-4}}}} = 0.973 + 0.45\dot{\epsilon}^{0.33}$$
$$\sigma_{Y_{2\times10^{-4}}}^{av} = 300\,MPa$$

$$\frac{\sigma_Y}{\sigma_{Y_{2.8\times10^{-4}}}} = 0.971 + 0.20\dot{\epsilon}^{0.235}$$
$$\sigma_{Y_{2.8\times10^{-4}}}^{av} = 162\,MPa$$

Figure 3   Comparison of stress-strain hysteresis loops.

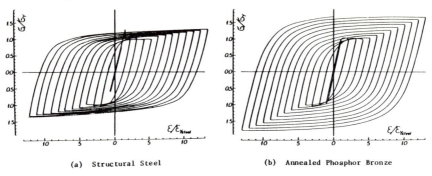

(a)   Structural Steel                    (b)   Annealed Phosphor Bronze

Figure 4   Skeleton curves for various load histories.

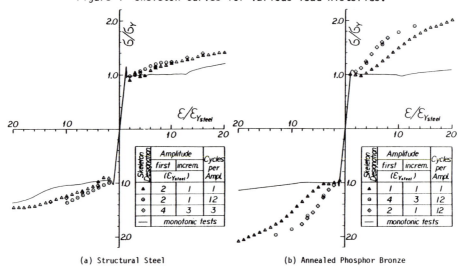

(a) Structural Steel                    (b) Annealed Phosphor Bronze

Figure 5   Comparison of cantilever load
displacement responses
(pseudo-prototype domain).

Figure 6   Shear-displacement response of
one column to 177% El Centro
motion (prototype domain).

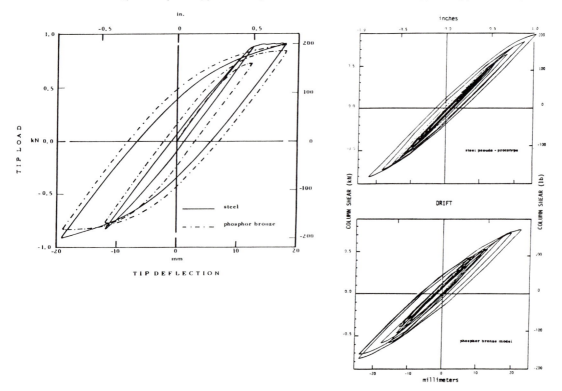

Figure 7 Microconcrete compression test.

Figure 8 Stress-strain curves for microconcrete compressed under various strain rates.

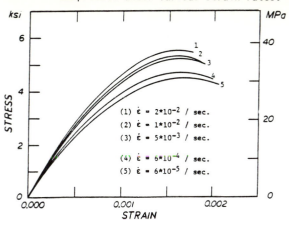

Figure 9 Effect of strain rates on compressive strength of microconcretes of different strengths.

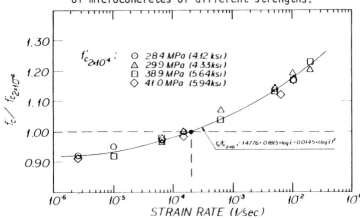

Figure 10 Reinforced microconcrete test specimen.

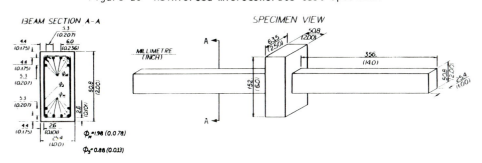

Figure 11 Hysteresis loops for prototype and model of reinforced concrete cantilever beam.

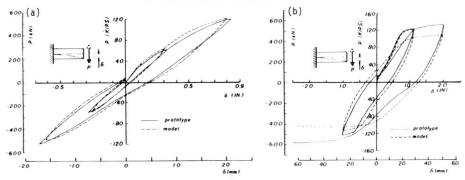

167

Figure 12  Load-displacement response from
high shear test.

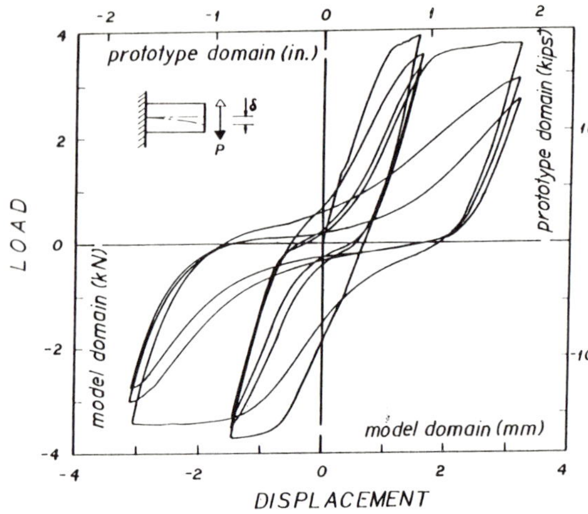

Figure 13  Crack pattern from
high shear test.

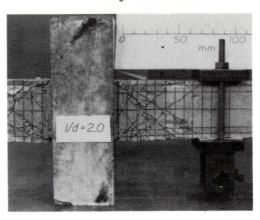

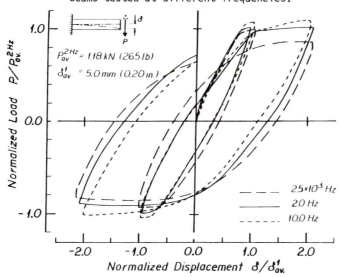

Figure 14  Hysteresis loops for models of R.C. cantilever
beams tested at different frequencies.

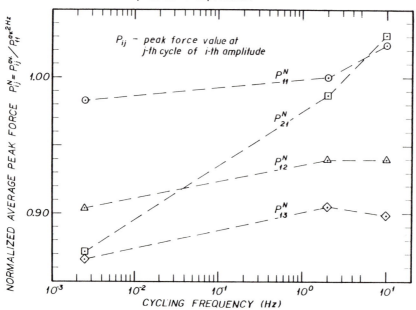

Figure 15  Influence of cycling frequency on the peak load at
a displacement amplitude.

168

# Contribution to discussion of Paper No. 18

Piotr D Moncarz
*Failure Analysis Associates, Palo Alto, California, USA*
Helmut Krawinkler
*Stanford University, Stanford, California, USA*

The purpose of this supplement is to expand on those aspects of the paper which initiated discussions with other participants of the seminar.

The basic similitude requirements presented in Table 1 pertain to three types of models: 1) True replica models (column 1), i.e., models in which all the pertinent physical quantities and material properties are scaled according to similitude laws. 2) Adequate models (columns 2, 3 and 4), i.e., models in which the distortion of one physical quantity does not require an adjustment of other physical quantities and of the prediction equation. Models with artificial mass simulation can be divided into lumped mass systems[2] used in the modelling of structures in which the masses can be concentrated at the floor levels, and distributed mass systems in which the distribution of the mass along the individual members is necessary. Models without gravity force simulation can be used in cases where stresses induced by gravity loads are small and may be viewed as negligible in comparison to inertia force effects. 3) Strain distorted models in which the nondimensional quantity of strain is subjected to scaling. This type of distortion seems promising for small scale model studies provided that no geometrical nonlinearities are expected.

The cold rolling process used for surface deformation of model reinforcement is accepted as the most reliable and suitable way of assurance of concrete-reinforcement bond simulation. It is, however, not always feasible to deform also the smaller wire used for the fabrication of stirrups. In this case chemical wire surface deformation might be used as a means to increase the bond properties of the stirrups. Figure 16 presents an example of surface roughening with hydrochloride and sulphur acids. This method allowed fabrication of very small shear reinforcement (stirrups) which were used in short beam specimens and resulted in a good simulation of shear behavior (Fig. 12) and crack pattern (Fig. 13).

Instrumentation for small-scale models poses particular problems, not as much for the measurement of forces and displacements but the measurement of localized deformations. Standard instruments can rarely be used for this purpose. However, commercially available displacement transducers composed of silastic tubing filled with mercury, together with in-house fabricated attachments were found to be well suited for the measurement of relative displacement over a short gage length. The gages are accurate and very inexpensive, but lose accuracy rapidly after a life of about two months. Applications for these gages are shown in Fig. 7 (extensometers for microconcrete cylinder tests), Fig. 17 (beam rotation measurement) and in Fig. 4 of Ref. 2 (shear distortion measurement in joint panel zones).

References

2.   Mills, R. M., "Small Scale Modelling of the Nonlinear Response
     of Steel Framed Buildings to Earthquakes," Proceedings,
     Dynamic Modelling of Structures, Building Research Station,
     Garston, Watford, England, 19th   20th November 1981.

Acknowledgements

This research was supported by the U.S.A. National Science
Foundation.  The cooperation of Dr. Russell M. Mills in this
project during his graduate studies is highly appreciated.

**Figure 16.  Surface of Stirrup Wire before and after chemical treatment**

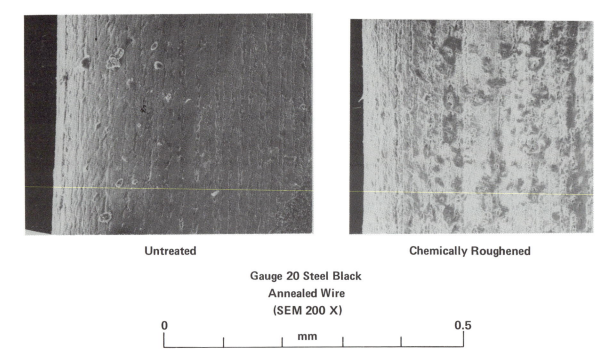

**Untreated**                                    **Chemically Roughened**

**Gauge 20 Steel Black
Annealed Wire
(SEM 200 X)**

0                                                            0.5
                              mm

**Figure 17.  Instrumentation for Beam Specimens.**

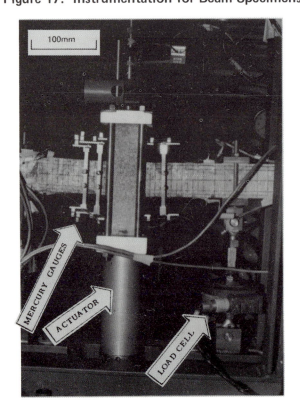

170

# Small-scale modeling of the nonlinear response of steel-framed buildings to earthquakes

19

Russell S Mills
*Project Research Engineer, URS/John A Blume & Associates, Engineers, San Francisco, California, USA*

SUMMARY

For certain structures, nonlinear dynamic response to earthquake loading may be accurately reproduced by testing small-scale models on earthquake simulators.  Artificial mass simulation is an applicable modeling technique provided that the mass can be effectively isolated from the structural material.  Steel-framed buildings often satisfy this condition, and building mass may be assumed to be concentrated at floor levels.

A comprehensive experimental study of a model of a previously tested prototype was undertaken to investigate this form of testing.  Correlations between prototype and model tests indicated that the model provided accurate simulation of the prototype and duplicated the primary energy-dissipating mechanism.  Minor discrepancies in correlation were attributed to fabrication techniques that limit the applicability of this modeling method and to anomalous motion of the earthquake simulator.

INTRODUCTION

Of primary importance to the performance of a building during a major earthquake is its ability to dissipate response energy without sustaining catastrophic structural collapse.  For steel-framed buildings, this energy is commonly dissipated through inelastic action of the structural system.  When coupled with dynamic considerations, this nonlinear response behavior produces a complex problem of analysis that may possibly be investigated by testing small-scale models on earthquake simulators.  The primary objective of such a study would be to replicate all aspects of the prototype structure that contribute to the response characteristics of interest.  To study earthquake-induced response, the energy-dissipating mechanisms and the level of strength and ductility must be reproduced.

When strength and post-yield response character are important and gravity effects cannot be neglected, dynamic similitude theory dictates strict physical requirements that the model must satisfy.  Fortunately, for steel-framed buildings the mass can often be assumed to be concentrated at floor levels, permitting a simplification of the modeling constraints through artificial mass simulation (AMS).  Modeling by AMS involves the addition of structurally uncoupled mass to augment the density of the model and permits selection of a model structural material without regard for mass density scaling.  The similitude relationships that must be satisfied by these models are summarized in Table 1.

Table 1   Modeling laws for artificial mass simulation

| Scaling Parameters*† | | Artificial Mass Simulation | |
|---|---|---|---|
| | | Any Material | Prototype Material |
| Length | $\ell_r$ | $\ell_r$ | $\ell_r$ |
| Time | $t_r$ | $\ell_r^{1/2}$ | $\ell_r^{1/2}$ |
| Frequency | $\omega_r$ | $\ell_r^{-1/2}$ | $\ell_r^{-1/2}$ |
| Velocity | $v_r$ | $\ell_r^{1/2}$ | $\ell_r^{1/2}$ |
| Gravitational acceleration | $g_r$ | 1 | 1 |
| Acceleration | $a_r$ | 1 | 1 |
| Structure mass | $M_r$ | $E_r \ell_r^2$ | $\ell_r^2$ |
| Strain | $\epsilon_r$ | 1 | 1 |
| Stress | $\sigma_r$ | $E_r$ | 1 |
| Modulus of elasticity | $E_r$ | $E_r$ | 1 |
| Displacement | $\delta_r$ | $\ell_r$ | $\ell_r$ |
| Force | $F_r$ | $E_r \ell_r^2$ | $\ell_r^2$ |
| Energy | $(EN)_r$ | $E_r \ell_r^3$ | $\ell_r^3$ |

*Subscript "r" refers to a ratio of model to prototype parameter (e.g., the model scale, $\ell_r = \ell_{model}/\ell_{prototype}$).

†Underlined scale ratios are chosen by the investigator.

With AMS modeling, the prototype structural material may be used as the modeling material to reduce the complexity of material property simulation.   Strain rate effects are an obvious source of concern since the prototype and model have different time scales (i.e., $t_r = \ell_r^{1/2}$).   However, research reported by Moncarz and Krawinkler (1) has shown that strain rate effects are generally insignificant when compared to other sources of experimental uncertainty.   Another potential problem is produced by the scaling requirements for mass and force. If the prototype structural material is used in the model, the model mass and force are scaled in the ratios $M_r = F_r = \ell_r^2$, which may exceed the capacity of the earthquake simulator.   This limitation can be overcome by using a model material that accurately replicates the properties of the prototype material and reduces the requirements on mass and force scaling to $M_r = F_r = E_r \ell_r^2$.   Moncarz and Krawinkler have found that suitably treated copper alloys, $E_r \approx 0.5$, can satisfactorily simulate the hysteritic properties of common structural steels and reduce mass and force requirements by 50% when used as the modeling material.

## PROTOTYPE AND MODEL DESCRIPTIONS

In order to explore the feasibility and limitations of this modeling technique, we undertook to fabricate and test a scale model at a Stanford University dynamic testing facility (2).   A well-defined prototype structure was necessary to enable an accurate evaluation of the applicability and accuracy of a small-scale model in replicating seismically induced response.   A three-story, single-bay structure previously tested on the shake table at the University of California, Berkeley (3), was selected as a suitable prototype.   This structure was designed to sustain yielding of the panel zone formed by the girder-column connections (Fig. 1) in shear prior to yielding of girder or column elements.   Thus, the primary objective of the small-scale model was to duplicate this behavior.

The prototype consisted of two parallel steel frames fabricated from standard rolled shapes of ASTM A36 grade steel (4) with moment-resisting welded connections.   The primary frames were bolted at floor levels to cross-beams and diagonal bracing, thus approximating floor diaphragms rigid in their own plane.  Bay bracing in the column's weak direction resisted motion transverse to the excitation axis.  To simulate gravity loads and to provide dynamic inertia, blocks of concrete weighing approximately 3,600 kg were added to each floor.

The configuration of the prototype was particularly suited for modeling by AMS.  A length scale, $\ell_r$, of 1:6 was selected with the resulting model configuration shown in Fig. 2.  The prototype material, steel, was selected as the modeling material to minimize simulation problems.  To satisfy the mass requirement, blocks weighing 114 kg were placed at each floor level.  Though the prototype structure was treated as full-scale for this study, it was actually smaller in size than a true structure.  Thus, the 1:6 scale chosen for the AMS model would have been comparable to a 1:10 to 1:20 scale if an actual structure had been used for the prototype.

Because the region near the joint panel zone was critical to the inelastic behavior of the structure, the detailing of the prototype girder-column and primary beam-column connections was faithfully reproduced in the model (Fig. 3).  However, some variations from prototype design were adopted to reduce the cost and difficulty of fabrication, but only when no significant deviations in response were expected as a consequence.

## MODEL FABRICATION

The model's wide-flange sections were produced by milling from ASTM A36 bar stock, the same base material used for the prototype.  The geometric properties of the sections that were produced are summarized in Table 2 and are compared to standard mill tolerances (4) reduced in scale to correspond to the model length ratio of $\ell_r$ = 1:6.  As can be seen, this technique of fabrication produced model elements that adequately satisfied geometric requirements.  All other auxiliary structural elements were also produced by milling.

Table 2   Machined elements for the model

| Dimension | Model Girder (W6x12 Prototype) | | | | Model Column (W5x16 Prototype) | | | |
|---|---|---|---|---|---|---|---|---|
| | Specified | Actual Average | Tolerance* | Actual Maximum | Specified | Actual Average | Tolerance* | Actual Maximum |
| Width, b (mm) | 16.99 | 17.02 | +1.0/-0.8 | ±0.10 | 21.16 | 21.16 | +1.0/-0.8 | ±0.13 |
| Depth, d (mm) | 25.53 | 25.55 | ±0.5 | ±0.18 | 21.03 | 21.06 | ±0.5 | ±0.20 |
| Web thickness, $t_w$ (mm) | 1.04 | 1.02 | -- | ±0.10 | 1.04 | 1.04 | -- | ±0.10 |
| Flange thickness, $t_f$ (mm) | 1.19 | 1.19 | -- | ±0.18 | 1.55 | 1.57 | -- | ±0.15 |
| Area, A (mm²) | 64.6 | 66.3 | ±2.0 | ±3.2 | 84.3 | 87.5 | ±2.0 | ±3.2 |
| Moment of inertia, $I_x$ (mm⁴) | 7,130. | 7,360. | -- | -- | 6,750. | 7,000. | -- | -- |
| Camber (mm) | -- | -- | 0.6 | 0.5 | -- | -- | 1.0 | 0.3† |
| Sweep (mm) | -- | -- | 1.3 | 0.3 | -- | -- | 1.0 | 1.0 |
| Flange out of square (mm) | -- | -- | 1.0 | 0.3 | -- | -- | 1.0 | 0.3 |

*Tolerances are scaled values from standard mill practice, Manual of Steel Construction (4).
†Column camber increased to 1.3 mm after stiffeners were welded and heat treatment was provided.

All primary structural elements were welded using the tungsten-inert gas (TIG) process with argon as the shielding gas.  Welding at such a reduced scale required great skill, and precautions were taken to minimize distortions caused by concentrated heating.  The model frames were clamped to a heat-sink during welding to restrict adverse motion yet still permit longitudinal thermal expansion.  Nevertheless, some distortion of the model elements, primarily column camber, was unavoidable, and the joint welds were larger than was required to produce exact prototype simulation.  The finished frames were positioned and heat treated at 595°C for one hour, then allowed to cool in-furnace

173

to ambient temperature to remove the distortions produced by high initial stresses without altering the properties of the base material. As a consequence, however, the model did not replicate the initial stress state of the prototype and would not be suitable for studying any response phenomenon related to the initial state of stress. In addition, the oversized welds produced joint panels somewhat smaller in size than those existing in the prototype. A detail of a girder-column joint (Fig. 4) illustrates the welded connection and the instrumentation, consisting of a strain-gage rosette and strain-sensitive mercury-filled silastic gages, used to detect deformation produced by inelastic yielding of the panel zone.

## TEST RESULTS

A comprehensive experimental study encompassing material, subassemblage, and earthquake simulation tests was used to define the adequacy of prototype replication by the model. The material and subassemblage tests served to define mechanical properties, to verify the adequacy of fabrication techniques, and to refine the instrumentation system. The results indicated that yield levels between 10% and 20% higher could be expected for the model as a result of a slightly higher yield stress and the smaller joint panel zones produced by the oversized welds. Initial static and low-amplitude dynamic tests of the completed model were performed on the earthquake simulator to determine the initial state of stress and to define fundamental dynamic properties (Fig. 5).

Earthquake simulation tests utilized actual and artificial earthquake records to produce both elastic and inelastic structural response. After model response had been converted to the prototype domain by the appropriate scaling laws, macroscopic and microscopic response quantities were used to compare prototype and model behavior. Typical results of an inelastic test using earthquake input motion are shown in Figs. 6 and 7. In Fig. 6a the relationship between energy input by ground motion to the structure and energy dissipated through inelastic action of the joint panel zones illustrates nonlinear inelastic response. Ductility demand and energy dissipation are illustrated at a microscopic level by moment-deformation behavior (Fig. 7), where the girder flexural moment, measured at equivalent locations for the prototype and model, is shown versus the diagonal strain measured in the center of the joint panel zone.

## TEST SUMMARY

Accurate simulation of the prototype was demonstrated by the similar energy-dissipation mechanisms and dynamic properties of the model. The yielding of the joint panel zones in shear was duplicated by the model, and agreement was observed between the two structures in terms of the amplitude and frequency content of response and the ductility demand and number of inelastic excursions.

Minor discrepancies in correlation were produced by the inadequate modeling of the initial stress state, the oversized welds of the model, and the influence of anomalous motion of the earthquake simulator on test-structure response. Relieving the stress of the finished model frames reduced the distortion produced by high initial stresses in the model but limited application of the study to topics where initial stress is insignificant. The model's oversized welds contributed to an increase of approximately 10% in yield strength and a proportional increase in inelastic stiffness. However, this result did not significantly alter the correlation and was anticipated.

Of significance is the influence of the earthquake simulator's reproduction capability on the observed response. This influence occurs because a lightly damped elastic structure is extremely sensitive to fluctuations within a narrow frequency interval of the input motion spectrum. This problem is of less importance for high-intensity inelastic tests, but sufficient energy must still be transferred to the structure at the elastic level before the yield threshold is reached. Since all earthquake simulators possess some reproduction inadequacies, various input motions should be used to provide a basis for prototype-model correlations.

## CONCLUSIONS

Within limitations, it is possible to accurately reproduce the nonlinear dynamic response of structures to earthquake motions by testing small-scale models on earthquake simulators. Artificial mass simulation is an applicable modeling technique provided that the mass can be effectively isolated from the structural load-resisting material. Steel-framed buildings often satisfy this condition, and building mass may be assumed to be concentrated at the floor levels.

For small-scale models of actual buildings, fabrication techniques somewhat different from those presented here may become desirable. Instead of using steel as the model material, other materials with lower elastic modulii may be used to reduce the resulting mass and force scaling requirements, provided that other prototype material properties are properly simulated. With these materials, a more efficient means of element fabrication than milling may be desirable (extrusion, for example), especially when many members of similar geometric properties are necessary.

Replicating certain aspects of prototype behavior by small-scale modeling is not currently feasible. For example, it was not possible to duplicate the initial stress state of the prototype for this study. Thus, phenomena that are a function of initial stress, such as buckling and initial yielding, could not be duplicated by the model. Also, distortions in the model's weld sizes would preclude the application of these techniques to studies of weld fracture.

## ACKNOWLEDGMENT

This research was supported by the National Science Foundation through grants ENV75-20036 and ENV77-14444. The author performed this work as a graduate student and research assistant at Stanford University, under the guidance of Professor Helmut Krawinkler and in cooperation with then fellow graduate student Dr. Piotr D. Moncarz.

## REFERENCES

(1)  MONCARZ, P D, and KRAWINKLER, H, 'Modelling of steel and reinforced concrete structures for seismic response simulation,' Proceedings, Dynamic Modelling of Structures, Building Research Station, Garston, Watford, England (1981).

(2)  MILLS, R S, KRAWINKLER, H, and GERE, J M, Model Tests on Earthquake Simulators -- Development and Implementation of Experimental Procedures, Blume Earthquake Engineering Center Report No. 39, Stanford University, California (1979).

(3)  CLOUGH, R W, and TANG, D T, Earthquake Simulator Study of a Steel Frame Structure, Vol. I: Experimental Results, Earthquake Engineering Research Center (EERC) Report No. 75-6, University of California, Berkeley (1975).

(4)  Manual of Steel Construction, 7th edition, American Institute of Steel Construction, New York (1973).

Figure 1  Joint panel zone.          Figure 2  Plans and elevations of the model.

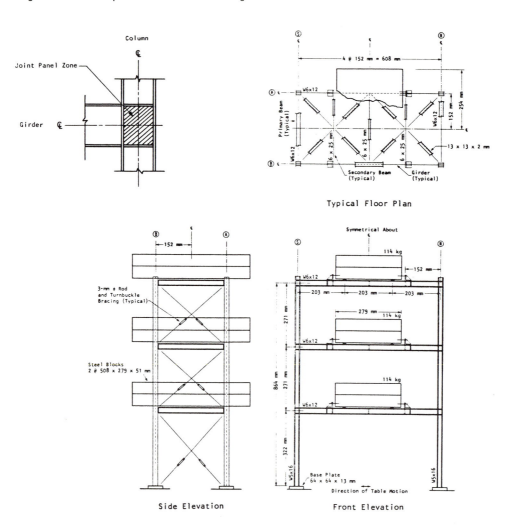

Side Elevation          Front Elevation

Figure 3  Girder and primary beam connections to column.

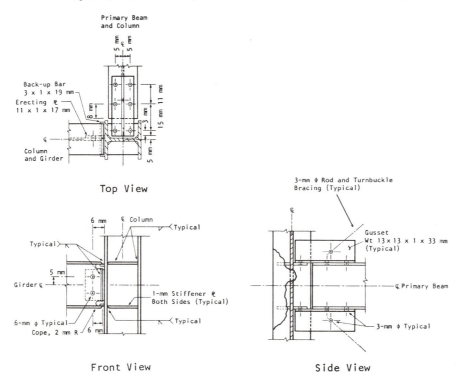

Top View

Front View          Side View

176

Figure 5  Comparison of prototype and model linear-dynamic properties.

Figure 4  Detail of the girder-column joint.

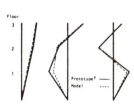

| Property | Mode | | |
|---|---|---|---|
| | 1st | 2nd | 3rd |
| Frequency,* Hz | | | |
| Prototype | 2.3 | 7.8 | 15.2 |
| Model | 2.4 | 8.3 | 15.5 |
| Damping, % | | | |
| Prototype | 0.11 | 0.08 | 0.57 |
| Model | 0.28 | 0.16 | 0.29 |

*Prototype time reference

†Prototype structure with 300-kg shaker at the first floor

Figure 6  Macroscopic inelastic response to earthquake.

a)  Energy distribution

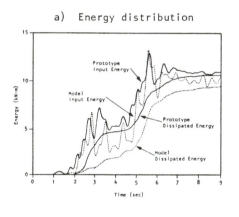

b)  Base shear

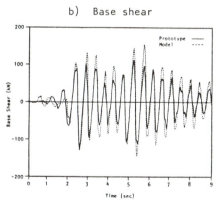

Figure 7  Microscopic inelastic response to earthquake girder moment versus 45° panel strain at center of joint.

a)  Prototype

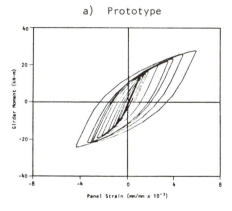

b)  Model

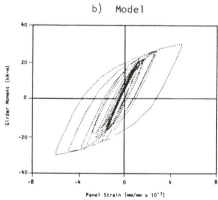

177

## 20

# The reduction of degrees of freedom in the dynamic analysis of complex three-dimensional structures

V M Trbojevic, R R Kunar and D C White
*Principia Mechanica Limited, Newton House, 50 Vineyard Path, London SW14 8ET, UK*

## SUMMARY

The reduction of degrees of freedom in dynamic analysis of complex structures is discussed. The reduction procedure which blends Guyan's reduction with concepts of modal synthesis is presented and applied to a three-dimensional structure. The results for full three-dimensional and the reduced model are compared.

## INTRODUCTION

The dynamic analysis of structures subjected to a specific earthquake input is often performed in the time domain using a direct integration scheme. The solution of the equations of motion is evaluated successively at many different time intervals. This procedure can be quite expensive for a large number of degrees of freedom as generally encountered in three-dimensional analysis. Therefore it is necessary to keep the degrees of freedom as few as possible while maintaining the proper vibrational characteristics of the structure.

There are essentially two methods by which a suitable model for dynamic analysis may be set up: the empirical generation of beam cantilever and lumped mass model, or the finite element idealisation. The former method leads to fewer degrees of freedom but relies heavily on the experience and expertise of the analyst. The finite element approach on the other hand replaces some of the engineering experience by a rigorous calculation of stiffness and inertial properties of a structure. The implementation of the finite element idealisation of complex three-dimensional structures usually results in a large number of degrees of freedom, the handling of which by a direct integration scheme becomes prohibitively expensive. To use this method therefore, the reduction of the number of degrees of freedom is important and desirable. It is also vitally important that the vibrational characteristics of the three-dimensional model are not significantly affected by any reduction procedure. One reduction method has been suggested by Guyan [1]. This method may introduce inaccuracies in the results because the choice of degrees of freedom to be retained is governed by physical arguments which are not always sufficient to preserve the inertial properties of a structure. In order to overcome this difficulty and make the choice of degrees of freedom to be retained in the analysis as simple as possible, a procedure is utilised which combines Guyan's reduction (static condensation) and concepts of modal synthesis. This dynamic reduction procedure retains important selected physical degrees of freedom and replaces the others by a few generalised degrees of freedom which represent the important modes of vibrations in the structure. This procedure is not new as it has been suggested in ref.[2] and the similar version has been implemented in the NASTRAN computer code. However, its implications in three-dimensional dynamic analysis are not fully appreciated and deserve public discussion. For example, in seismic soil-structure interaction problems in which floor response spectra are to be computed, this method can be used in the structure to improve analysis time considerably without a loss of accuracy [3].

The purpose of this paper is to present a theoretical basis for the reduction of the number of degrees of freedom in a structural model without loss of accuracy. This dynamic reduction procedure is applied to a structure typical of the nuclear reprocessing industry. Analyses are performed for the reduced and unreduced models and the results of the two are compared to demonstrate the accuracy of the method.

## Theoretical Basis of the Reduction Procedure

For the better understanding of this dynamic reduction procedure, it is best to start from the so-called 'static condensation' or Guyan's reduction [1]. Starting from the static equilibrium equation

$$\mathbf{K} \, \mathbf{r} = \mathbf{R} \tag{1}$$

where

$$\mathbf{r} = \text{displacement vector}$$
$$\mathbf{R} = \text{corresponding external forces}$$
$$\mathbf{K} = \text{structural stiffness matrix}$$

The condensation is achieved by partitioning the degrees of freedom into a set to be eliminated $\mathbf{r}_s$ (slaves) and the set $\mathbf{r}_m$ (masters) to be retained. Note that the forces $\mathbf{R}_s$ corresponding to $\mathbf{r}_s$ are zero. The equation (1) may now be written in the partitioned form as follows

$$\begin{bmatrix} \mathbf{K}_{ss} & \mathbf{K}_{ms} \\ \mathbf{K}_{sm} & \mathbf{K}_{mm} \end{bmatrix} \begin{bmatrix} \mathbf{r}_s \\ \mathbf{r}_m \end{bmatrix} = \begin{bmatrix} 0 \\ \mathbf{R}_m \end{bmatrix} \tag{2}$$

Eliminating slave degrees of freedom results in

$$\overset{*}{\mathbf{K}} \, \mathbf{r}_m = \mathbf{R}_m \tag{3}$$

where the reduced stiffness matrix is of the form

$$\overset{*}{\mathbf{K}} = \mathbf{K}_{mm} - \mathbf{K}_{ms} \, \mathbf{K}_{ss}^{-1} \, \mathbf{K}_{sm} \tag{4}$$

The same result may be obtained by the following

transformation

$$\mathbf{r} = \mathbf{T} \ \mathbf{r}_m \qquad\qquad (5)$$

where the transformation matrix $\mathbf{T}$ is given by

$$\mathbf{T} = \begin{bmatrix} - \ \mathbf{K}_{ss}^{-1} \ \mathbf{K}_{sm} \\ \mathbf{I}_{mm} \end{bmatrix} \qquad\qquad (6)$$

and $\mathbf{I}_{mm}$ is m x m unit matrix

This transformation gives reduced stiffness and mass matrices

$$\overset{*}{\mathbf{K}} = \mathbf{T}^t \ \mathbf{K} \ \mathbf{T}$$
$$\overset{*}{\mathbf{M}} = \mathbf{T}^t \ \mathbf{M} \ \mathbf{T} \qquad\qquad (7)$$

where

$\mathbf{M}$ = structural mass matrix
$\overset{*}{\mathbf{M}}$ = reduced mass matrix

If the mass matrix is partitioned in the same fashion as the stiffness matrix in the eq.(2), the full expression for $\overset{*}{\mathbf{M}}$ yields

$$\overset{*}{\mathbf{M}} = \mathbf{M}_{mm} - \mathbf{M}_{ms} \ \mathbf{K}_{ss}^{-1} \ \mathbf{K}_{sm} - \mathbf{K}_{ms} \ \mathbf{K}_{ss}^{-1} \ \mathbf{M}_{sm}$$
$$+ \ \mathbf{K}_{ms} \ \mathbf{K}_{ss}^{-1} \ \mathbf{M}_{ss} \ \mathbf{K}_{ss}^{-1} \ \mathbf{K}_{sm} \qquad\qquad (8)$$

From equation (8) it is evident that the accuracy of the analysis depends on the combination of mass and stiffness terms, basically on the transformation (6) and the mass associated with the slave degrees of freedom $\mathbf{r}_s$. Clearly, if no mass is lumped to the slave degrees of freedom, the reduced model is exact. For a general three-dimensional structure this is often not the case and to obtain a reliable reduced model it is necessary to account for the inertial effects of all degrees of freedom which should be eliminated. This is achieved by expressing the appropriate part of the structural degrees of freedom in terms of a few selected modes of vibration. The resulting relationship is in the form

$$\mathbf{r}_s = \boldsymbol{\phi}_n \ \mathbf{r}_n - \mathbf{K}_{ss}^{-1} \ \mathbf{K}_{sm} \ \mathbf{r}_m \qquad\qquad (9)$$

where

$\boldsymbol{\phi}_n$ = set of n modal vectors
$\mathbf{r}_n$ = corresponding modal amplitudes (generalised degrees of freedom)

Note that if all modal vectors are taken into account, the number of generalised degrees of freedom $\mathbf{r}_n$ would be the same as the number of slaves $\mathbf{r}_s$, and the model would be exact.

If however the response in the lowest modes is dominant, then only a few of the important modes are taken into account which ensures a reliable reduced model.

The transformation between the original displacements $\mathbf{r}$ and the new set consisting of $\mathbf{r}_n$ and $\mathbf{r}_m$ is now written

180

as

$$r = T \begin{bmatrix} r_n \\ r_m \end{bmatrix} \qquad (10)$$

where

$$T = \begin{bmatrix} \emptyset_n & -K_{ss}^{-1} K_{sm} \\ 0 & I_{mm} \end{bmatrix} \qquad (11)$$

The reduced stiffness and mass matrices are of the form

$$\hat{K} = \begin{bmatrix} \emptyset_n^t K_{ss} \emptyset_n & \vdots \\ & \vdots & \overset{*}{K} \end{bmatrix} \qquad (12)$$

$$\hat{M} = \begin{bmatrix} \emptyset_n^t M_{ss} \emptyset_n & \vdots & \emptyset_n^t (M_{sm} - M_{ss} K_{ss}^{-1} K_{sm}) \\ symm. & \vdots & \overset{*}{M} \end{bmatrix} \qquad (13)$$

A point to note at this stage is that $\emptyset_n$ are modes of vibration of the structure with all master degrees of freedom constrained. The number of modes should be equal to the number of frequencies of interest.

Example of Application

A structure used in this example is shown in Fig.1. It consists of a basemat of thickness 1.5m and dimensions 8m x 18m (Fig.2) and two towers, the first of which is 18m high with a base 6m x 6m, wall thickness 1m, and the second is 15m high with a base 2m x 3m. The heights of both towers are measured from the top of the basemat. The material is concrete with the following properties: Young's modulus E = 2.411 x $10^{10}$ N/m$^2$, Poisson's ratio $\mu$ = 0.15, mass density $\rho$ = 2.32 x $10^3$ kg/m$^3$ and damping is 5% of critical (Rayleigh damping was assumed).

The full three-dimensional model consists of 117 8-noded brick elements, with 272 nodes and 816 degrees of freedom. The reduced model consists of nodes marked by dots (Figs.1 and 2), with 51 master degrees of freedom and 7 generalised ones, i.e. 58 degrees of freedom altogether. Seven modes included in the reduction span all frequencies up to 48.2Hz for the structure which consists of slave degrees of freedom. The eigenvalue analysis is now performed on both models for a free-free structure. The results of this test are given in Table 1.

181

Table 1     Frequencies for reduced and unreduced
            structure

| No. | Frequency [Hz] | |
| --- | --- | --- |
| | 3D Model | Reduced Model |
| 1 | 2.365 | 2.365 |
| 2 | 3.507 | 3.508 |
| 3 | 7.311 | 7.337 |
| 4 | 14.510 | 14.619 |
| 5 | 17.975 | 18.119 |
| 6 | 22.453 | 22.703 |
| 7 | 24.376 | 25.585 |
| 8 | 26.102 | 27.363 |
| 9 | 34.063 | 35.690 |

The typical earthquake motion in the horizontal
direction is then applied to the base of the models.
Peak input acceleration was 0.2g and the duration 5
seconds. The explicit time integration scheme is used
for the reduced model while the three-dimensional model
was analysed by SAPIV [4] utilising implicit Wilson θ
algorithm. The time histories of the horizontal
accelerations at points A and B (Fig.1) are used for the
calculation of the corresponding response spectra for 5%
damping. The results are presented in Figs.3 and 4.
The agreement between two models is very good which
proves the effectiveness of this reduction procedure.

## REFERENCES

[1] GUYAN, R J, 'Reduction of stiffness and mass
matrices', AIAA Journal, Vol.3, No.2 (1965) 380.

[2] CLOUGH, R W, WILSON, E L, 'Dynamic analysis of
large structural systems with local non-linearities',
Comp. Meth. Appl. Mech. Engng., 17/18 (1979) 107-129.

[3] KUNAR, R R, WHITE, D C, ASHDOWN, M J, WAKER, C H,
DAINTITH, D, 'A sensitivity study for soil-structure
interaction', Proc. 6th Int. Conf. on Struct. Mech. in
Reactor Tech., paper M7/5 (1981).

[4] BATHE, K-J, WILSON, E L, PETERSON, F E, 'SAPIV - A
structural analysis program for static and dynamic
response of linear systems', Rep.No.EERC73-11,
University of California, Berkeley, California (1974).

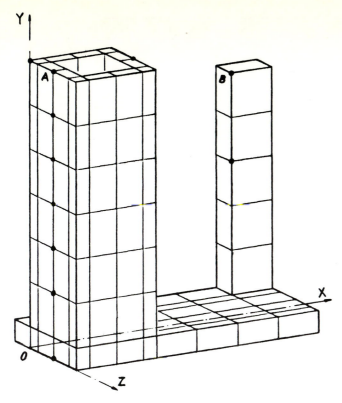

Figure 1   3D Structural Model

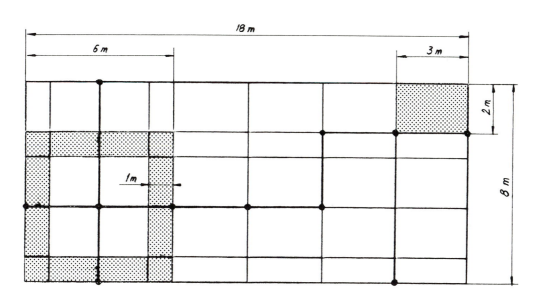

Figure 2   Basemat Idealisation

183

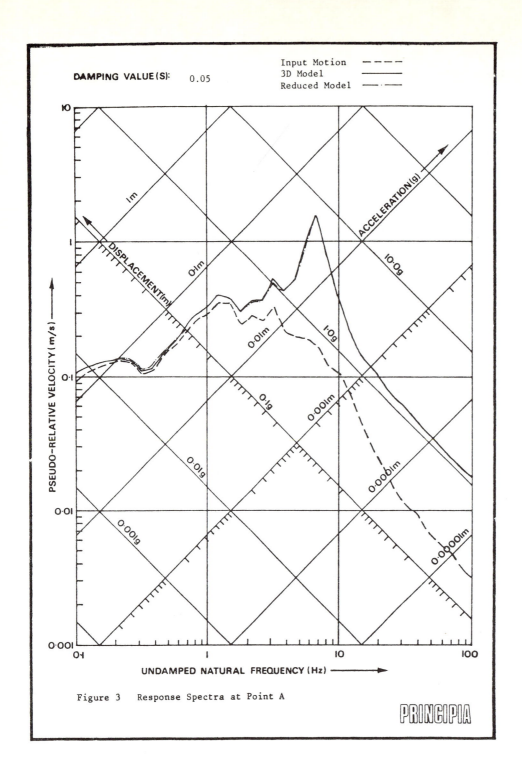

Figure 3    Response Spectra at Point A

184

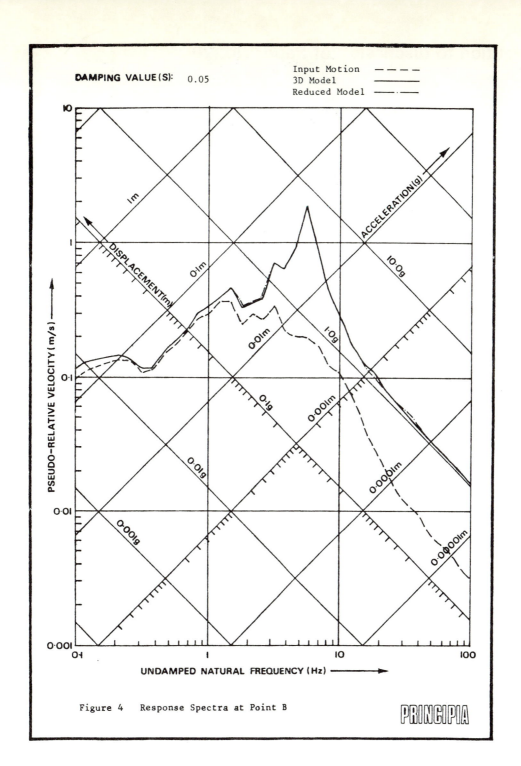

Figure 4    Response Spectra at Point B

PRINCIPIA

185

# Contribution to discussion of Paper No. 20

M A Murray
*Department of Civil Engineering, The Queen's University of Belfast, UK*

It was pleasant to find in the midst of a number of papers describing elegant experimental models of dynamic structural behaviour, one addressing itself to an approximate numerical technique. Although I find myself at variance with the authors on several points, they are to be commended for their attempt to seek a public discussion on the mass condensation technique and in particular the improvement to it deriving from the inclusion of some constrained structural models.

It is clear that the eliminated variables should be associated with areas of high stiffness or low mass, if the effect of the neglected inertia upon engineering intuition is questionable when complex structures are encountered. It is for this reason that I would have liked their Table 1 to have included the frequencies predicted by the Guyan reduction alone. The choice of seven generalised mode shapes to augment the master freedoms appears rather arbitrary and I would be interested to find how sensitive the results presented in Table 1 are to increasing the number of modal shapes. I suspect that the inclusion of some key mode shapes will improve accuracy dramatically. Let us assume, using the authors' notation, that the interior eigenvalue problem comprising the slave nodes, can be written as

$$K_{ss} r_s = \mu M_{ss} r_s$$

with $\phi_i$ the nodal vector associated with each eigenvalue $\mu_i$. From the orthogonality condition condition,

$$\phi_i^T M_{ss} \phi_i = m_i \text{ and } \phi_i^T K_{ss} \phi_i = k_i$$

then $m_i r_n = \mu_i k_i r_n$

Combining several of these lower nodes in the rectangular modal matrix $\phi_n$ and denoting the generalised vector of modal amplitudes $r_n$ we obtain the authors' equation (9). It is useful to write,

$$\phi_n^T M_{ss} \phi_n = M_D \text{ and } \phi_n^T K_{ss} \phi_n = K_D \text{ hence}$$

$$M_D r_n = \mu K_D r_n \quad \phi r,$$

$$M_{ss} \phi_n r_n = \mu K_{ss} \phi_n r_n$$

i.e., $\quad \phi_n r_n = \mu M_{ss}^{-1} K_{ss} \phi_n r_n$

The off diagonal mass sub matrix from equation (13) has the form

$$M^*_{ms} = M^{*T}_{sm} = (M_{ms} - K_{ms}K_{ss}^{-1}M_{ss}) \, \phi_n r_n$$

hence on substituting for $r_n$

i.e. $M^*_{ms} = \mu(M_{ms} - K_{ms}K_{ss}^{-1}M_{ss}) \, M_{ss}^{-1}K_{ss}\phi_n r_n$ we obtain,

$$M^*_{ms} = \mu \, (M_{ms}M_{ss}^{-1} - K_{ms}K_{ss}^{-1}) \, K_{ss}\phi_n r_n$$

Let $R = (M_{ms} M_{ss}^{-1} - K_{ms}K_{ss}^{-1}) \, K_{ss}\phi_n$

$$\therefore \quad M^*_{ms} = \mu R \, r_n$$

The equation of motion can then be written as,

$$\begin{bmatrix} M_D & \mu R^T \\ \mu R & M^* \end{bmatrix} \begin{Bmatrix} r_n \\ r_m \end{Bmatrix} = \lambda \begin{bmatrix} K_D & 0 \\ 0 & K^* \end{bmatrix} \begin{Bmatrix} r_n \\ r_m \end{Bmatrix}$$

i.e.,

$$(M^* - \lambda K^*)r_m + \mu R r_n = 0$$

and $\quad (M_D - \lambda K_D)r_n + \mu R^T r_m = 0$

note that $M_D r_n = \mu K_D r_n$

i.e., $(\mu - \lambda)K_D r_n + \mu R^T r_m = 0$

or $r_n = (-\mu \, K_D^{-1} \, (\mu - \lambda)^{-1} R^T) \, r_m$

enabling the equation of motion to be written as

$$(M^* - \lambda \, K^*) - \frac{\mu^2}{(\mu - \lambda)} \cdot R \, K_D^{-1} R^T \, r_m = 0$$

Note it is a physical impossibility for $\lambda_i = u_i$.

The format of the above equation clearly shows the improvement on the Guyan reduction due to modal superposition. More importantly it illustrates its sensitivity since the effect of additional modes $q_i$ diminish approximately at the rate $\mu_i 2/\lambda_i - \mu_i$. In the paper the seventh frequency of the interior problem is stated to be 48.2Hz which from the above reasoning means that its inclusion effects the fundamental frequency only in its fourth significant figure. It is questionable of course whether any dynamic problem requires that sort of accuracy. Some guidance is therefore required on the number of superimposed modes required to obtain reasonable accuracy.

# Contribution to discussion of Paper No. 20

V M Trbojevic, R R Kunar and D C White
*Principia Mechanica Limited, UK*

The writers would like to thank Eatock Taylor for pointing out that the reduction procedure dates back to 1971. The method of dynamic substructuring has been developed in the work of Hurty (1), and also presented by Eatock Taylor (2), and as mentioned in the paper (3), by Clough and Wilson (4). It is also a feature of NASTRAN program.

In this presentation the authors do not claim that the dynamic reduction procedure is a novel concept. The intention was clearly to present the technique as a particularly good way of obtaining structural models for soil-structure interaction analysis. While dynamic reduction has been used in various engineering fields, the authors feel that the concept is not sufficiently known and appreciated in areas concerned with seismic soil-structure inter-action. Its application in this field was deserving of public discussion.

## References

1.  HURTY, W.C., et al "Dynamic analysis of large structures by modal synthesis techniques",
    Computers and Structures, 1971, 1, 535-563.

2.  EATOCK TAYLOR, R, "Structural dynamics of offshore platforms", paper no.10, Conference on Offshore Structures, Instn. Civil Eng., London, 7-8 October 1974.

3.  TRBOJEVIC, V.M., KUNAR, R.R. and WHITE, D.C., "The reduction of degrees of freedom in the dynamic analysis of complex three-dimensional structures", Paper no. 18, Joint I. Struct. E./ BRE Seminar: Dynamic Modelling of Structures, BRE, Garston, Watford, England, 19-20 November 1981.

4.  CLOUGH, R.W. and WILSON, E.L., "Dynamic analysis of large structural systems with local non-linearities", Comp. Meth. Appl. Mech. Engng., 17/18 (1979), 107-129.

# Seismic response of reactor building for different soil conditions

## 21

Mohamed Sobaih
*Structural Engineering Department, Cairo University, Giza, Egypt*

## SUMMARY

A recently developed mathematical model for the seismic response of nuclear reactor building has proven its adequacy. The model takes into account the soil-structure interaction. Till now no data has been reported that compares the seismic response of a typical reactor building when supported on different soil conditions. Evaluation of characteristic seismic values for this problem using the mathematical model are presented.

## INTRODUCTION

Safety requirements dictate that nuclear power plants should be designed conservatively to withstand seismic disturbances. The most important part of the dynamic analysis of the reactor building is to devise a mathematical model that will satisfactorily represent the dynamic behaviour of the structure.

An important aspect in the evaluation of seismic response of reactor buildings is to consider soil-structure interaction. For this purpose two general approaches are commonly used to idealize the reactor building and its supporting soil. These are the finite element and the lumped-mass idealization.

Most of lumped-mass models, as appear in the literature, are quite simplified. Recently, a more elaborate lumped-mass model has been developed by the author (1). It considers the interaction between the containment, the interior structure, and the supporting soil.

A survey of the published literature reveals that no results have been reported that compares the seismic response of a reactor building when supported on different soil conditions. Therefore, the objectives of this paper are as follows:
1- To demonstrate the suitability of the model developed by the author to investigate the effect of different soil conditions on the seismic response of a typical reactor building.
2- Characteristic seismic values are reported, for the first time, as functions of soil properties.

## MATHEMATICAL MODEL

A mathematical model for the complex multi-degree of freedom
system of a reactor building has been developed by the author
(1). The model is based on lumped-mass idealization for each
of the containment building and the interior structure. Soil
has been represented by springs whose properties are evaluated
from the theory of a rigid base resting on an elastic half
space (2), (3).

A typical CANDU reactor is chosen for the purpose of illustration. Dimensions and layout of such a reactor are shown in
Figs. 1 and 2.

The mathematical model devised for this typical reactor building
is shown in Fig. 3. It consists of 64 nodal points, 114 elements,
and 64 dynamic degrees of freedom. More details about this
model can be found in reference (1).

Figure 1   Containment building with mat foundation

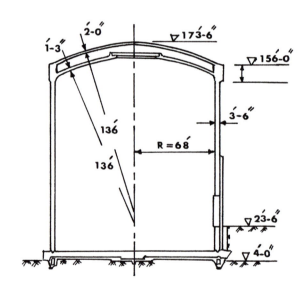

## SEISMIC ANALYSIS METHOD

Three-dimensional finite element general purpose programs have
been proven to be very efficient to perform dynamic analysis
for such complex systems (4), (5). The ANSYS program has been
chosen to analyse the mathematical model of Fig. 3 under the
effect of a horizontal response spectrum shown in Fig. 4.

Figure 2 Layout of different floors of interior structure of reactor building.

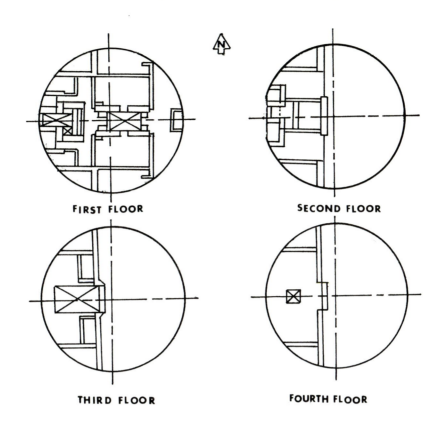

FIRST FLOOR

SECOND FLOOR

THIRD FLOOR

FOURTH FLOOR

Figure 3 Mathematical model for reactor building.

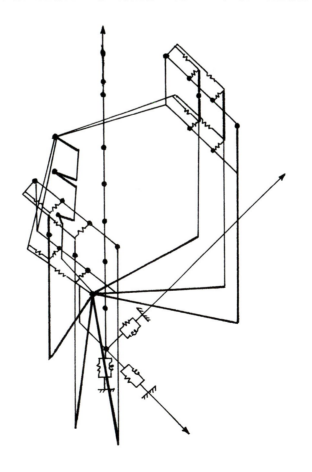

Figure 4　Horizontal response spectrum for
the design basis earthquake.

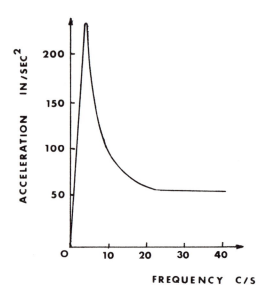

FREQUENCY C/S

## EFFECT OF SOIL PROPERTIES

The devised model together with the suggested method of analysis are adaptable to any change in the properties of the
supporting soil. Nori, et al. (6) have presented results for
the accelerations and displacements of different floor levels
of the internal structure of a typical reactor building for
different arbitrary soil conditions. The soil properties
corresponded to hard rock, dense sand, and loose sand. However,
the mathematical model used by Nori et al. was two-dimensional
and the outer wall, dome, and deep reinforced concrete walls
were assumed to be infinitely rigid. Besides, only horizontal
and rocking stiffnesses of the soil were considered.

In order to investigate the effect of soil properties on the
seismic behaviour of the CANDU reactor, three different actual
sites were considered. Soil properties for these three sites
were available through complete soil investigation. These are
shown in Table 1 as cases 1, 2, and 4. Two additional arbitrary cases have been considered and appear in Table 1 as cases
3 and 5. Case 5 corresponds to the condition of fixed base for
the reactor building.

Table 1　Soil properties for different sites.

| Case | Dynamic G　psi |
|---|---|
| (1) First site | $0.65 \times 10^5$ |
| (2) Second site | $1.15 \times 10^5$ |
| (3) Case study | $5.00 \times 10^5$ |
| (4) Third site | $19.00 \times 10^5$ |
| (5) Fixed base | $\infty$ |

192

## RESULTS

Computations were made for the five cases of soil conditions on a CDC 6600. The results of the analysis are plotted as functions of the dynamic shear modulus of the soil. Fig. 5 shows the variation of the frequency f and the maximum spectral lateral displacement δ with respect to the soil dynamic shear modulus G. Fig. 6 shows the variation of the spectral base shear Q and the base moment M.

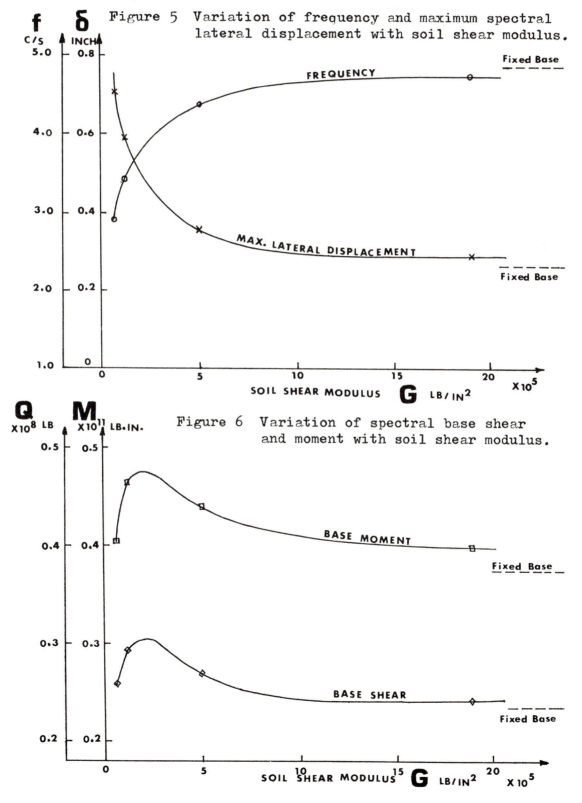

Figure 5  Variation of frequency and maximum spectral lateral displacement with soil shear modulus.

Figure 6  Variation of spectral base shear and moment with soil shear modulus.

The model was also analyzed for a vertical response spectrum which equals two-thirds of the horizontal spectrum. In both horizontal and vertical spectrums damping was assumed to be 2 %. Fig. 7 shows the variation of frequency f and spectral vertical reaction R with soil shear modulus G.

Figure 7 Variation of frequency and spectral vertical reaction with soil shear modulus due to vertical excitation.

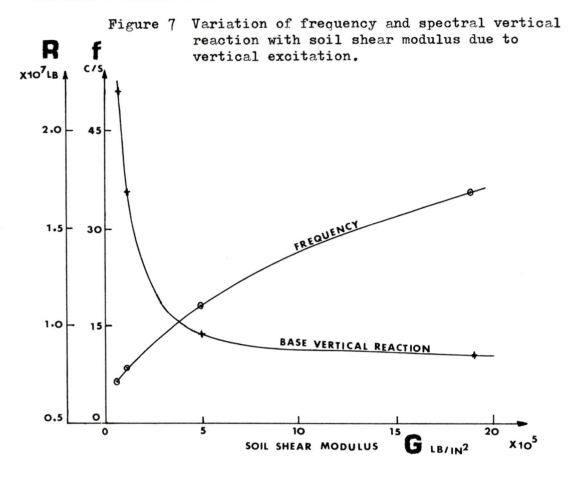

CONCLUSIONS

The three-dimensional mathematical model developed by the author (1) yields a convenient tool to investigate the modification of seismic behaviour of a typical reactor building when supported on different soil conditions. The results presented in this paper demonstrate the impact of soil conditions on the seismic response and consequently on the total cost of the reactor building. It is believed that this approach can be helpful in the process of site selection of nuclear power plants especially in developing countries.

REFERENCES

(1)   SOBAIH, M, 'Seismic analysis methodology for nuclear reactor building', Bulletin of the International Institute of Seismology and Earthquake Engineering, Vol. 18, (1980).

(2)    RICHART, F E JR, HALL, J R JR, and WOODS, R D, Vibration
       of Soils and Foundations, Prentice Hall Inc., New Jersey,
       (1970).

(3)    WHITMAN, R V, and RICHART F E JR, 'Design procedures for
       dynamically loaded foundations', Journal of the Soil
       Mechanics and Foundation Division, ASCE, Vol. 93, No.
       SM6, (November 1967).

(4)    DE SALVO, G J, and SWANSON, J A, ANSYS: Engineering Ana-
       lysis System User's Manual, Swanson Analysis Systems,
       Inc., Pennsylvania, (1975).

(5)    SOBAIH, M, SAP IV DOS-Version, AMAC Center, Cairo, Egypt,
       (1979).

(6)    NORI, V V, et AL, 'Response of a typical reactor building
       to strong motion earthquake', Fifth Symposium on Earth-
       quake Engineering, Roorkee, India, (1974).

# Chairman's summing up

G Somerville
*Director of Research, Cement and Concrete Association, UK*

In the short time available, I am not going to attempt to summarize the 11 papers which we have had this afternoon. This would be unfair to the authors, and, in any case, I feel that the papers speak for themselves.

However, I would like to take this opportunity to make some general remarks about models - to give a sense of perspective to the seminar as a whole. These remarks relate particularly to the accuracy of models (whether physical or mathematical) and to their role in engineering.

It must be realized that everything we do in research or design involves modelling of one type or another. Activities under these headings can vary appreciably, and therefore the demands made of the 'model' can also vary. As an illustration, within the context of this seminar, we have had examples of models being used for:-

- representing actual structural performance "as accurately as possible";
- checking that performance was better than a stated minimum requirement;
- advancing knowledge (in R & D terms) by a combination of physical and numerical methods.

Each of these uses has quite <u>different requirements in terms of accuracy</u>. I have been encouraged in this seminar to see so many people developing different types of models to represent dynamic behaviour and doing it very well in most cases. However, even at this stage, it is essential to keep the final fields of application in mind, since this can influence both technical development and the accuracy required. This general problem was acknowledged and analysed a number of years ago in the structural field[1] and perhaps it is time that it is reviewed once again, to give a sense of purpose and perspective to all the present activity.

Turning next to the present state-of-the-art in modelling, I still believe that more work is necessary to correlate results from physical models of different sizes (including full-scale) with those from mathematical models - all aimed at a better understanding of the problems involved in these new fields of application, but within the defined limits required for design (where, for example, errors in analysis are routinely allowed for in the general treatment of safety).

Problems of scale, similitude and absolute size still remain - in terms of materials, loads, deformations and, especially in the dynamics field, of time and boundary conditions. In some cases, these problems have not been recognised, never mind solved.

In summary, we must always ask ourselves the question:- "How accurate need we be?" in our modelling. Extreme precision in the model itself is pointless, it the assumptions on which it is based are doubtful, or if our knowledge of the input factors (e.g. loads) is vague at best (frequently the case with dynamic problems).

In case these remarks should be taken as being highly critical of modelling as a useful tool, let me hasten to say that I am a firm believer in the technique, having been involved in this field off and on for over 20 years. However, this experience has also taught me to be clinically realistic, and to take the broad view.

Finally, I wish to record my view that seminars like this can do nothing but good. Apart from the information obtained, new contacts are made, and outlooks are broadened. This last point is particularly important, since research workers can be insular in their approach - for example, a study of the references to the papers presented reveals that, mostly, these are taken from only one country, and, occasionally, from only one institute.

Reference

1.    Little, W.A., Cohen, E. and Somerville, G. Accuracy of structural models. Paper SP24.4. American Concrete Institute publication number 24. 'Models for Concrete Structures', pp 65-124. Detroit 1979.

# Studies of the response of reinforced concrete structures to short duration dynamic loads

## 22

I Ll Davies
*Taylor Woodrow Construction Ltd, UK*

## INTRODUCTION

Undoubtedly our first experience with the effects of short duration dynamic loads will have been as very young children when we found we could lie on the floor (static load) with no discomfort but a fall (dynamic load) could be very painful. Later, when as under-graduate engineers, we began to quantify such dynamic effects we learnt that for elastic conditions a suddenly applied load would double the deflections compared to the same load when applied slowly (statically).

Further, these quantifications showed that as a weight is dropped from increasing heights the dynamic effects increase so that when the drop height is forty times greater than the static deflection the dynamic deflections are ten times greater than the static ones.

Throughout these assessments we assumed that the structural system remains elastic. However, when structural materials are driven to exceed elastic limits the problem becomes increasingly complex and when reinforced concrete structures are considered further complexities occur due to composite behaviour and concrete cracking.

The study of such effects is becoming increasingly important for both safety and economic - protection of investment - reasons. With modern plant, in which equipment and materials are used under evermore onerous conditions and where ever greater capital sums are being invested, it becomes increasingly important that some minor event cannot escalate into a catastrophic and total failure with obvious safety and financial implications. This leads to a growing requirement for the introduction of systems which will provide people and equipment with adequate protection but which need not remain undamaged after an incident of design magnitude. Such protection must be provided at minimum cost and in minimum space.

The size of the events and structures which are of concern make it impractical to study the effects experimentally at full scale so extensive use of models is essential, whilst analytical assessments need information against which they can be validated and and calibrated. This requires that detailed consideration be given to similarity methods before any practical work can be undertaken. Notwithstanding, the ultimate aim of all such studies is the development and validation of computational techniques to reduce to a minimum the need for expensive experiments.

## SIMILITUDE

The use of models in the study of structural behaviour is not new nor are the controversies which the approach always generates. Vitruvius is reported by Leonardo da Vinci to have considered and discarded the technique while da Vinci himself, as outlined in his notebooks in use in 1500, set out to disprove the Vitruvian conclusion that models do not behave as the larger prototype. Further, although the formal presentation of the basic concepts of dimensional analysis has been made only within the last 70 years or so, it has been suggested that the basic ideas date from Galileo (1564 - 1642). Nonetheless it was not until 1915 when Buckingham proposed his $\pi$ theorems that the approach was put on a completely formal basis. Coincidentally, it was in the same year that Hopkinson propounded his "rules" to the Ordnance Board and made what is believed to be the first statements on the applicability of the concept to the assessment of structural response to dynamic loads. That it is possible even to contemplate the prediction of prototype performance from observations of model behaviour depends completely on recognition of the fact that locations, times and forces in a model system may be homologous with locations, times and forces in the prototype (homologous - at corresponding but not necessarily equal values of a variable). In addition in such studies it is essential that strict adherence to the concepts of similarity be maintained if such similitude in behaviour is to be achieved. However, it has long been realised that small scale experiments do not always give a good indication of prototype performance but the aim of model theory is to predict where lack of correspondence arises and to determine conditions under which results from model tests can be used to forecast prototype behaviour. Where uncertainties, scale effects and incompatibilities arise precise corrections can be determined by repeating some experiments at enlarged scales. It must be stressed, however, that there is no unique scaling law for any particular situation; all the parameters involved in the process are listed; these are then converted into a set of dimensionless parameters and assembled into equations which constitute one model law. By setting down different sets of dimensionless parameters it should be possible to define other model laws which are equally valid.

Hopkinson presented his "rules" to the Ordnance Board with the following statement:

"If two structures be made to the same drawings and of the same materials, but to different scales, and if charges of guncotton be detonated in similar positions in relation to them, the effects will be similar in the two cases if the weights of the charges are in proportion to the cubes of the ratio of the linear dimensions."

He explained that similar blast waves, with identical peak pressures, are produced at scaled distances when explosive charges of the same explosive and similar geometrical shape are fired in the same atmosphere. This is often reduced to the Hopkinson parameters for:

i.    scaled distance          $z = {}^R/_{W^{\frac{1}{3}}}$

ii.   scaled time              $\bar{t} = {}^t/_{W^{\frac{1}{3}}}$

iii.  scaled impulse           $\xi = {}^I/_{W^{\frac{1}{3}}}$

(Note:  $W^{\frac{1}{3}} \propto L$)

and implies that all quantities with the dimensions of density $\{ML^{-3}\}$, pressure (or stress) $\{ML^{-1}T^{-2}\}$ and velocity $\{LT^{-1}\}$ must be invariant between the model and the prototype systems.  The dimensional, structural amd material similitude of the structure, whose response under blast loading is being assessed, are very concisely specified.

These are the first recorded statements made relating to the dynamic response of what is now called "replica" modelling.

It is possible to remove some of the Hopkinson restrictions to produce a rather more general set of relationships which may be used in studying structural behaviour generally under dynamic loads but since our studies have been executed with "replica" modelling it is not proposed here to proceed further with such a generalisation.

Other short duration dynamic loads may nevertheless be similarly treated.  Consideration of the blast wave loading shows that although this has been specified by Hopkinson by reference to:

1.    W - the weight of the explosive charge (energy in
                                             charge)

2.    R - the distance from the source

3.    $\alpha_0$, $\gamma$, $p_0$ - the ambient state quantities describing the atmosphere

these may be replaced by the three quantities:

1.    $p_i$ - peak incident pressure

2.    p  - a function describing the wave shape

3.    $\tau$  - wave duration

which define the blast wave loading variations explicitly with respect to time.  Thus, if on the model scale, it is possible to produce a loading pulse which has identical pressures to those of the prototype loading but with relevant times scaled by the scaling factor for linear dimensions then a "replica" model approach may be used.

The replication of the materials with which the structure

is made may be similarly achieved without using the
same materials so long as absolute invariance exists
between the densities and the stress-strain
characteristics of all the materials in both prototype
and model systems.  However, with such a versatile and
complex material as reinforced concrete it is a fruitless
and unnecessary exercise to attempt a search for
substitutes which are normally only used to reduce the
cost of tests.  In the work described later in this
paper the concrete has been represented in the model
by a microconcrete in which the mix constituents and
proportions have been adjusted to yield the identity
of those stress-strain curves for both tensile and
compressive loads which may be required to achieve
similarity of response whilst the reinforcing has been
specially annealed to ensure that its tensile
characteristics are identical to those specified for the
prototype structure.

No discussion of scaling can be complete without some
mention being made of those parameters which may not
scale and which can affect the efficacy of the model
study.  Tables 1 and 2 list the parameters which are
believed to be important in a study involving short
duration loads arising from blast waves and impact or
collision situations;  similar lists may be made for
other impulsive loading conditions but these can, in
general, be reduced to the form of Table 1 where the
pressure time curve is defined explicitly.  When we
consider these tables we note that Table 1 really over-
defines the situation because not only do we include the
charge weight, source distance and atmospheric state
parameters but we have included information for the
complete description of the blast wave itself.  Table 2
includes a number of parameters related to temperature
and thermal energy considerations;  it is known that
high velocity impact thermal effects are very important
but for the lower velocity impact ($v_i$ < 200 m/s) the
effects are less significant so that temperature and
thermal effects have been ignored in the studies I will
be describing later in this paper.  From these two
tables we may rearrange our list to form Table 3 which
now collects variables under broad parameter headings.

This table indicates first how a completely general model
law might be developed (Col. 4) in which every significant
parameter may be given its own scaling relationship;
such an approach would be extremely complex to undertake
experimentally but it does allow us to define our two
systems (model and prototype) to suit our own purposes,
we may for instance, introduce these into computer
assessments to reduce running times.  Column 5 shows how
in "replica" modelling in which the scaling factors
$S_\rho$, $S_\delta$, $S_\sigma$, $S_\upsilon$ will all be set to equal unity the
scaling law is enormously simplified.

Up to this point the structural loading parameters which
we have considered are only those related to effects
arising from external events with no thought given to
the intrinsic dead loads which the structure will apply
to itself.  According to Table 3 acceleration should
scale as $^1$/S so that dead weight forces in the model

should be mg/s.  However, in most of our test
the gravitational acceleration (g) is an invariant
quantity which means that dead weight and dead weight
effects are much lower than they should be in an ideally
executed experiment which should be mounted under a
gravitational acceleration of $g/s$.  The significance of
this deviation from complete similitude has to be very
carefully assessed.  However, for the short duration
loads I am to describe it was found that dead weight
loads were insignificant when compared to the forces
applied by the blast wave or collision loading.  One
final non-scaling parameter which needs to be considered
is the way in which the speed of distortion of our
structure may modify the material strength parameters.
Under dynamic loads movement can be very rapid and the
rate at which strains develop may produce significant
changes in material behaviour. When we consider typical
rate of strain curves we find that the one for concrete
typically follows the form given in Fig.1 while that
for many metals has been found to follow similar curves
of which two are also plotted in Fig. 1.  When one
notes that typically a static test with a cylinder or
a cube of concrete is carried out at a strain rate of
the order of $2 \times 10^{-5}$ per sec it is evident that
distortion has to be extremely fast before significant
changes occur in the structural behaviour of concrete
under dynamic loads.  Nonetheless, attention must be
directed at this problem to ensure that corrections
are introduced or the effects quantified when they
are seen to be significant.  What we find, in fact, is
that with scaling ratios which are normally of the order
of 10 and never larger than 100 the model material
appears to be stronger but the effect is usually small
except where local behaviour is critical.  Furthermore,
if we know the magnitude of the effect we can introduce
compensating solutions which may be made either in
the analysis of the results or alternatively, if we
are to produce visual evidence for extrapolation to the
larger scale, minor changes can be made to the strength
of materials in the model so that at the distortion
rate which will develop the strength parameters will
more nearly equal those in the prototype under similar
conditions.

MATHEMATICAL MODELLING

The behaviour of a structure subjected to loading which
varies with time may be represented by the equation
of motion which may be written in the form:

$$P(t) \quad = \quad m\ddot{x} \quad + \quad c\dot{x} \quad + \quad kx$$

(Loading) =  Resistances
(Inertia) + (Damping) + (Intrinsic)

When we consider dynamic conditions our interest must be
directed at the peak values of the response parameters.
For a short duration single pulse load, where resonance
effects cannot develop, the peak response almost
invariably occurs at the first maximum value of the
fundamental mode in which the structure vibrates.  In
the time to reach this first peak the effects introduced
by damping are usually found to be small so that
damping may be ignored and the equation of motion

rewritten as:

$$P(t) = m\ddot{x} + kx$$

Such an omission tends to make the analysis conservative because one resistive mechanism which will, in fact, develop is ignored requiring an increased contribution for the inertial and intrinsic resistance functions and this can only be achieved by increases in deflection and acceleration.

Perhaps the greatest problem in the mathematical modelling arises from a true determination of the ultimate capacity of our structure whether this be as an assessment of local or global response. For studies related to a "once in a lifetime" loading we will need to know the true ultimate capacity of our structure and the route by which this ultimate state is reached. The ultimate strength derived using methods outlined in Codes of Practice etc. are not able to give a true strength value for many simple structural forms, load support mechanisms develop due to the composite action of our structures which almost invariably make these codified solutions extremely conservative for the ultimate state and thus yield a very pessimistic answer in the solution of our equations of motion. When we are looking at the behaviour of structures under extreme conditions it is important to know the true strength characteristics so that a reasonable assessment of dynamic behaviour can be made.

Another approach which has been extensively used may be termed the "energy method" in which an attempt is made to determine how much energy is implanted into the "target" structure so that an assessment of distortions may be made. This method has some very serious limitations in that it is extremely difficult, if not impossible, to deduce how the energy is partitioned or to determine how much of the energy is lost in irreversible processes.

DESIGN PROBLEMS

Two design problems have been of particular interest in making these studies of the behaviour of reinforced concrete structures to short duration dynamic loads. In the first case it was required to design a heavy dual purpose shelter structure capable of withstanding the loading which would be experienced close to the ground zero of a large airburst nuclear weapon. The second study concerned experimental work related to the behaviour of reinforced concrete structures subjected to loads generated by the impact of heavy objects moving at velocities up to 250 m/s. Although these impact tests dealt specifically with test and analysis work which has been done in the consideration of the effects that develop when a heavy military aeroplane impacts against the containment building surrounding some nuclear system, the techniques which are being evolved may be applied to the collision of a ship against an offshore oil producing platform and, with some development, to the impact of any high speed missile such as those resulting from some disrupting pressure

203

vessel or rotating machinery. In all of these problems
the structures are allowed to experience extensive damage;
so long as they retain their protective requirements, i.e.
so long as people being sheltered can escape from the
shelter structure without heavy casualties or injury it
does not matter how far the shelter is damaged it is
highly unlikely that an enemy will want to repeat a
strike at the same place even if he has the capacity
to do so; in impact situations, so long as the missile
cannot escalate the accident situation, the object of
the exercise will have been achieved.

For the shelter complex a composite structure of columns,
beams and slabs was selected (Fig.2&3) as giving the
most effective system to deal with the hazards (blast
loads and ionising radiation) which would result from
a weapon attack. It was argued that a balanced design
was required because there was little point in failure
of different components occurring at different pressure
levels. To achieve this balance a series of preliminary
model studies was executed so that when a 1/12th scale
composite of part of the roof system was tested all
components would develop the same degree of distress
as the load was increased. In these static tests the
peak pressure reached before a snap through failure of
one of the roof panels occurred was ≃60 p.s.i. all
components were showing considerable distress and final
failure might have resulted from collapse of any one of
them.

Having achieved this balance for a structure designed
to have an ultimate static capacity of 50 p.s.i. a model
was tested under static loading to check the design
calculations and to obtain an accurate definition of
the "kx" term of the equation of motion. Applying
this information some dynamic assessments using the
simplified equations of motion were undertaken to
estimate the structural response to blast loadings of
various peak pressures and thence determine the
positions at which six similar structures should be
placed relative to the centre of a hemispherical stack
of 100 tons of explosive. The calculations indicated
that at 50 p.s.i. peak pressure the structure should
be cracked but no plastic yielding should occur while
at 100 p.s.i. peak pressure there would be very
extensive damage and probable "snap through" so that the
structure would be almost in a condition of collapse.
Models were placed in the field where peak pressures
of 40, 50, 60, 70, 80 and 100 p.s.i. were expected.
Actual peak pressures measured in the test were 42, 52,
61, 71, 85 and 110 p.s.i. In this first test the two
foremost models, at 85 and 110 p.s.i. were mildly
cracked whilst the remaining four models were apparently
undamaged although maximum panel deflections equal
to about 1/5 of the thickness were measured during the
passage of the blast wave. These four models were
later exposed to the blast from a 500 ton explosive
charge at positions where the measured peak pressures
were 145, 125, 110 and 95 p.s.i. In this trial the
leading target collapsed completely whilst the other
structures at the lower pressures fitted into the
pattern of damage established in the first test.

Computations were continued and, although acceptable agreement was obtained for the lower pressures and distortions, the assessments for the higher pressures, where heavy damage and collapse might be close, have always indicated that the structure has a lower potential than has been observed in the tests, i.e. the computations were conservative. Quite clearly, at that stage in the study the mathematical modelling was inadequate for those situations where extensive damage might occur and heavy reliance had to be placed upon the model tests. Nonetheless the tests did indicate that a shelter configuration of the form tested would be completely adequate for peak pressures up to 100 p.s.i. and that it could probably be used for pressures up to 125 p.s.i. Although scaling with this particular complex structure was never tested experimentally under dynamic conditions, tests with simpler structures showed that for scaling factors up to about 1/10 the behaviour of model and prototype structures, subjected to correctly scaled blast waves were similar when all the similitude requirements were met.

The impact studies had two objectives:

1.   to produce experimental data at two scales to verify that "replica" scaling could be used in impact studies,

2.   to obtain information on target response under impact loads which could be used in assessments to calibrate and validate a computer program which is being developed to deal with problems related to short duration dynamic loads.

The study involved the firing of long thin walled cylindrical missiles to impact end on against reinforced concrete targets. Various impact velocities between 200 and 250 m/s were used. The tests of the study were divided into three major phases:

i.    firings of missiles against a Kistler load cell to measure the way in which the force between the missile and target varied with time during the period in which the missile is in contact with the target;

ii.   firings of these same sized missiles at R.C. targets with measurements taken at a number of positions to determine how target deflections varied with time; afterwards targets were examined for damage and in some cases they were sectioned using a diamond saw to enable internal cracking and damage to be assessed;

iii.  firings similar to those described in ii. above but with all dimensions increased by a factor of 5.6 times; close identity of material character-istics for the concrete and reinforcing was sought and, in the main, was obtained, after a first apparently disastrous result.

Phase i. of these tests, in which a model missile (Fig. 4) was fired against a suitable load cell produced the measured load function given in Fig. 5. It may be

205

noted that this function has some 25 major peaks while
the missile was found to have a similar number of buckle
convolutions after the firing (Fig. 4). It is
interesting to note also that the missile used in the
larger scale tests (Phase iii.) had a similar number
of buckle convolutions and looked very similar to the
model after the firings. To achieve this degree of
similitude it was necessary to select the materials
for the two missiles so that there was identity of
material parameters and to ensure that missile mass and
mass distribution were similarly and accurately modelled.
As has already been noted the impact velocity at the
two sizes was invariant to achieve the similarity in
the loading function.

For the tests at the smaller scale (Phase ii.) the
target structure was a concrete plate 1.16 m wide x
1.07 m high x 0.125 m thick having $\simeq\frac{1}{2}$% reinforcing each
way each face and with various quantities of shear
reinforcement (as stirrups). The plates were supported
upon a "ring" of 48 points lying on a 0.96 m square and
held back against a heavy abutment by tensioned
prestressing wires; this effectively produced a square
of simply supported edges for the target. The tests at
the larger scale (Phase iii.) were made against targets
which were increased in size by 5.6 times being 6.5 m x
6.0 m x 0.7 m and also having $\simeq\frac{1}{2}$% ewef, with various
quantities of shear steel, also supported at 48 points
and stressed against a very massive abutment.

The tests were arranged so that a number, usually three,
of small scale experiments were made with the missile
fired at a different velocity in each test to determine
the velocity at which the firing should be made at the
larger scale to give the optimum results in this
larger more expensive experiment. The results of the
first series in this comparison were, at first sight,
most discouraging because in the small scale experiments
the deflection at a point close to the centre when scaled
to full size averaged only about 20 mm and the target was
quite lightly damaged by the formation of a shear plug
in the zone adjacent to the impact. At the larger scale
the missile almost perforated the target, damage was
extensive and the deflection at a similar point to that
noted for the small target test reached over 100 mm.
At first it was postulated that strain rate effects
were far more important than had been originally
envisaged and that this would be extremely difficult to
compensate for. However, prior analysis of the small
scale tests had indicated that, scaled to full size, the
maximum deflection at the reference point would be about
25 mm (cf 20 mm measured) and this demanded detailed
consideration to disclose the reason(s) for the
differences between model and prototype.

Closer study of the details of the experiments showed
some unexpected differences. First it was noted that
whereas the compressive strength of the concrete at the
time of test in the small models was about 38 N/mm$^2$ this
was some 15% higher than the concrete strength in the
prototype structure (33 N/mm$^2$). Later, comparisons of
the strengths of the respective reinforcing steels showed

that again the material in the larger target was of
lower strength (by about 25%) than the reinforcing
in the model targets.  Finally, when more precise
recorded data on impact velocities were determined from
the field equipment it was found that the impact
velocity upon the larger structure was 236 m/s some
8% higher than the average value in the smaller tests
(218 m/s).

The prefiring computations which were made for comparison
and planning purposes indicated, as already noted, that
the reference point deflection maximum would be about
25 mm;  the measured deflection was found to be about
20 mm (Fig.6).  Further computations were undertaken
in which material strengths were reduced and impact
velocities increased.  Typical of these are shown in
the upper curves of Fig.6 which demonstrate how a
small increase in velocity combined with a reduction
in concrete strength have pushed the maximum deflections
up by almost three times;  the effect of different
steel strengths has not been included.

The suggestion that this divergence of results between
the tests at the two scales was caused by these quite
small inaccuracies in the identity of material
parameters and the small differences in impact velocity
were, at first, considered to be unacceptable by some
people:  once again the controversies which have for
so long persisted with respect to the use of models,
began to surface.  However, two further model tests in
which the material properties and impact velocities
were arranged to be almost identical to those of the
larger prototype test were carried out and gave good
agreement between the results from the two scales.

Similar good agreement has been obtained throughout the
whole series of comparisons in which the quantity of
shear steel was varied.  One test in which the target
thickness was also modified also showed a remarkable
level of agreement.  In one further test a set of
targets similar to but not identical with the sets
(prototype and model) detailed above were subjected
to impact by a "hard" non deformable missile, a similar
scale factor was employed, again there was very close
agreement between deflections, etc. but there appeared
to be somewhat more damage, as indicated by the extent
of the backface spalling, in the prototype than there
was in the model structures.

CONCLUSIONS

i.      The use of models to study the response of
        reinforced concrete structures to short duration
        loads has been demonstrated to be an effective
        design tool.

ii.     For reinforced concrete "replica" modelling appears
        to be well suited to these studies.

iii.    In such model tests it is essential to obtain close
        identity of the material parameters used in the
        prototype and model systems.

iv.   The invariance of velocities from the model scale
      to the prototype is also seen to be important.

v.    Computational techniques are available by which
      it is possible to consider parametric variations
      and to assess the effect of such variations upon
      response.

SHORT BIBLIOGRAPHY

(1)   DAVIES, I Ll,  Model Analysis  'Proceedings of
      the Symposium on Protective Structures for
      Civilian Population', National Academy of
      Sciences, National Research Council, Washington
      DC (1965).

(2)   DAVIES, I Ll, 'The Design and Testing of a
      Reinforced Concrete Roof System for an Underground
      Garage/Shelter to Resist a 50 p.s.i. Blast Wave
      from a Nuclear Weapon', MICE Thesis, U.K.A.E.A.
      (1967).

(3)   CHRISTOPHERSON, D G, PHIL, D, 'Structural Defence',
      (1945), Ministry of Home Security, Research and
      Experiments Department.

(4)   DAVIES, I Ll, CARLTON, D, O'BRIEN, T P, 'Scaling
      Laws Applied to Impact Testing and Computer
      Assessments made to Compare Tests at Two Scales',
      Paper J8/1 Transactions 5th SMIRT Conference,
      Berlin (1979).

(5)   DAVIES, I Ll,  'The Use of Models in Studying the
      Response to Dynamic Loading'  Paper No. 15
      Seminar on Reinforced and Prestressed Micro-
      concrete Models, BRE Garston (1978).

TABLE 1    LIST OF PARAMETERS RELATED TO BLAST WAVES

| PARAMETER | VALUE | DIMENSIONS | EFFECTS OF HOPKINSON RESTRICTIONS | |
|---|---|---|---|---|
| **Source Parameters** | | | | |
| Energy in explosive | $W$ | $ML^2T^{-2}$ | $\pi_1 = r_i$ $\qquad Sr_i = 1$ | Geometric similarity of whole experiment |
| Typical source dimension | $r$ | $L$ | $\pi_2 = \dfrac{R}{r}$ $\qquad S_R = S_r = S_L$ | |
| Shape factor ratios | $r_i$ | | | |
| Detonation velocity | $v$ | $LT^{-1}$ | $\pi_3 = U\sqrt{\dfrac{\rho_o}{\gamma P_o}}$ $\qquad S_u = S_{\alpha_o} = 1$ | Invariance of velocities |
| Distance from source | $R$ | $L$ | $\pi_4 = \dfrac{U}{u}$ $\qquad S_U = S_u = 1$ | |
| **Blast wave Parameters** | | | | |
| Peak incident pressure | $p_i$ | $ML^{-1}T^{-2}$ | $\pi_5 = \dfrac{v}{u}$ $\qquad S_v = S_u = 1$ | |
| Blast wave shape factor | $p$ | | $\pi_6 = \dfrac{p_i}{p_o}$ $\qquad Sp_i = Sp_o = 1$ | Invariance of pressures and pulse shape |
| Shock wave velocity | $U$ | $LT^{-1}$ | $\pi_7 = p$ $\qquad S_p = p$ | |
| Particle velocity behind shock | $u$ | $LT^{-1}$ | $\pi_8 = \dfrac{p_i r^3}{w}$ $\qquad S_w = S_r^3$ | Scaling of blast energy |
| Gas density behind shock | $\rho$ | $ML^{-3}$ | | |
| Shock duration | $\tau$ | $T$ | $\pi_9 = \dfrac{\rho u^2}{p_i}$ $\qquad S_\rho = 1$ | Invariance of shock density |
| Time for sound wave to arrive | $t$ | $T$ | $\pi_{10} = \dfrac{t\alpha_o}{r}$ $\qquad S_t = S_r = S_L$ | Equivalence of time and Space scaling |
| **Environment Parameters** | | | | |
| Ambient pressure | $p_o$ | $ML^{-1}T^{-2}$ | $\pi_{11} = \dfrac{t}{r}\sqrt{\dfrac{\rho \rho_o}{\gamma P_o}}$ $\qquad S_\tau = S_r = S_L$ | |
| Sound speed in ambient air | $\alpha_o$ | $LT^{-1}$ | | |
| Ratio of specific heats | $\gamma$ | | $\pi_{12} = \gamma$ $\qquad S_\gamma = 1$ | Atmospheric invariance |

We may write

$$\alpha_o = \sqrt{\frac{\gamma P_o}{\rho_o}} \qquad \text{and} \qquad t = \frac{R}{\alpha_o} = R\sqrt{\frac{\rho_o}{\gamma P_o}}$$

$S$ = scale factors

Hopkinson restrictions:   i. identical explosives
   ii. identical materials
   iii. identical atmospheres

209

## TABLE 2

LIST OF PARAMETERS FOR IMPACT STUDIES

| Parameter | Value | Dimensions |
|---|---|---|
| Missile diameter | $d$ | $L$ |
| Shape factor ratios | $r_i$ | |
| Angle of attack | $\beta$ | |
| Density of missile material | $\rho_m$ | $ML^{-3}$ |
| Density of target material | $\rho_t$ | $ML^{-3}$ |
| Impact velocity | $v$ | $LT^{-1}$ |
| Target thickness | $h$ | $L$ |
| Missile temperature | $\theta_m$ | $\theta$ |
| Target temperature | $\theta_t$ | $\theta$ |
| Specific heat of missile material | $c_m$ | $L^2 \theta^{-1} T^{-2}$ |
| Specific heat of target material | $c_t$ | $L^2 \theta^{-1} T^{-2}$ |
| Heat of fusion of missile material | $n_m$ | $L^2 T^{-2}$ |
| Heat of fusion of target material | $n_t$ | $L^2 T^{-2}$ |
| Ultimate strength of missile material | $\sigma$ | $ML^{-1} T^{-2}$ |
| Ultimate strength of target material | $s$ | $ML^{-1} T^{-2}$ |
| Other stress ratios | $\sigma_i$ | |
| Strain | $\varepsilon$ | |

## TABLE 3

SCALING RELATIONSHIPS ASSOCIATED WITH SHORT DURATION LOADS

| Parameter | Dimensions | Full Scale Value | General Model Value | Model value using REPLICA modelling | Typical Quantities |
|---|---|---|---|---|---|
| Length | $L$ | $d$ | $s_L d$ | $sd$ | Missile dimensions, Target dimensions, Geometry |
| Mass* | $M$ | $m$ | $s_m m^*$ | $s^3 m$ | Missile mass, Target weight |
| Time | $T$ | $t$ | $s_t t$ | $st$ | Duration of Load, Natural period of target |
| Density | $ML^{-3}$ | $\rho$ | $s_\rho \rho$ | $\rho$ | Densities of materials in structure and missile, Density behind shock |
| Strain | — | $\varepsilon$ | $s_\varepsilon \varepsilon$ | $\varepsilon$ | |
| Stress | $ML^{-1} T^{-2}$ | $\sigma$ | $s_\sigma \sigma$ | $\sigma$ | Includes pressure and material moduli |
| Velocity* | $LT^{-1}$ | $v$ | $s_v v^*$ | $v$ | Missile velocity, Stress wave velocities, Ejecta velocity, Shock and sound velocities (Except gravity) |
| Acceleration | $LT^{-2}$ | $a$ | $s_a a$ | $s^{-1} a$ | |
| Rate of Strain | $T^{-1}$ | $r$ | $s_r r$ | $s^{-1} r$ | |
| Force* | $MLT^{-2}$ | $w$ | $s_w w^*$ | $s^2 w$ | Applied load, Structural reactions |
| Energy* | $ML^2 T^{-2}$ | $E$ | $s_E E^*$ | $s^3 E$ | Kinetic energy of missile, Energy in explosive |

The scale factor s is defined as $s = f_m/f_p$

Parameters marked * are <u>not</u> independent quantities

210

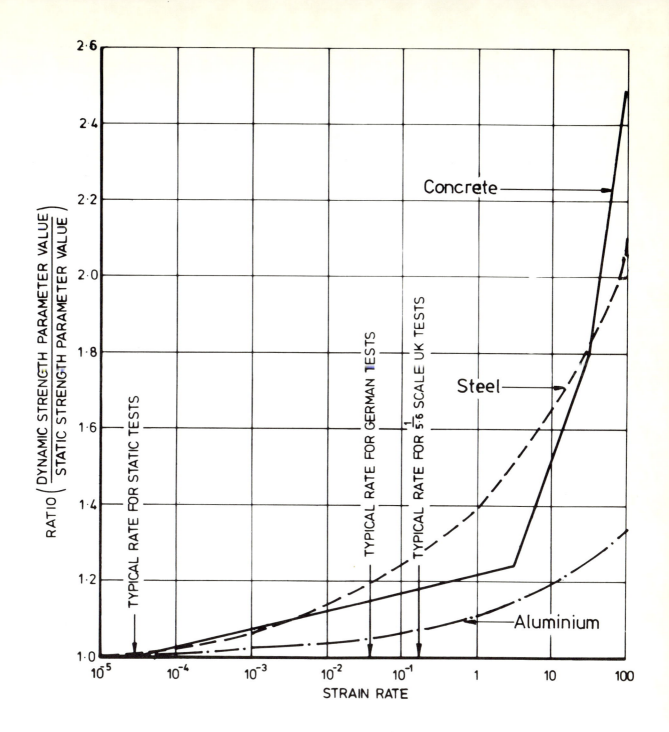

FIGURE 1. EFFECT OF STRAIN RATE ON STRENGTH OF
CONRETE, STEEL & ALUMINIUM

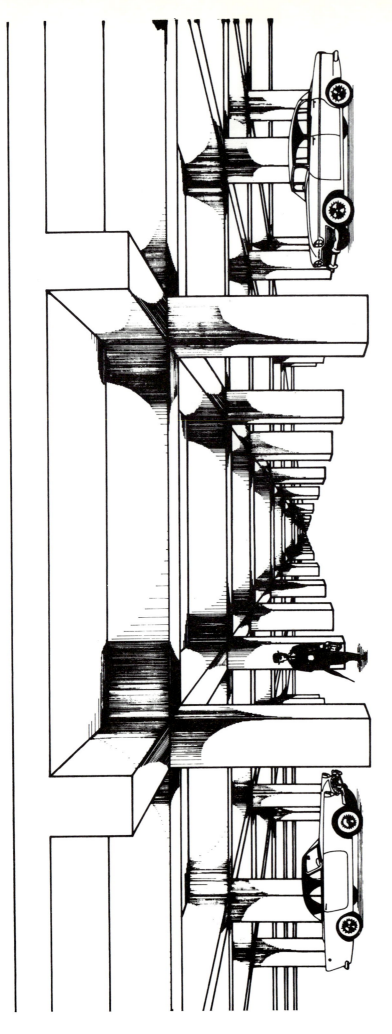

Figure 2    SKETCH OF PROPOSED GARAGE SHELTER

212

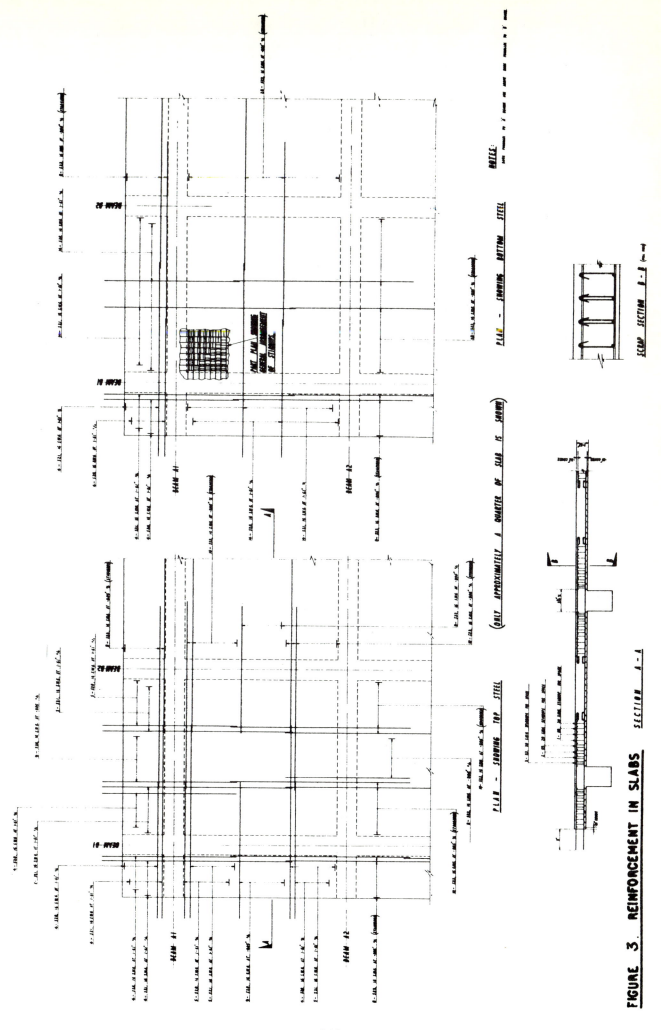

FIGURE 3. REINFORCEMENT IN SLABS

MISSILE BEFORE TEST

MISSILE AFTER IMPACT AT 233ms$^{-1}$

FIGURE 4 $\frac{1}{5 \cdot 6}$ SCALE MODEL OF GERMAN MISSILE

214

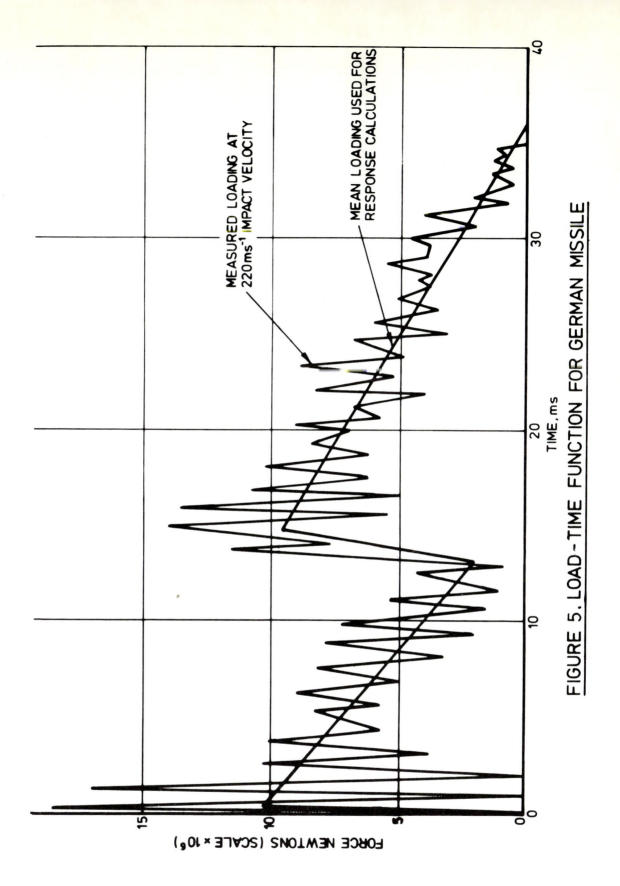

MEASURED LOADING AT 220ms⁻¹ IMPACT VELOCITY

MEAN LOADING USED FOR RESPONSE CALCULATIONS

TIME, ms

FORCE NEWTONS (SCALE × 10⁶)

FIGURE 5. LOAD-TIME FUNCTION FOR GERMAN MISSILE

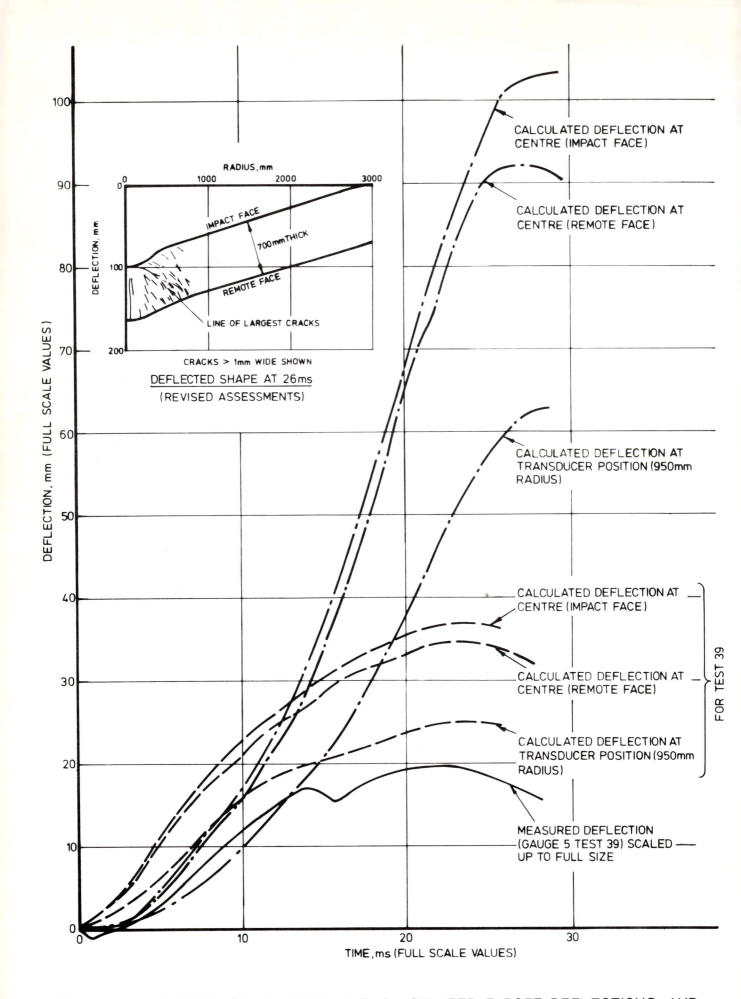

Inset graph labels:
RADIUS, mm
0   1000   2000   3000
DEFLECTION, mm
0
IMPACT FACE
700mm THICK
100
REMOTE FACE
200
LINE OF LARGEST CRACKS
CRACKS > 1mm WIDE SHOWN
DEFLECTED SHAPE AT 26ms
(REVISED ASSESSMENTS)

Main graph labels:
DEFLECTION, mm (FULL SCALE VALUES)
TIME, ms (FULL SCALE VALUES)

CALCULATED DEFLECTION AT CENTRE (IMPACT FACE)
CALCULATED DEFLECTION AT CENTRE (REMOTE FACE)
CALCULATED DEFLECTION AT TRANSDUCER POSITION (950mm RADIUS)
CALCULATED DEFLECTION AT CENTRE (IMPACT FACE)
CALCULATED DEFLECTION AT CENTRE (REMOTE FACE)
CALCULATED DEFLECTION AT TRANSDUCER POSITION (950mm RADIUS)
MEASURED DEFLECTION (GAUGE 5 TEST 39) SCALED UP TO FULL SIZE
FOR TEST 39

FIGURE 6. COMPARISON OF MEASURED AND CALCULATED TARGET DEFLECTIONS AND EFFECT OF CHANGE IN MISSILE VELOCITY AND TARGET STRENGTH PARAMETERS

216

# Contribution to discussion of Paper No. 22

Norman Jones
*Department of Mechanical Engineering, The University of Liverpool, UK*

The writer wondered if the author was aware of Duffey's work on scaling laws in which he considers the dynamic plastic behaviour of structures and identifies 19 independent $\pi$ variables (T. A. Duffey, Scaling Laws for Fuel Capsules Subjected to Blast, Impact and Thermal Loading, SAE reprint number 719107, proceedings Intersociety Energy Conversion Engineering Conference, 1971)?

The damage to the missile illustrated in Figure 4 is characteristic of dynamic plastic buckling. Various theoretical and experimental studies have been conducted on cylindrical shells impacted axially, principally by a research group at Stanford. In particular, the theoretical study presented by Vaughan (The Response of a Plastic Cylindrical Shell to Axial Impact, Z.A.M.P., Vol. 20, 321-328, 1969) could be used to predict the buckling wave number and other information about the response including the threshold impulse as described in a more recent publication (Jones, N. and E.A. Papageorgiou, Dynamic Axial Plastic Buckling of Stringer Stiffened Cylindrical Shells, Dept. Mech. Eng. Rep. ES/01/80, University of Liverpool and Int. J. Mech. Sci., In Press).

# Contribution to discussion of Paper No. 22

R K Muller
*Institut for Modellstatik, Stuttgart University, West Germany*

Several contributors at the seminar speaking about small scale
replica models for reinforced concrete structures, have stated
that similarity cannot be achieved because of remarkable scaling
effects on bond and shear mechanisms.

It seems necessary for us to express that with a modelling technique presently used at the IMS these mechanisms can be simulated
accurately to a scale factor of at least 1:10.  The model material
is a specially designed microconcrete and deformed steel wires are
used as reinforcement.

The results of a recent test on simply supported circular slabs are
shown in Fig. 1 (Ref. 1).  This test indicates the similarity of
bondage behaviour and shear failure between small scale models and
their prototype.

Initial investigations on material properties have been carried out
in cooperation with the Technische Hogeschool, Delft, concerning
the response characteristics of such models under dynamic loadings.
Measurements on microconcrete specimens under high tensile strain
rates indicate good resemblance to prototype concrete (see Fig. 2).

## Reference

1.  Institution of Structural Engineers Informal Study Group on
    Model Analysis as a Design Tool - Newsletter No. 19

Fig. 1   Impact Load of Hopkinson-bar

| Experiment No. | Loading Speed o N/mm² s | Tensile Strength N/mm² | Stress % |
|---|---|---|---|
| 1 | 19 600 | 5.66 | 0.27 |
| 2 | 26 400 | 6.18 | 0.35 |
| 3 | 27 700 | 5.97 | 0.33 |
| 4 | 28 600 | 6.54 | 0.35 |
| 5 | 29 900 | 6.04 | 0.26 |
| 6 | 31 500 | 6.59 | 0.26 |

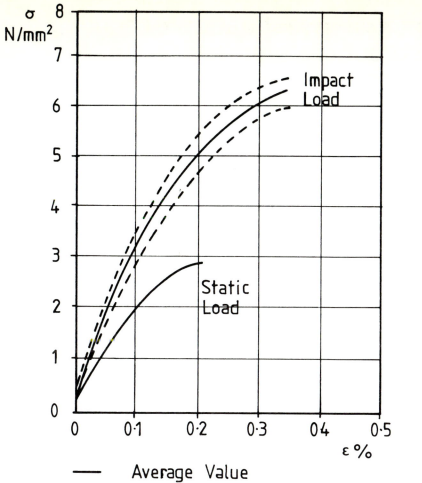

—— Average Value

- - - Dispersion Area of the Experiment Valves

Fig 2.   Punching Shear Failure
         Failure Cone of a 1:6.45 Scale Model

# An experimental investigation of scaling of reinforced concrete structures under impact loading

## 23

P Barr, W D Howe, A J Neilson and P G Carter
*UKAEA, AEE Winfrith, UK*
W Nachtsheim
*Zerna Schnellenbach und Partner, Bochum, West Germany*
H Riech and E Rudiger
*Hochtief A G, Frankfurt, West Germany*

INTRODUCTION

The behaviour of reactor structures under impact loading is being studied experimentally and theoretically in both Germany and the United Kingdom. Present interest in Germany, originating from the aircraft crash load case, is largely focused on the study of the effects of deformable missiles with impact velocities between 200 m/s and 300 m/s. Work in the UK covers both rigid and deformable missiles. A formal technical collaboration agreement covering the exchange of data on experimental and theoretical studies and the implementation of co-ordinated experimental programmes in the missile impact field is in force between the two countries.

The ultimate objective of workers both in Germany and in the UK is the development of a well validated calculation technique to describe the effects of missile impacts both locally and globally on reinforced concrete structures. Experimental validation of any calculation method is essential, and a validation of scaling even over a limited range of applicability, has potentially great direct benefit since

(a) If a general computation technique proves difficult to apply to the necessarily complex features of actual reactor structures, experimental qualification of such structures using models would be possible.

(b) The economies accruing from small scale investigation of minor parametric variations are considerable.

The present German programme is managed for BMFT by GRS Cologne with technical participation by Hochtief AG, Zerna Schnellenbach und Partner, Bundesanstalt fur Materialprufung and Bundeswehr. The programme in the UK is carried out and funded by the UKAEA, with technical linking to practical conditions through co-ordination with the CEGB and the National Nuclear Corporation.

In this paper we describe a series of experiments in the collaborative programme which examine the response of reinforced concrete panels to the impact of deformable missiles.

EXPERIMENTAL TECHNIQUES

General

The experimental facilities in both Germany and the UK are similar in that flat reinforced concrete targets are impacted by missiles

projected by a launcher, and target response is appropriately
monitored, but the apparatus in the 2 test sites differs markedly in
size. At Meppen, in Germany, the launcher is capable of projecting
a missile of about 1T mass and 600 mm diameter at velocities up to
300 m/s (45 MJ kinetic energy). At Winfrith, in the UK, some 1 MJ
of kinetic energy can be imparted to missiles up to 150 mm diameter.
So, in essence, the Winfrith facility is capable of operating
experiments at up to one $\frac{1}{4}$ scale of those conducted at Meppen.

Replica scaling implies that the dimensions of prototype and model
scale geometrically and that the material properties (density and
stress-strain relationship) are identical in prototype and model.
With these requirements satisfied, identical velocities will
provide deformations of missile and target which scale geometrically
and reaction loadings which scale according to the square of the
length scale factor. Most materials of engineering interest display
some effect of the loading rate on the stress-strain relationship
and such effects may produce differences in behaviour of model and
prototype structures.

True replica scaling is difficult to achieve and experiments at
different geometric scales can be used to determine the most
significant parameters that must scale from prototype to model. The
collaboration agreement between UK and Germany provides the UK with
information on the validity of the scaling rules used and provides
Germany with guidance on the impact velocities to be used in their
large scale tests.

With these aims in view, a series of 4 experiments involving the
impact of deformable missiles on concrete targets with two levels
of shear reinforcement were carried out at Winfrith. Prior to these
experiments with concrete targets, a series of experiments involving
the impact of the deformable missiles against a load cell was
mounted to assess the loading to be imposed on the concrete targets.

Targets

The overall dimensions of the Meppen targets were 6.5 x 6 x 0.7 m
and the gross weight 70T. The targets were supported against a
massive concrete abutment weighing some 6000T on 48 identical
steel pillars, equispaced around a square of 5.4 m side (Figure 1).
Eight of these pillars were strain gauged so that they acted as load
cells. These load cells were precompressed by 38 tension rods to
maintain contact between target and load cells at all stages of the
impact process.

The targets used at Winfrith were geometrically similar and $\frac{1}{4}$ of the
Meppen size (1.625 x 1.5 x 0.175 m). They were suspended against
the abutment in a similar manner, but 16 of the support pillars
acted as load cells. The Winfrith abutment weighed about 37T
and was braced through an intervening sandbag layer to a building
structure weighing some thousands of tonnes.

The sizes and positioning of reinforcement were replicated in model
and prototype, but the surface deformation of the prototype reinforcing bar was not replicated in the model. The properties of the
reinforcement in the model (yield strength, ultimate strength and
elongation at break) corresponded to the German norm for the
prototype reinforcement.

The aggregate size fractions for the model and prototype panels are

given in Table 1.  Some differences existed at the extremes of the
size distribution,and these arose from practicalities of supply of
the model aggregate.  The microconcrete mix was designed to produce
compressive and tensile strengths which were as close as possible
to those expected from the prototype concrete mix.

Table 1   Typical Meppen Target Constituents Per $M^3$ Concrete
(as poured)

| Aggregate | Weight fraction % | Weight (dry) kg | Density kg m$^{-3}$ |
|-----------|-------------------|-----------------|---------------------|
| 0 -  2 mm | 36 | 666 | 2600 |
| 2 -  8 mm | 12 | 223 | 2620 |
| 8 - 32 mm | 52 | 969 | 2620 |
| Cement Water | | 310 175 | |

Nominal 28d   UCS   35 MPa on Ø150 x 300 cylinders

Typical Winfrith Target Constituents (as poured) for a Complete
Target

| Aggregate | Weight fraction % | Weight (dry) kg |
|-----------|-------------------|-----------------|
| 0.15 - 0.3 mm | 16 | 190.4 |
| 0.3  - 0.5 mm | 17 | 202.3 |
| 0.5  - 1.5 mm | 17 | 202.3 |
| 1.7  - 4.9 mm | 25 | 297.5 |
| 5  - 10  mm | 25 | 297.5 |
| Cement Water | - - | 238 133 |

Nominal 28 d UCS   35 MPa on Ø150 x 300 cylinders

A nominal compressive strength of 35 MPa (on 150 mm diameter by
300 mm long cylinders), and a nominal tensile strength of 3 MPa
(Brazilian) were required and achieved within the limits of
experimentation.  Bending reinforcement was 0.29% ew on the impact
face, and 0.57% ew on the rear face.  Shear stirrup reinforcement
in the first series of panels was included at 0.38%, and 0.51%
plan area in the subsequent experiments.

Missiles

The missiles used at Meppen, designated type 11, weighed 1T and
comprised a tubular body of 600 mm diameter,with a part spherical
nose and flat rear end,all of mild steel.  Their overall length was
6 M, and the tubular body was of 2 thicknesses.  The forward
section 2.5 M long was of 7 mm wall thickness and the rear section
of 10 mm wall thickness.  An accelerometer system and telemetry
transmitter were housed in the rearmost section.

The $\frac{1}{4}$ scale deformable missiles comprised essentially 2 lengths of mild steel tube manufactured from sheet thicknesses of 1.6 mm and 2.5 mm, welded together such that, with backing plate and nose piece, the overall length was 1.5 m, the overall diameter 151 mm, and the total weight about 15.6 kg. Exact $\frac{1}{4}$ scaling would have needed 1.76 mm and 2.52 mm thick sheet but for practical expediency standard sheet thicknesses were employed.

The material properties of these sheets were also different in detail from those quoted for the material to be used for the full scale missiles at Meppen. Table 2 summarises the expected effects of these differences in size and material properties on the scaled crushing strengths of the $\frac{1}{4}$ scale and prototype missiles, based on the assumption that $P \propto \sigma_y D^2 \left(\dfrac{t}{D}\right)^{\frac{3}{2}}$ where P = crushing load,

$\sigma_y$ = yield stress, D = missile diameter, t = tube thickness.

The reaction load from the crushing missile has two components, (i) the destruction of the momentum of the undamaged portion and (ii) the crushing of the missile at the interface between damaged and undamaged zones. The dominant component at the velocities of interest comes from the momentum destruction. To a first approximation the reaction load is proportional to the mass per unit length of the missile sections. Calculations based on a momentum balance suggested that the reaction load in the period during which the first section is being crushed will be 10% too low in the actual model, the peak load in the collapse of the second section will be 2% too low and the total period of load application will be 5% too long. Experimental measurements of the reaction load of the model missile confirmed the validity of these calculations. (see Section on Missile Collapse Loads).

## Instrumentation

Measurements were made of missile impact velocity, transient target deformation, and transient force applied to the abutment by the target during impact. Also high speed cine photography of the impact was carried out at both laboratories.

Additionally at Meppen missile deceleration during impact was measured by on-board accelerometers and radio telemetry, and transient strain measurements were made for both bending and shear reinforcement steel. At Winfrith, missile exit velocity (in the case of perforating missiles) and ejecta velocity were measured, and the rear face of the panel was monitored by high speed cinephotography.

Load cells capable of registering the direct transient load of deformable missiles were used at Winfrith and Meppen to establish the imposed loading of targets. These loadings were measured separately from target panel impacts as the inclusion of load cells in the impact area would markedly influence panel behaviour.

## EXPERIMENTAL RESULTS

## Missile Collapse Loads

Measurements of the missile collapse load have been made at both Winfrith and Meppen.

At Winfrith a piezoelectric load cell (1 MN capacity) was mounted between a steel impact anvil and a large concrete block bearing on to the abutment used to support the target panels.

At Meppen a load cell with a capacity of 60 MN was constructed. In this load cell the impact anvil forms an integral unit with an annular support skirt which bears through a large concrete block on to the backing structure of the abutment. The reaction load is measured as the reduction of prestress in twelve prestressed bolts set in recessed holes in the support skirt.

The results of one test firing at Winfrith and one test firing at Meppen are shown in Figure 3 in which the Winfrith results have been rescaled (by x 16 for the load and x 4 for the timescale) for direct comparison with the Meppen results. There was a small difference in the test velocities; at Winfrith a velocity of 235 m/s was obtained compared with a velocity of 241 m/s at Meppen. The load results confirm the conclusions reached from the calculations described previously. The mean load level during the collapse of the first section was some 10% lower in the Winfrith test but very close agreement was achieved for the load generated by the collapse of the second section. The contact time deduced from the Winfrith test was some 10% greater than that obtained at Meppen. The final lengths of the missiles showed very close agreement; at Winfrith the length of the undamaged portion was 300 mm (scaling to 1200 mm) compared with a final undamaged length of 1140 mm obtained at Meppen.

## Impacts on Concrete Panels

Four tests on concrete targets were performed at Winfrith prior to the two tests at Meppen. The first three of these ¼ scale tests used targets with a shear reinforcement level of 0.37%, and the fourth test a level of 0.51%. One test at each shear reinforcement level was performed at prototype scale at Meppen.

The objectives of the ¼ scale tests were to discover (a) the velocity at which the missile just perforated the target (b) the velocity at which significant rear face scabbing occurred, and (c) the effects of increasing the shear reinforcement quantity.

The results of the initial test giving perforation of the panel, together with calculations made at Hochtief A.G., were used as a guide for selecting the velocities in the second and third test, and the overall results of the Winfrith tests were used to provide a guide for selecting experimental conditions at Meppen.

Results are summarised in Table 3.

Perforation Test (Winfrith M125)

The impact velocity was 278 m/s and the measured exit velocity 41 m/s. The missile was considerably deformed and had an overall length of 420 mm after the test. The impact attitude of the missile was at right angles to the target surface, but, asymmetry in its collapse mode or in the target perforation process, caused some yawing of the missile, resulting in the formation of a slightly elliptical entry hole. On the basis of an energy balance, the velocity of a missile of this type which would just perforate a target is about 272 m/s.

"No Scabbing" Tests (Winfrith M126, Meppen III1)

The impact velocity of test M126 was 237 m/s. No scabbing of the rear face concrete resulted, but fine radial cracks were detected extending to the corners of the target, and some light cracking within a circle of about 300 mm diameter centred on the impact point. Crack widths were fractions of a millimetre, and a magnifying glass was necessary for their detection. The concrete cover over the impact face reinforcement bars was destroyed over a diameter of about 180 mm, and there was slight deformation of the reinforcing bars in this area.

Meppen test II11 was intended to repeat the impact conditions of Winfrith test M126. As a result of unusually low ambient temperature conditions the launcher produced an impact velocity of 232 m/s instead of 237 m/s.

The concrete cover on the impact face was removed by the impact to expose the reinforcing bars over an area of about 1000 mm diameter.

A clear mark o.d. 710 mm, i.d. 530 mm corresponding to the collapse imprint of the missile was visible. Only slight deformation of the reinforcing bars was apparent. The rear face of the target exhibited a region of fine cracks within an area of about 1.5 metres square in the centre of the target, and fine diagonal cracks from the central region towards the target corners.

"Significant Scabbing" Test (Winfrith M127)

An impact velocity of ∿252 m/s was achieved in this test. The missile penetrated the impact face of the target to a depth of 100 mm, shearing the bending reinforcing bars over a diameter of about 200 mm, and forming a shear plug of concrete. Scabbing on the target rear face extended over an area of about 600 mm diameter, and a permanent rear face heave of about 100 mm was measured. Radial cracking from the scabbed area, essentially along the diagonals to each of the 4 corners, was also evident. The 4 displacement transducers, used for observing transient target deformation nearest to the target centre, were within the scabbed area, so although an initial 20 mm of movement was recorded, thereafter no meaningful displacement was observed.

Increased Shear Reinforcement Experiments (Winfrith M129, Meppen II12)

The fourth experiment at Winfrith and the second experiment at Meppen were carried out on targets having the same bending reinforcement quantities but with the shear reinforcement increased to 0.51% of the plan area. The velocities for both tests were chosen to be the same as in the test giving significant scabbing on the 0.37% shear Winfrith panel (M127).

The missile impact velocity in the Winfrith test was 247 m/s. Concrete cover on the impact face was removed over an area under the missile of about 180 mm diameter together with some deformation of the reinforcing bars. The zone of missile collapse OD 160 mm ID 130 mm was sharply delineated. Rear face cracking within an area of approximately 300 mm square, with slight radial cracks extending out towards the target corners, was evident. Concrete had been ejected, as fairly finely divided particles, in small quantities from the cracks in the central area, and significant scabbing seemed to be incipient. No evidence of shear plug formation was apparent, but a permanent heave of 5 mm was produced on the rear face of the target. The overall missile length after impact was 420 mm.

The missile impact velocity for the comparative Meppen test II12 was 241.5 m/s. The missile overall length after impact was 1960 mm. Concrete cover of the reinforcing bars on the front face had been removed by the impact over an area of about 900 mm diameter, and there was evident deformation, but no shearing, of the exposed reinforcing bars. The target rear face showed an area of cracking in the central region approximately 800 x 950 mm, with diagonal cracking extending to the corners. Some cracks were relatively wide but no scabbing had occurred.

## DISCUSSION OF THE TESTS ON CONCRETE TARGETS

The measured penetrations of the missile are plotted against impact velocity in Figure 4. Results from both Winfrith and Meppen are included. In addition to showing the close agreement obtained for tests at the same impact velocity, two other points are apparent. Firstly there was an abrupt change in the damage to the panel as the impact velocity increased. For the panels with 0.38% shear reinforcement at impact velocities below 240 m/s the missile only uncovered the front reinforcement, thus penetrations of approximately 10% panel thickness were observed. An increase in velocity to 252 m/s, i.e. 4%, was sufficient to produce penetration of some 50% of the panel thickness and to displace the central shear cone sufficiently to allow one to see through the ¼ scale panel. A further increase of 8% in the missile velocity produced complete perforation. Thus in this critical region, a velocity increase of some 12% was sufficient to alter the barrier behaviour from a principally 'elastic' response to complete failure. The abrupt change is ascribed to the formation of a complete shear cone failure. Up to critical velocity the cracks defining this cone were forming and approaching the impact face. Above this critical velocity the cone had formed completely and was restrained by the rear face reinforcement and the shear stirrups. Because of the relatively small quantities of reinforcement the damage to the target increased rapidly with velocity once this critical velocity had been exceeded. The second point to emerge from the results is that the damage can be reduced by increasing the amount of shear reinforcement. An increase of shear reinforcement by some 30% reduced the penetration of the missile impact at ∿250 m/s by a factor of 4.

The reaction load measured by the load cell array on one of the Winfrith panels is shown in Figure 5. In this particular test the impact velocity was 237 m/s. The computed reaction load for impact of the missile on a rigid surface at 235 m/s is included for comparison. This computed reaction load produced good agreement with an experimental result obtained when a model missile was fired against an anvil and load cell mounted on the massive abutment (Figure 3). There were some differences between the computed reaction load on a rigid surface and the total reaction load measured on the panel. Four potential causes could be identified. Firstly the total reaction load was not measured directly. At Winfrith, signals from only 16 of the 48 load cells were recorded dynamically, and sufficient information exists to show systematic variations in reaction load along the sides of the panels. As might be expected, the load was greatest in the mid-side region and decreased rapidly as the corners were approached. Also some of the reaction load is carried in the pretensioning tie bolts. Secondly the presence of the massive panel acted to reduce the natural frequency of the load cells. From the estimated mass of the panel and the stiffness of the load cell array, signals with rise times less than 0.4 ms were expected to experience some attenuation. Thus errors in the overall shape of the curve were to be expected by simply adding the curves together and scaling the results so that the total impulse was conserved. A third reason for the disparity in the loading and reaction forces is that although the panel was not significantly damaged some cracking had occurred and this would lead to the modification of the forces transmitted through the panel.

The overall panel damage and rear face crack pattern corresponded very well in the Meppen and Winfrith tests. Detailed crack pattern differences were discernible but were to be expected.

CONCLUSIONS

In the description of the model tests attention has been drawn to the parameters which do not scale exactly with the prototype experiments. While true replica scaling has not been achieved, the prototype panels have been scaled geometrically,and the prototype missiles have been scaled to the nearest practical dimensions. The strength parameters of the reinforcement and missile have been matched as closely as possible, and the strengths of the concrete were 'identical' within experimental error.

The results of the experiments confirm that the modelling techniques employed provide good representations of the overall prototype behaviour. Additional experiments in the UK and German programmes, not described in this paper, provide additional evidence to support these conclusions and extend the scale range, for soft missile impacts, to 1/5.6.

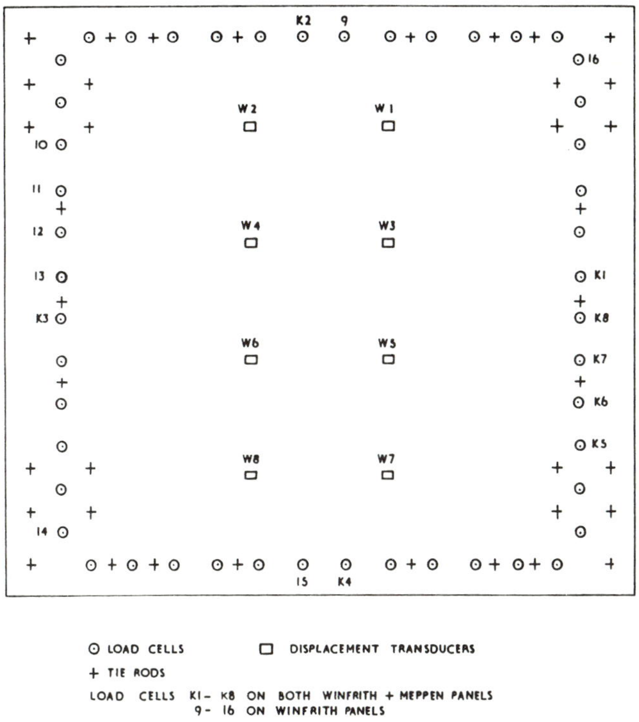

FIG. 1 CONCRETE PANEL INSTRUMENTATION.

Table 2  Comparison of model and prototype missile strengths

| Section Thickness in Model, mm | | Ratio of Crushing Strengths Actual/Replica from Differences in | | |
|---|---|---|---|---|
| Actual | Replica | Geometry | Material | Overall |
| 1.6 | 1.76 | 0.87 | 1.06 | 0.92 |
| 2.5 | 2.52 | 0.99 | 0.93 | 0.92 |

Table 3  Summary of test results

| Test Site | Meppen Slab Type | Impact Velocity | Missile Length After Impact mm | Target Face Conditions | |
|---|---|---|---|---|---|
| | | | | Front | Rear |
| Winfrith (M125) | B16 | 278 | 420 | Perforation | Exit Hole |
| Winfrith (M126) | B16 | 237 | 440 | Concrete Cover Removed ∅ 180 mm | Light Radial Cracks |
| Meppen (II.11) | B16 | 223 | 2035 | Concrete Cover Removed ∿∅ 1000 mm | Light Radial Cracks |
| Winfrith (M127) | B16 | 252 | 440 | Penetration 100 mm | Scabbing over ∅ 600 Radial Cracks in Corners |
| Winfrith (M129) | B13 | 247 | 420 | Concrete Cover Removed ∅ 180 mm | Light Cracking over 300 x 300 mm |
| Meppen (II.12) | B13 | 244 | 1960 | Concrete Cover Removed ∿∅900 mm | Cracking over region ∿950 x 950 mm |

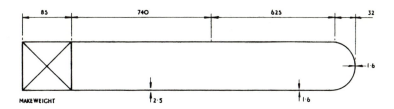

85 | 740 | 625 | 32

MAKEWEIGHT

2·5    1·6    1·6

TOTAL LENGTH 1·45 m EXCLUDING NOSE CONE

FIG.2  DIMENSIONS OF COLLAPSING MISSILE  (WINFRITH SCALE)

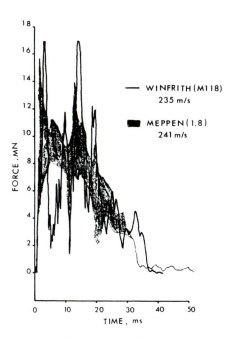

WINFRITH (M118)
235 m/s

MEPPEN (1.8)
241 m/s

FIG.3   MISSILE COLLAPSE LOADS

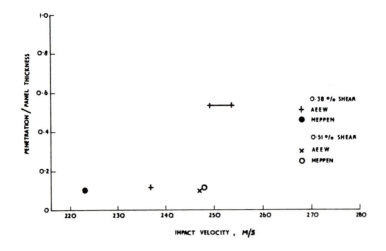

FIG. 4    MEASURED  PENETRATIONS  OF  MISSILES  IN  TESTS.

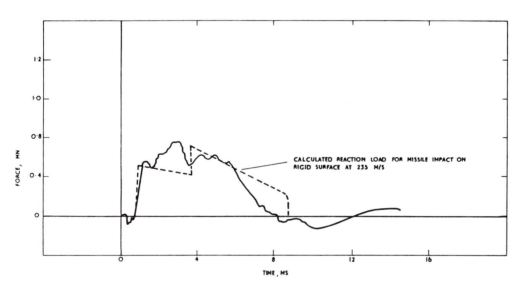

FIG. 5  PANEL REACTION LOAD AT 237 M/S (TEST M126)

# Postscript to Paper No. 23

Since publishing the paper additional experimental data has become available from Meppen, notably target transient deflection measurements, target reaction loads, and a target has been sectioned transversely to observe internal cracks (Fig 8).

A comparison of the measured reaction loads in corresponding load cells for Winfrith and Meppen tests is shown in Fig 6 (see Fig 1) for load cell location). The Winfrith measurements have been scaled by a factor of 16 to account for the difference in scales of the experiments. It can be seen that very similar loading functions have been observed in this particular load cell position, and similarly good agreement has been found in other corresponding load cell locations.

Fig 7 shows transient target displacements in corresponding locations on a Meppen and on a Winfrith target panel. An apparent disparity is observed in all corresponding measurement locations. Two possible reasons for the disparity are postulated. Firstly a calculation (Nachtsheim's) taking account of the admitted differences in the Meppen and Winfrith missiles indicate that the Meppen displacements would be expected to be 50% greater than those of the Winfrith panels. Secondly the method of attachment of the displacement transducers varied between Meppen and Winfrith. At Meppen the transducers were embedded in the concrete cover over the reinforcement, whereas at Winfrith the transducers were connected directly to the target rear face bending reinforcement mesh. The transverse cross section of the Meppen target (Fig 8) shows a crack between the rear face reinforcing bars and the concrete cover. This crack might have allowed the cover to move more than the main body of the target during the impact process and thereby give an apparently larger deflection.

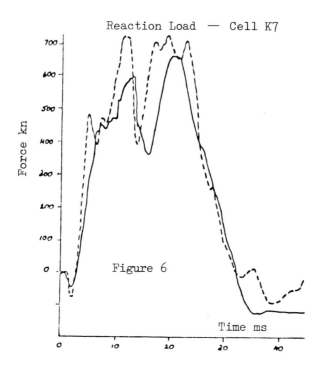

Reaction Load — Cell K7

Force kn

Figure 6

Time ms

Type II Missiles Impacting B13 Targets

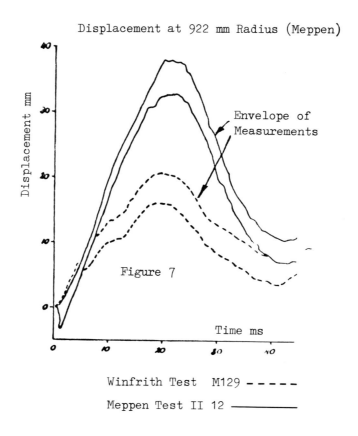

Displacement at 922 mm Radius (Meppen)

Displacement mm

Envelope of
Measurements

Figure 7

Time ms

Winfrith Test M129 - - - - -
Meppen Test II 12 ───────

232

Figure 8          Cross Section of B13 Target          Meppen Test II 12

# Allowing for strain-rate sensitivity in impulsively loaded structures

24

Brian Hobbs
*Department of Civil and Structural Engineering, University of Sheffield, UK*

## SUMMARY

The influence of strain-rate sensitivity on the behaviour of steel and reinforced concrete structures subjected to high rates of loading is discussed. Attention is drawn to the importance of allowing for strain-rate effects when the results from small scale model tests are used to predict the behaviour of full size structures. It is suggested that for many practical structures subjected to high intensity, short duration loading, the influence of strain-rate sensitivity may be considerable and may suppress the travelling hinge deformation mechanisms predicted by rigid-plastic analyses. It is therefore suggested that in many cases the use of simple, single degree of freedom idealization for the deformation modes is justified. A simple analytical approach for impulsively loaded structures is outlined. The method is based on energy consider-ations and involves a two stage procedure, in which the initial elastic phase of the behaviour and the subsequent plastic deformations are considered separately. The case of simply supported beams subjected to uniformly distributed impulsive pressure loading is used as an example of the application of this approach and detailed equations are developed for this case. The analytical results obtained are compared with existing experimental data for small scale model steel beams. The analysis predicts that the effect or strain-rate sensitivity in these model beams would have been to increase the yield stress by up to 60%, with a consequent reduction in plastic deformation. The agreement between the experimental and analytical results is shown to be good and it is suggested that the proposed method of analysis is capable of being extended to cover more complex structures and, despite the many simplifications involved, may be of sufficient accuracy for many practical purposes.

## INTRODUCTION

The response of structures to high intensity impulsive loads has received a considerable amount of attention in the applied mechanics literature. Reviews presented by Symonds (1), and more recently by Jones (2), (3), show that most analyses have been based on the assumption of rigid-plastic structural response, although some recent work on elastic-plastic behaviour has been reported (3). This paper is concerned with the problem of estimating the overall response of building and civil engineering structures to high intensity, short duration loading and, in particular, with the influence of rate of straining on that response. The principal materials of interest, therefore, are reinforced concrete and structural steel. The primary mode of deformation considered is flexure and behaviour past the elastic limit is considered, the principal parameter of interest being the resulting permanent deformation.

There is considerable uncertainty surrounding the response of large scale structures to heavy dynamic loads, particularly in the case of those constructed in reinforced concrete. The use of suitable approximate methods of calculation is thus frequently appropriate and the development of such methods has been pioneered by Newmark (4). The application of simplified methods to the solution of a range of problems is described by Biggs (5). This basic approach is developed herein to take into account the effects of strain-rate sensitivity.

## INFLUENCE OF RATE OF STRAINING

### Material Behaviour

It is known that the yield stress of steel may be considerably increased at high rates of straining and it is generally accepted (3) that a reasonable estimate of the dynamic yield stress, $f_{yd}$, may be obtained by use of the relation

$$f_{yd} = \{1 + (\frac{\dot{\varepsilon}}{D})^{1/p}\} f_y \qquad (1)$$

in which $f_y$ is the static yield stress, $\dot{\varepsilon}$ is the strain-rate and D and p are constants. Suitable values for mild steel are D = 40 and p = 5 (3). Equation (1) then predicts a 30% increase in yield stress at a strain rate of 0.1 sec$^{-1}$ and a 75% increase at 10 sec$^{-1}$.

Test data for concrete subjected to high rates of straining has been summarised by Mainstone (6). It is seen that both the compressive strength and the modulus of elasticity may be increased at high strain-rates but that the influence of strain-rate reduces as the static strength of the concrete increases. For structural concrete with an average static compressive strength in the region of 40 N/mm$^2$ the reported test results indicate increases in strength of the order of 30% to 50% for strain-rates in the range 0.1 sec$^{-1}$ to 10 sec$^{-1}$. There is a considerable amount of variability in the results, however, and more recently published work by Hughes and Watson (7) indicates no effect of strain-rates below about 8 sec$^{-1}$ and an 11% increase in compressive strength at 10 sec$^{-1}$. There is thus, as yet, no generally accepted relationship for concrete equivalent to equation (1). It does appear, however, that strain-rate effects may be rather smaller in concrete than in mild steel.

### Structural Response

In the absence of premature failure due to instability the strength of a steel structure is obviously directly related to the yield stress of the steel. The situation with reinforced concrete is more complex, however, since the strength depends upon the properties of two materials. In many practical structures, however, the flexural response is governed primarily by the reinforcement, and concrete strength has a smaller influence. For very many structures of interest, therefore, the overall response to large dynamic load is governed primarily by the behaviour of steel. This is fortunate, since the available data on strain rate effects in concrete appears to be insufficient for incorporation into an analytical procedure.

The influence of strain rate is thus of considerable importance since the behaviour of steel is so sensitive to it. Experimental work on impulsively loaded structures is generally carried out at a small scale due to the difficulties involved in producing the high intensity loadings required. Methods of analysis are checked against these small scale test results. If an analysis is to be used to predict the behaviour of large scale structures, it is essential that due account should be taken of strain-rate effects, since rates of straining in these structures may be very different from those experienced by a small test structure.

PROPOSED ANALYTICAL APPROACH

Idealization of the Structure

The flexural response of most structures of interest may be reasonably accurately represented by an elastic-perfectly plastic idealization. The rigid-plastic idealization is a further simplification and analyses on this basis predict the occurrence of complex deformation mechanisms involving travelling plastic hinges. It has been noted by Bodner (8), however, that strain rate sensitivity in a structure tends to suppress these travelling hinge mechanisms and to give rise to simpler modes of deformation with fixed hinge positions. The use of these simpler mechanisms is thus felt to be justified for structures of interest here.

For many structures and loading conditions it is possible to predict the mode of deformation and to characterize the response in terms of a single deflection parameter. This allows a structure to be idealized as an equivalent single degree of freedom system. Such a system and its force-displacement characteristic are illustrated in Fig. 1. The initial slope of this characteristic may be estimated from the elastic response of the structure to the given distribution of loading, and the plastic collapse load, $F_{em}$, from the plastic deformation mechanism. General procedures, and some standard transformation factors, for reducing structures to equivalent single degree of freedom systems are given by Biggs (5). It should be noted that, in general, different transformation factors apply to the elastic and plastic phases of the motion.

Figure 1  Equivalent single degree of freedom system and force displacement relation.

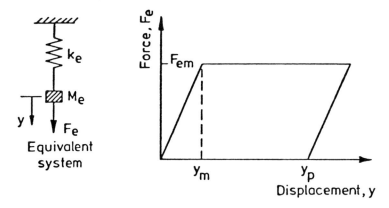

Idealization of the Loading

The loading is treated as an ideal impulse. Only the magnitude of the impulse, i.e. the area under the load-time curve, is thus significant and the precise magnitudes of the peak load and duration of loading need not be known. As noted by Biggs (5), this idealization is satisfactory provided that the duration of the load pulse is less than about 1/5 of the natural period of the structure. As a result of this criterion, significant permanent deformations will only take place if the peak load is considerably higher than the static collapse load of the structure.

## Basis of the Analysis

The basic assumption underlying the analysis is that the initial kinetic energy imparted to the structure by the impulsive load is dissipated during the plastic deformation phase. The load is applied at the beginning of the elastic phase of the motion and the energy imparted is therefore a function of the elastic properties of the structure. It is in this respect that the current approach differs from the rigid-plastic mode approximation techniques.

## Inclusion of Strain-Rate Sensitivity

The time to reach the yield deflection, $Y_m$ (Fig. 1), is estimated and used, in conjuction with the yield strain for the material, to calculate a rate of straining. The dynamic yield stress is then calculated using equation (1). This is only a very approximate procedure, but the form of equation (1) is such that, over a range of strain-rates from 0.1 sec$^{-1}$ to 10 sec$^{-1}$, a 50% error in the strain-rate estimate results in an error of < 5% in the dynamic yield stress for mild steel. The yield stress so calculated is assumed to give rise to a constant enhanced resistance over the whole of the plastic phase of the motion. Perrone (9) has shown that the use of this type of approach, in which an overall correction factor for dynamic yield stress is applied, leads to small errors in rigid-plastic analyses.

## APPLICATION TO SIMPLY SUPPORTED BEAMS UNDER UNIFORMLY DISTRIBUTED IMPULSES

### Derivation of Equations

In the initial elastic phase of the motion the transformation factors are 0.50 for mass and 0.64 for loading. Thus, for a beam of span L and mass per unit length m, the equivalent mass, $M_e$, is given by

$$M_e = 0.50 \, m \, L \qquad (2)$$

and if the beam has a breadth B and is subjected to an impulse i per unit area, the total equivalent impulse, $I_e$, is given by

$$I_e = 0.64 \, i \, B \, L \qquad (3)$$

The initial impulsively generated velocity, $v_i$, is then given by

$$v_i = \frac{I_e}{M_e} = 1.28 \frac{i \, B}{m} \qquad (4)$$

and the maximum elastic displacement, $Y_m$, by

$$Y_m = \frac{8 \, M_{pd}}{k \, L} \qquad (5)$$

in which $k = 384EI/5L^3$ and $M_{pd}$ is the dynamic plastic moment of resistance. An estimate of the time to reach yield, $t_y$, is then given by

$$t_y \doteqdot \frac{Y_m}{v_i} = \frac{6.25 \, m \, M_{pd}}{k \, L \, i \, B} \qquad (6)$$

and hence the strain rate for the material for the beam is

$$\dot{\varepsilon} \doteqdot \frac{f_{yd}}{Et_y} = \frac{f_{yd} \, k \, L \, i \, B}{6.25 \, E \, m \, M_{pd}} \qquad (7)$$

and since $M_{pd} = \dfrac{f_{yd} \, M_p}{f_y}$ we have

$$\dot{\varepsilon} \doteqdot \frac{f_y \, k \, L \, i \, B}{6.25 \, E \, m \, M_p} \qquad (8)$$

The kinetic energy, $U_k$, imparted to the system is given by

$$U_k = \frac{1}{2} M_e v_i^2 = 0.41 \frac{i^2 B^2 L}{m} \qquad (9)$$

When the motion enters the plastic range, giving a permanent midspan deflection $y_p$, the energy absorbed by plastic work, $U_p$, is given by

$$U_p = 4 M_{pd} \frac{y_p}{L} \qquad (10)$$

and by equating $U_k$ and $U_p$, we obtain

$$\frac{y_p}{L} \doteq \frac{i^2 B^2 L}{10 \, m \, M_{pd}} \qquad (11)$$

In order to estimate the critical impulse $i_c$, below which no plastic deformation will take place, the maximum elastic strain energy, $U_m$, is determined from

$$U_m = 0.64 \frac{8 \, M_{pd}}{L} \frac{y_m}{2}$$

$$= 20.5 \frac{M_{pd}^2}{k \, L^2} \qquad (12)$$

Equating $U_m$ and $U_k$, we obtain from equations (9) and (12)

$$i_c = \frac{M_{pd}}{BL} \sqrt{\frac{m}{k \, L}} \qquad (13)$$

Comparison with Test Results

Tests on 457mm span beams of rectangular section have been reported by Florence and Firth (10). Their results for beams of 1018 steel, both in the as-received condition and after annealing, are reproduced in Fig. 2. Results based on the present analysis are given for comparison. Since no specific data is available for the 1018 steel used, values of the constants in equation (1) appropriate to mild steel (i.e. D = 4, p = 5) have been used for both sets of calculations. It is seen that, for impulses significantly greater than the elastic critical values, agreement between the experimental and analytical results is good. The calculated effect of the strain-rates in these tests is to increase the yield stress by between 45% and 60% and this effect is thus of considerable importance in the interpretation of these model tests.

The discrepancies between the analytical and test results for the untreated beams at low impulses may be due to a number of factors. The assumption of a sudden change from purely elastic to purely plastic behaviour as the motion passes the critical displacement $y_p$ is obviously a gross simplification. For beams of solid rectangular section the first yield moment is only 2/3 of the fully plastic moment and the threshold impulse, at which some permanent deformation first takes place, might be more accurately assessed on this basis. This would reduce the relevant values of $i_c$ to approximately 0.24 kN sec/m$^2$ and 0.48 kN sec/m$^2$. This effect would be less serious in most structures of interest, since the yield moment is generally rather closer to the plastic moment. In addition, some of the initial kinetic energy imparted to the beam will be dissipated in elastic vibrations. At low impulses this may form a significant part of the total energy and the analysis will therefore overestimate the permanent deformation.

Figure 2    Comparison between present analysis and test results of
Florence and Firth (10)

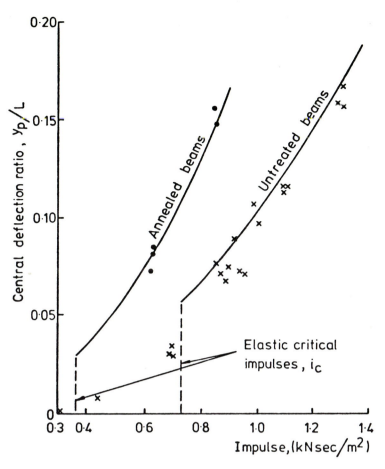

—— Analytical results

• Annealed beam tests

× Untreated beam tests

CONCLUSIONS

The strain-rates experienced by small model structures may have a
considerable influence on the resulting deformations.  It is therefore essential that analytical techniques developed from small scale
tests should take account of any material strain-rate sensitivity.
The proposed analytical approach, involving the use of a simple
two stage mode approximation technqiue, combined with a single
overall strain rate correction, is based on considerable
simplifications of the loading and structural response. Nevertheless, for impulses significantly (approx. 20% or more) greater
than the calculated elastic critical values, the results are in
good agreement with existing test data for small scale beams.
Greater accuracy in the region of the elastic critical impulses
might be obtainable by modifying the form of Fig. 1 to include a
representation of the region between the first yield and fully
plastic loads, though this may be of less significance in many
structures of interest.  The application of the approach to more
complex structures may be achieved by use of the procedures outlined by Newmark (4) and Biggs (5) for deriving suitable single
degree of freedom idealizations.

It must be emphasised, however, that, as presented here, this approach deals only with the overall flexural response of the structure. The possible effects of local damage and shear failure are not considered. It is known that dynamic loading conditions may tend to change the mode of failure from flexure to shear and this may be a particular problem in reinforced concrete structures. There is, however, very little data available on which to base an extension of this approach to take into account shear failure in either structural steelwork or reinforced concrete.

## ACKNOWLEDGEMENT

The approach described in this paper was developed as part of a research project sponsored by the Ministry of Defence, Procurement Executive under contract no. AT/2031/062/RAR.

## REFERENCES

(1)   SYMONDS, P S, 'Survey of methods of analysis for plastic deformation of structures under dynamic loadings', Brown University Report BU/NSRDC/1-67, (1967).

(2)   JONES, N, 'A literature review of the dynamic plastic response of structures', M.I.T. Department of Ocean Engineering Report 74-9, (1974).

(3)   JONES, N, 'Response of structures to dynamic loading', Mechanical Properties at High Rates of Strain, Ed. J. Harding, Institute of Physics, London, (1980), pp. 254-276.

(4)   NEWMARK, N M, 'An engineering approach to blast resistant design', Transactions of the American Society of Civil Engineers, Vol. 120, (1955), pp. 45-61.

(5)   BIGGS, J M, Introduction to Structural Dynamics, McGraw-Hill, New York, (1964).

(6)   MAINSTONE, R J, 'Properties of materials at high rates of straining or loading', Materials and Structures, Vol. 8, No. 44, (1975), pp. 102-115.

(7)   HUGHES, B P, and WATSON, A J, 'Compressive strength and ultimate strain of concrete under impact loading', Magazine of Concrete Research, Vol. 30, No. 105, (1978), pp. 189-199.

(8)   BODNER, S R, 'Deformation of rate-sensitive structures under impulsive loading', Engineering Plasticity, Eds. J. Heyman and F.A. Leckie, Cambridge University Press, Cambridge, (1968), pp. 77-91.

(9)   PERRONE, N, 'On a simplified method for solving impulsively loaded structures of rate-sensitive materials', Journal of Applied Mechanics, Vol. 32, (1965), pp. 489-492.

(10)  FLORENCE, A L, and FIRTH, R D, 'Rigid-plastic beams under uniformly distributed impulses', Journal of Applied Mechanics, Vol. 32, (1965), pp. 481-488.

# Contribution to discussion of Paper No. 24

Norman Jones
*Department of Mechanical Engineering, The University of Liverpool, UK*

As the author remarked, mode approximation methods have been used widely in dynamic plasticity since, when appropriate, they can lead to quite simple, yet surprisingly accurate, predictions for metal structures (e.g., Jones, Int. J. Solids and Structures, Vol. 7, 1007-1029, 1971). Martin and Symonds (Proc. A.S.C.E., J. Eng. Mechs. Div., Vol. 92, No. EM5, 43-66, 1966) have examined the behaviour of rigid-plastic structures responding in a modal form and developed criteria for optimum modes. It appears that the coefficients in equations (2) and (3) are valid for elastic behaviour. The criteria of Martin and Symonds could be used to obtain the corresponding coefficients for     plastic behaviour.

# Impact resistance of reinforced concrete structures

## 25

A J Watson and T H Ang
*Civil and Structural Engineering Department, University of Sheffield, UK*

## 1. Summary

The paper describes the behaviour of $\frac{1}{8}$th scale, reinforced microconcrete beams 44mm wide x 65mm deep on a simply supported span of
442mm under static and impact load. The impact load is produced by
a 360mm long steel rod of 1.78kg mass which is fired from an air
gun at a velocity of 16m/s to impact a 1 metre long cylindrical steel
pressure bar which is placed in contact with the beam. A maximum
force of 191kN is applied for a max duration of 158µs. The aim of
the tests is to compare the load-deflection and cracking response of
the beam under static and impact loading and to determine the
residual strength of the beam after it has been damaged by impact
loads. The paper also describes tests carried out to determine the
static and dynamic bond stresses and bond slip characteristics for
wire reinforcement in microconcrete and describes the device
developed for these tests.

## 2. Introduction

There are many instances where reinforced concrete structures
designed for static loads are subjected to accidental or deliberate
impact loads because of industrial or transportation accidents,
military or terrorist activity. Even when such loading can be
forseen, it is frequently uneconomic or even technically impossible
to design the structure to resist impact loads. For such instances
it may be more feasible to attempt to calculate the ultimate impact
resistance of the structure and relate that to the probability level
of impact loading. Such calculations might also be valuable in
planning and carrying out the demolition of structures.

Small scale models have very great advantages of economy and
practicability, especially in the study of dynamic loading on
structures. It has already been demonstrated (1) that models as
small as 1/10th scale can be used to predict the moment-rotation
characteristics and the displacement of prototype reinforced concrete
structures, although a difference in the tensile strength:compressive
strength ratio has been observed between model and prototype concrete.

This paper describes the equipment and reports the results of impact
and static point load tests on $\frac{1}{8}$th scale reinforced microconcrete
beams with rectangular cross section and simply supported on a span

242

of 442mm with a span:overall depth ratio of 6.8. In every test the load was applied at mid span and all the impact load beams were subsequently tested to collapse under static load. Some of these beams had been impacted once and some impacted twice.

During all the beam tests the deflection-time history at $\frac{span}{2}$, $\frac{span}{3}$, $\frac{span}{6}$ was measured as well as the load-time history at mid span. The independent test variables were rate of loading (impact or static); number of blows; percentage of bottom longitudinal reinforcement, top longitudinal reinforcement and the spacing and type of transverse stirrup reinforcement. All other independent variables such as beam dimensions, span and material properties remained the same. After the test, measurements were made for impact punch width and depth, defined in fig 4; number, location and length of cracks in the beams and residual deflection of the impacted beams. Calculations were then made for the load-deflection characteristics of all the beams; the energy absorption and ductility factor under static loading; the static reserve strength of impacted beams.

The paper also describes the equipment and reports some results for model scale impact and static pull-out tests with the same reinforcing bars used in the model beams. These bars were embedded in 50 x 100mm microconcrete clyinders. During the tests the load-time history and the bar slip-time history were measured. The independent variables were rate of loading (impact or static); bar diameter, type and embedment length.

## 3. Test Specimens

Dimensions of the model scale beams and details of the reinforcement are given in fig. 1 and the pull-out specimens in fig. 3. The microconcrete mix was designed to be of medium workability (34mm slump) and to have a 50mm dia x 100mm cylinder crushing strength of about $46N/mm^2$ at 14 days using Ferrocrete Rapid Hardening Portland Cement with a water-cement ratio = 0.6. The aggregate was river gravel sand with a maximum aggregate size B.S. sieve 7 and an aggregate-cement ratio of 3.0. The cylinder splitting strength using 50mm dia x 100mm cylinders was about 1/10th of the cylinder crushing strength.

The tension and compression reinforcement in the model was black annealed wire of 4mm, 3.15mm and 2mm diameter which, except the 2mm, had been cold worked and deformed on four sides by passing the wire through two pairs of perpendicular knurls. The knurling machine is similar to that used by Harris and others (2). The depth of the indentations produced by the knurls not only controls the amount of cold work hardening of the wire but improves the bond between wire and concrete. Sharp-Anvil micrometers were used to measure the depth of the indentations and a wire drawing machine was used to provide a uniform force to pull the wire through the knurls. The amount of cold working used was made just sufficient to raise the yield point from $155N/mm^2$ to that of prototype mild steel reinforcement $250N/mm^2$ and the local bond characteristics of the knurled wire were also closer to those of mild steel reinforcement. Plain black annealed wire of 2mm diameter was used to model the transverse stirrup reinforcement.

A single batch of concrete was used to cast 6 or 8 beams and 12 control cylinders or 18 cylinders for the pull-out tests. After 24 hours in moist conditions in the laboratory, the specimens were demoulded and kept at 22°C ± ½° and 95-99% RH for 6 days. The beams were then whitewashed and the position of all reinforcement marked on the side of the beam. The beams and cylinders remained in the laboratory until tested at 14 or 15 days old. When casting the pull-out specimens the embedment length was varied to 3, 6 or 12 bar diameters using nylon tubes as a debonding material. Each pull-out specimen had transverse spiral reinforcement of 2mm diameter at 7mm pitch, plain black annealed wire.

## 4. Test Equipment and Experimental Procedure

### 4.1 Beam tests

The beam was supported vertically and horizontally on simple supports and loaded at mid span in the horizontal plane as shown in fig. 2. The vertical support at each end of the beam was a 60mm diameter plate on rotary bearings with the axis vertical and the horizontal support was a 50mm diameter vertical peg. One support peg could rotate about its major axis but at the other support the peg was fixed. Both supports allowed free rotation of the beam. For the impact tests the end supports had a safety capping with soft rubber pads to limit but not prevent the beam from "lifting-off" the supports.

In the impact load beam tests a 28mm dia x 360mm long steel missile was fired by air pressure of $0.21N/mm^2$ down a 2.8m long, precision drawn steel barrel 28.3mm internal diameter. The missile impacted the end of a steel 25mm dia x 1 metre long stationary pressure bar mounted in linear bearings and placed in contact at mid span of the beam, at a velocity of $15.85 \pm 0.25$ m/s. The velocity was measured in each test using 2 photodiodes 100mm apart at the end of the steel barrel. The pressure bar had 2 sets of electrical resistance strain gauges to measure the amplitude of the strain pulse produced in the bar by the impact of the missile. This strain pulse was recorded electronically, and then analized and calibrated in the manner described by Hughes and Watson (3), to give the load-time history for the beam. Contact of the missile with the pressure bar was indicated by a microswitch and this was taken as the time origin. The deflection-time history was obtained using Linear Variable Differential transformer at half, third and sixth span positions and attached to the beam to record both push and pull movement. Typical force-time and deflection-time records are given in fig. 4. High speed photography was used to record deflection and crack development in the beam using a Hycam 16mm camera at a speed of 5300 p.p.s.

In the static load beam tests the load was applied through a socket containing a circular bearing by a screw jack against a load cell placed at mid span of the beam. Load was applied at 0.25kN increments until the first crack was observed, and then at 0.5kN increments until the maximum load and a mid span deflection of 30mm had both been reached. At each load level the deflection was measured by mechanical dial gauges at half, third and sixth span positions. Those beams which had been impact tested were subsequently loaded statically in the manner just described.

### 4.2 Bond tests

In the bond tests the specimen was arranged as shown in fig. 3 with a load cell placed between the concrete cylinder and the support to measure the applied load from the embedded wire. The load cell was a 38mm diameter aluminium tube, 600mm long x 0.6mm thick for the impact tests and 50mm long x 1mm thick for the static tests and gauged with electrical resistance strain gauges. The unloaded end of the embedded wire protruded 10mm from the concrete cylinder so that slip relative to the concrete could be measured using a linear variable differential transformer.

In the impact bond tests the embedded wire was fixed to a vertical rod on which a mass of 4.3kg was allowed to slide freely. The mass was released by a pull catch so that it dropped 1115mm onto a steel collar fixed at the bottom of the rod to impose an impact load on the embedded wire. The load-time history was measured on the load cell and the bond slip-time history with the L.V.D.T.

## 5. Test Results and Analysis

### 5.1 Beam tests

Fig. 4 shows the load-time and deflection-time records for a beam under an impact load. These results are typical although the impact peak-mid span and residual-mid span deflection were dependent upon the amount of reinforcement, both longitudinal and transverse. Fig. 4 also shows the dynamically deflected form of the beam during the impact, the static load-deflection curves of similar beams, with and without impact damage, and also the crack pattern produced which is typical for the impacted beams. Photographs of impact loaded and static loaded beams are given in plate 1.

### 5.1.1 Static load-deflection characteristics

Figs. 5, 6, 7 show the static load-deflection curves obtained at mid span for beams which differed only in the amount of tension reinforcement and in whether or not the beam had any previous impact damage. It is clear that all beams had initially a linear deflection response to load but that the initial stiffness decreased for those beams in proportion to the amount of impact damage. The limiting deflection for a linear response increased with the amount of impact damage but the limiting load decreased. After the initial linear response all beams had a rapid and continuing decrease in stiffness and the maximum load and, in general the deflection at which it occurred, decreased with the amount of impact damage. This decrease in deflection at maximum load is quantified by calculating a ductility factor as the ratio of deflection at maximum load to deflection at the linear response limit load, which was at about 70% of the maximum load, fig. 8.

The area beneath the load-deflection curve was used to calculate the static energy absorption capacity using a 25mm deflection limit. On average the energy absorption capacity for those beams given in Figs. 5, 6, 7 fell by 21% after a single impact test and by 55% after a double impact test.

A reserve strength was calculated for those beams with impact damage using the energy absorbed in a post impact static test divided by the energy absorbed by a similar but undamaged beam in a static test. For the impact damaged beams this ratio remained practically independent of the amount of reinforcement at about 80%-90% for the single impact and at about 50%-70% for the double impact damaged beams.

### 5.1.2 Impact load and deflection

The average impact load measured from the strain pulse in the pressure bar was a square pulse with a peak load of 184kN and an average duration of 154μs. For the 22 beams tested the sample standard deviation for load and duration was 4.37kN and 1.97μs.

On average the deflection of the beams did not commence until 1003μs after the missile had impacted the pressure bar. This is on average 806μs after the load pulse front reached the beam and 652μs after the end of the incident load pulse. From the high speed photographs it appeared that the cracking of the beam under the impact load also commenced on average at 1003μs after the missile had impacted the pressure bar, but with an interframe time of 189μs these times were not precisely defined.

The L.V.D.T. records show that after the impact load the beam moved through a single cycle of deformation with a period of oscillation of 31000μs on average and ended with a residual deflection. After the initial contact of the missile on the pressure bar, the two bars separated and did not make contact again until 10000μs later. This was about the time that the beam reformed to its original position so that the beam may have been undergoing a forced oscillation. The photographs show that the pressure bar separated from the beam at 13585μs after the missile impacted the pressure bar.

The impact peak mid span deflection of the beam for the first or second impact blow decreased significantly as the amount of bottom steel increased from 0.5 to 1.5%, fig. 9. It is also shown that beams with spiral reinforcement have a much lower peak deflection than beams with an equal amount of stirrup reinforcement. The amount of top steel had only a small influence on the peak impact deflection from the first impact blow but for the second blow the peak deflection decreased by 9% as the top steel was increased from 0.25 to 1.5%. The residual impact deflection at mid span also decreased most significantly as the amount of bottom steel was increased, fig. 10.

5.1.3 Crack patterns and deformation

Under static load the beam cracks initiated from the soffit of the beam as vertical cracks and later became inclined, plate 1. The length of span having cracks in the soffit depended upon the amount of reinforcement in the beam and varied from 0.30 to 0.68 of the span, and was approximately central about the load point.

Under the first impact load the crack pattern indicated a punching failure under the load point with the high speed film showing that inclined cracks were initiated from the load point and these propagated at velocities exceeding 500 m/s until they reached the soffit of the beam. The width of this punch zone at the beam soffit depended upon the amount of reinforcement in the beam and varied from 0.29 to 0.42 of the span and was symmetrical about the load point. Within the punch width there were between 6 and 11 other cracks which appeared about 200μs later and varied in inclination from vertical under the load to a maximum inclination of 40°-50° to the vertical at the sides of the punch zone, plate 1. In addition to the cracks within the punch zone under the load point, there were vertical cracks initiated from the top of the beam within a third of the span from the supports, plate 1. These cracks generally coincided with the position of the transverse reinforcement although not necessarily at every stirrup. Only a few of them extended through the full depth of the beam and were probably caused by flexure and produced by reverse curvature of the beam near the supports since the L.V.D.T. dynamic deformation records, fig. 4, indicated that deflection at mid span increased more rapidly than that at third or sixth span positions.

## 5.2 Bond tests

Results for the static pull out tests indicated that on average knurled wires had a 45% higher bond strength than plain wire in the static tests. The impact bond strength was greater than the static bond strength for the same embedment length thus indicating that bond strength is strain rate dependent. Bond slip did not commence until almost 1 ms after the falling weight struck the collar which is about double the time taken for the load pulse to reach the embedded wire.

## 6. Conclusions

The work described in this paper is continuing but from the results obtained so far on models some broad qualitative conclusions can be drawn which may be of value to the structural design engineer.

(1) An impact load up to thirty times larger than the static strength at the mid span of a simply supported beam with a low span:depth ratio, produced deformations and crack development which was quite distinct from that produced by a static load. Because the duration of the impact load was short it did not reduce the residual static load carrying capacity below that required for static service loads.

(2) Under repeated impact loads at mid span the residual strength of the simply supported beam decreased rapidly.

(3) Impact loads induced large shear forces and local damage near the impact zone and produced higher mode deformation than that produced by a static load. This resulted in tension cracking at the top of the simply supported beam near the supports.

(4) Beams damaged by an impact load retained an initial linear load-deflection response to static loads but with a reduced stiffness.

(5) The residual deflection at the mid span of the beams after impact depended largely upon the amount of longitudinal bottom reinforcement but the reserve strength of the damaged beam was practically independent of the amount of reinforcement and hence of the residual deflection.

## Acknowledgements

The authors are grateful to all the staff of the Civil and Structural Engineering Department, Sheffield University, who have helped in this project with advice and practical assistance.

## References

1. Chowdhury, A H, White, R N, Materials and MOdelling Techniques for Reinforced Concrete Frames, American Concrete Institute Journal, Proceedings Vol. 74, No. 11 (November 1977)

2. Watson, A J, Inkester, J E, Impact Loading of a Reinforced Concrete Beam to Column Joint, American Concrete Institute Special Publication, Dynamic Modelling of Concrete Structures (1980)

3.  Hughes, B P, Watson, A J, Compressive Strength and Ultimate
    Strain of Concrete under Impact Loading, Magazine of Concrete
    Research, Vol. 30, No. 105 (December 1978).

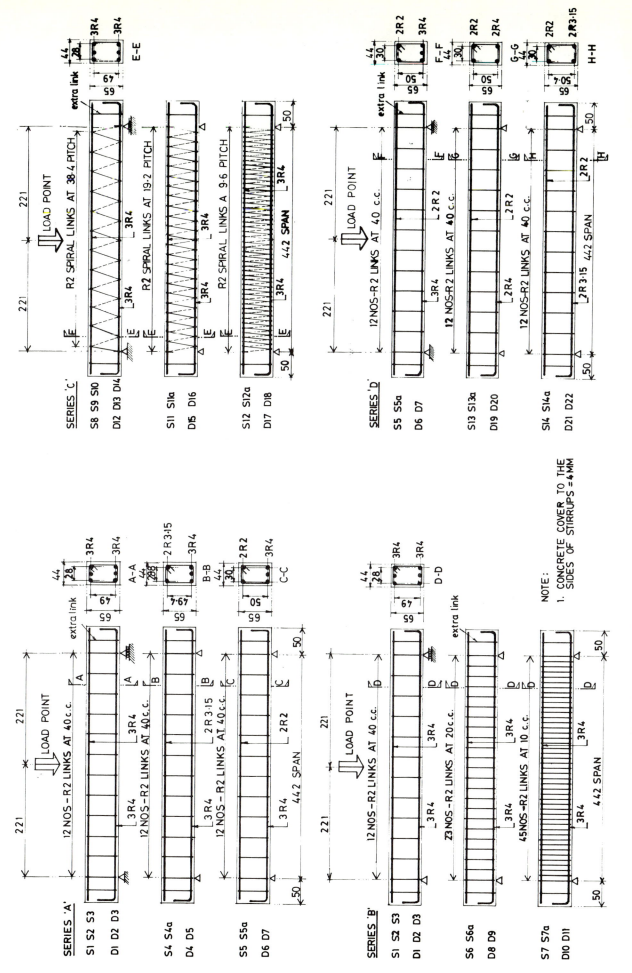

FIGURE 1: DETAILS OF REINFORCEMENT FOR SIMPLY SUPPORTED BEAMS

249

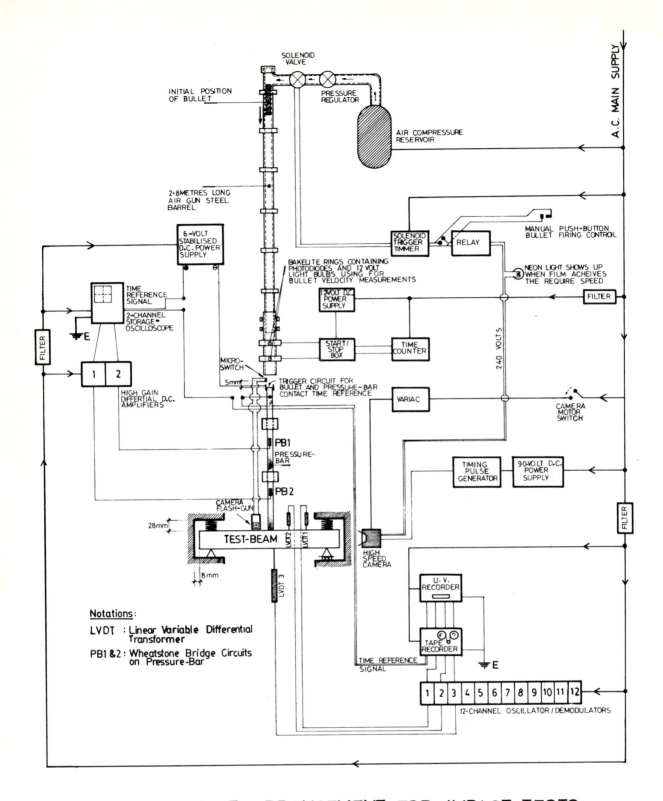

FIGURE 2:  TEST ARRANGEMENT FOR IMPACT TESTS
ON SIMPLY SUPPORTED BEAMS

250

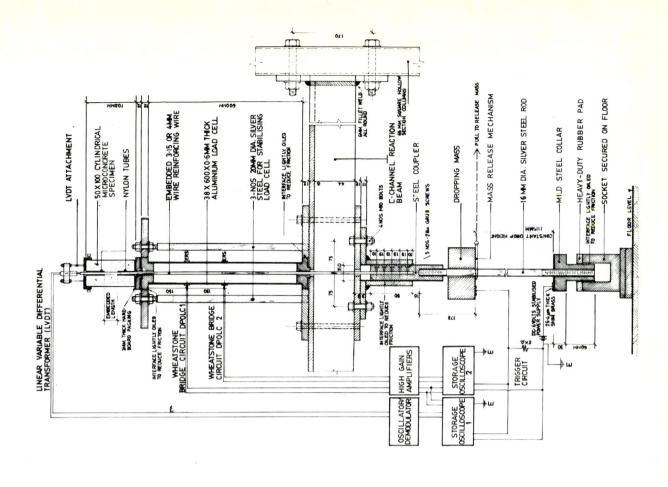

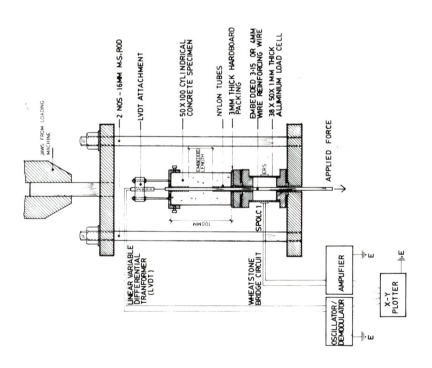

FIGURE 3: ELECTRIC CIRCUITRY AND RIG FOR STATIC AND IMPACT BOND TESTS

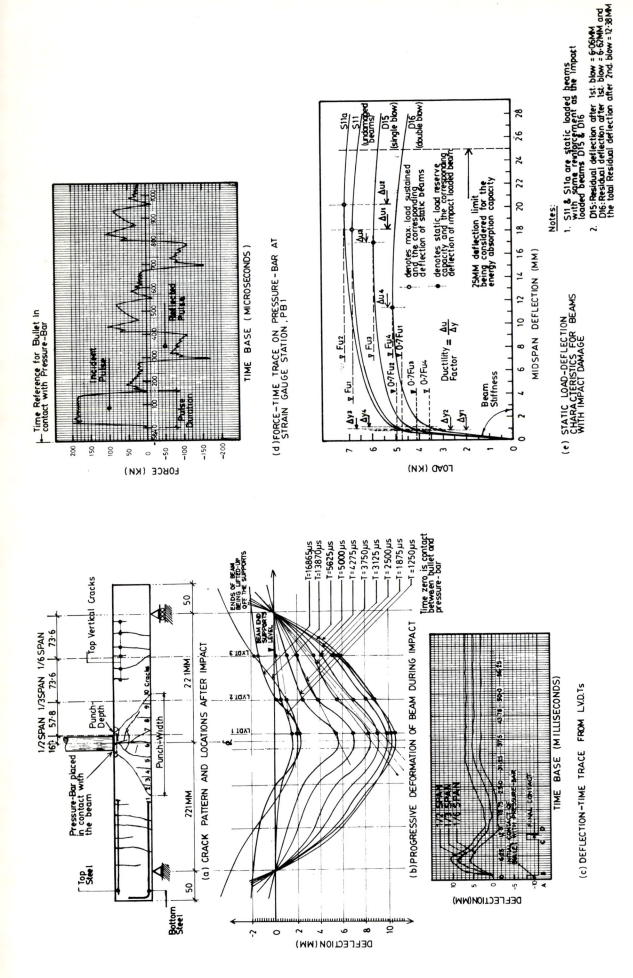

FIGURE 4: TYPICAL IMPACT TEST RESULTS

252

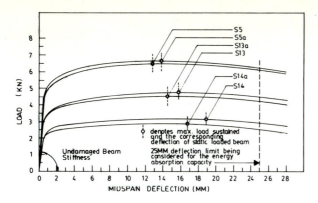

FIGURE 5: STATIC LOAD-DEFLECTION CHARACTERISTICS OF STATIC LOADED BEAMS S5, S5a, S13, S13a, S14 & S14a (TEST SERIES 'D')

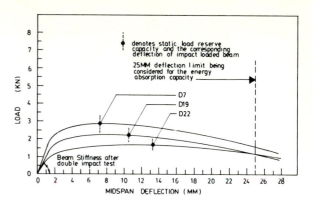

FIGURE 7: STATIC LOAD-DEFLECTION CHARACTERISTICS OF DOUBLE BLOW IMPACT BEAMS D7, D19 & D22 (TEST SERIES 'D')

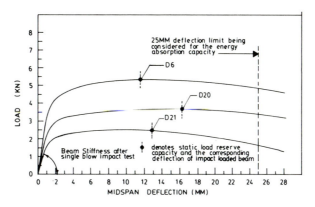

FIGURE 6: STATIC LOAD-DEFLECTION CHARACTERISTICS OF SINGLE BLOW IMPACT BEAMS D6, D20 &( D21 (TEST SERIES 'D')

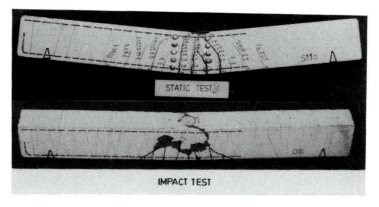

PLATE 1: CRACK PATTERN OF STATIC AND IMPACT LOADED BEAMS

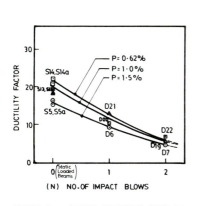

FIGURE 8: EFFECT OF BOTTOM STEEL ON DUCTILITY FACTOR OF IMPACT DAMAGED BEAMS

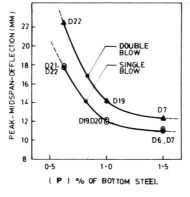

FIGURE 9: IMPACT PEAK-MIDSPAN-DEFLECTION Vs % OF BOTTOM STEEL

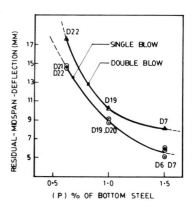

FIGURE 10: IMPACT RESIDUAL-MIDSPAN-DEFLECTION Vs % OF BOTTOM STEEL

253

# Model prestressed concrete slabs subjected to hard impact loading

## 26

S H Perry and I C Brown
*Department of Civil Engineering, Imperial College, London, UK*

SUMMARY

A research project which investigates the response of prestressed concrete slabs to hard impact loading is presented. After a general description of the tests, some aspects of impact modelling are discussed.

ARRANGEMENT OF TESTS

Introduction

A project is being carried out, in the Concrete Laboratories at Imperial College, to investigate the impact resistance of prestressed concrete. At present the test specimens are in the form of slabs but it is intended to extend the investigation to include cylindrical shells and domes, and, without major modification, it would be possible to test any model of width up to about 2m.

Impacting Body

The impacting body is dropped freely down the inside of a tube manufactured from two battened steel channels. The maximum height of drop which can be achieved depends to some extent on the length of the impacting body, but is about 5m, giving a maximum approach velocity of 10 m/s. The impacting body is a steel cylinder 1.2m long and 127mm diameter. At the front end of this cylinder is fixed a load cell to record the contact force during impact.

Model Slabs

Slab specimens are 1.5m square and 60mm thick. A normal concrete mix is used, with OPC and graded aggregate of 10mm maximum size. The mix achieves a cube strength of about 55 $N/mm^2$ at 28 days. Ordinary prestressing wires of 5mm or 7mm diameter are placed in straight ducts cast in the concrete. The ducts are formed by bright mild steel bar, 10mm diameter, around which is a PVC sleeve. About four hours after casting the slab, the mild steel bars and PVC sleeving are removed leaving ducts through the concrete. Prestressing is carried out in two stages. During the first stage, wires are stressed individually to 70% of their characteristic strength and locked off. In the second stage, losses caused by grip pull-in are mostly recovered by re-stressing each wire to 70% characteristic strength and placing shims between the barrel and stressing block. This second stage of stressing is especially important when short stressing wires are used because the strain loss caused by pull-in can be very considerable. After stressing, the ducts are fully grouted to provide a bonded structure.

## Instrumentation

Apart from recording the contact force by means of a load cell at the end of the impacting body, measurements are made of strains in the prestressing wires and on the concrete surface, and displacements of the slab. Difficulty was experienced in devising a suitable method for collecting the experimental data. Initially, it was intended to use a recording oscilloscope, but there are problems with accurate triggering of the oscilloscope, and the record length (one timebase sweep) is very short. The resulting traces are in analogue form and need digitising if further processing is required. Furthermore, the cost for several channels of information is high. Therefore it was decided to convert all data from analogue to digital form before storing the results. The scheme which has been adopted is to convert and store in static memory each individual channel of data. Subsequently, information from these transient recorders is read into a micro-computer for processing.

## IMPACT MODELLING

The stimulus for this research comes from concern about impact on maritime installations such as oil platforms and a lack of knowledge of the effects of such impact. Two main situations can be identified; firstly, accidental collision with a ship and, secondly, impact caused by rigid dropped objects. In the first case the impacting body deforms and yields, itself absorbing much of the impact energy. For this "soft" impact the loading time may be an order of magnitude greater than the fundamental periodic time of the structural member that has been hit. In the second case, a "hard" impact, the impacting body absorbs little energy and the loading time is much shorter. The present research is concerned with the latter situation, although, by suitable modifcations to the dropped mass, an investigation of soft impacts could be carried out.

It will be obvious from the description of the tests that no attempt has been made to consider a particular design impact condition; neither are the test slabs carefully modelled on any prototype. Instead, it is intended that the research should be used for a fundamental investigation of the problem. If the results of the tests can be properly understood and explained then it should be possible to predict the behaviour of a prototype structure. A justification for this approach is the difficulty in accurately modelling a hard impact. In such an impact, failure may be either of a "flexural" or of a "shear" type. In the first case, failure occurs after several reflections of stress waves from the supports and considerable deflection, approximately after one quarter cycle of the lowest frequency vibration of the element. In the second case, failure occurs very rapidly before significant stress wave reflections or deflections have occurred. The modelling techniques required for these two cases are different, and, as it is not possible to know before a test what the failure mode will be, it is also difficult to devise a satisfactory model for the structural element and impacting body for each test.

## ACKNOWLEDGEMENTS

Funds for this project have been provided by the Science Research Council as part of their Marine Technology Programme.

REFERENCES

(1)    BROWN I C and PERRY S H, 'Transverse Impact on Beams and
       Slabs', Proc. 2nd Int. Conf. on the Behaviour of Offshore
       Structures, <u>BHRA Fluid Engineering</u>, Cranfield, England, Vol. 2
       pp. 357 - 368 (1979).

(2)    OKAFOR H O, 'Bonded Prestressed Concrete Square Slabs
       Subjected to Concentrated Loading', Ph.D. Thesis, London
       (1980).

(3)    OKAFOR H O and PERRY S H, 'Post Yield Behaviour of Bonded
       Prestressed Concrete Square Slabs Subjected to Concentrated
       Loading', Publication pending, (1981).

Figure 1   General view of test rig showing support frame for slab
           and guide tube for impacting body.

Figure 2 Impacting body striking a test slab.

Figure 3 View of underside of slab after impact showing complete penetration of impacting body and fracture of prestressing wires.

# Contribution to Paper No. 26

S H Perry and I C Brown
*Department of Civil Engineering, Imperial College, London, UK*

ADDITIONAL CONTRIBUTION BY AUTHORS

Since writing the preprinted paper some preliminary tests have been carried out and the results warrant some discussion although only tentative conclusions are possible at this stage.

Fig. 1 shows the contact load that was recorded in three successive impacts carried out at progressively faster approach speeds on one prestressed slab. Fig. 2 shows a detailed trace of the initial part of the contact load during impact on another slab. On these traces periodic fluctuations of the load can be identified at three distinct frequencies. The fastest fluctuation, clearly visible on Fig. 2, with a period of about 200 microseconds, is caused by stress wave reflections within the load cell. This variation is unwanted, in that it is a function of the measuring system and occurs only in the model test, not in the prototype. However, because the impact duration is relatively long, these fast vibrations of the load cell are not too troublesome and are not even discernible on the traces of Fig. 1. The other vibration seen in Fig. 2, also just visible in Fig. 1, with a period of about 2 milliseconds, is caused by stress wave reflections in the steel billet which forms the dropped mass. This form of load variation will occur in both model and prototype, and emphasises the need for proper modelling of the impacting body as well as the impacted structure. Finally, the first trace of Fig. 1 shows the slowest fluctuation, with a period of about 6 milliseconds which is caused by transverse vibration of the slab.

Looking at Fig. 1, a suggested explanation for the sequence of events during an impact can be given. The initial sharp, peaked loading is the response of the concrete in the vicinity of the contact area. The form of this part of the loading is affected by local conditions in the slab including damage caused by previous impacts, but is insensitive to support conditions. It is during this period that cracks in the contact zone will be formed. In the first two traces on Fig. 1 contact is lost briefly after the initial impact and re-established as the slab undergoes transverse vibrations. The form of the subsequent part of the loading depends on how damaged the slab is. In the first case, with low approach velocity and little damage, the slab remains a complete structural unit and its transverse vibrations are indicated throughout the load trace. In the second case, at higher approach velocity, very considerate damage occurs during the initial impact and it is postulated that, subsequently, the major connection between the centre of the slab, supporting the steel billet, and the outside of the slab is by means of the prestressing wires acting as a net. Finally, in the third trace, at even higher approach velocity, complete penetration of the slab occurs with fracture of the prestressing wires and tearing of the wires through the concrete cover on the underside of the slab.

Figure 1   Contact load history for successive impacts on a
prestressed concrete slab

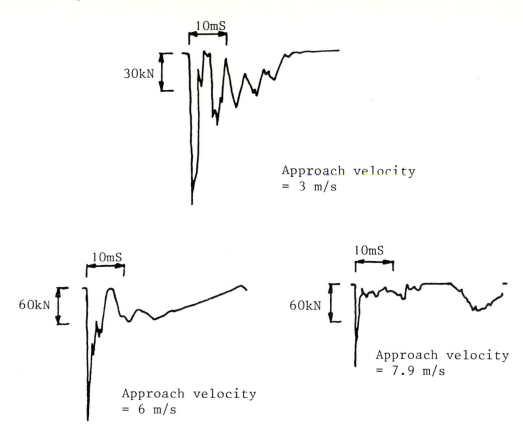

10mS

30kN

Approach velocity
= 3 m/s

10mS

60kN

Approach velocity
= 6 m/s

10mS

60kN

Approach velocity
= 7.9 m/s

Figure 2   Detail of initial contact load history for an impact
on a reinforced concrete slab

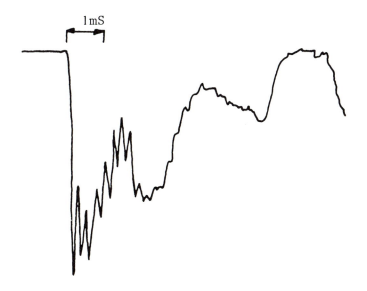

1mS

# Analysis of the beam impact problem    27

G Hughes
*Research & Development Division, Cement & Concrete Association, UK*

## SUMMARY

80 pin-ended beams were tested by dropping a nominally rigid striker
onto the beam at midspan.  For each test the impact force history and
the beam displacements (maximum and residual) were measured.   The
simple beam vibration model, which allows for strain energy of bending
and transverse inertia,is shown to be applicable over the test range.
The problem is amenable to parametric representation and the
importance of two parameters, the mass ratio and the pulse ratio, are
recognised.

## INTRODUCTION

There is a growing awareness that some structures must be designed
for dynamic as well as static loading.  The dynamic loading investigated is commonly known as the 'beam impact problem' : a rigid mass
falls on the midspan of a single span beam initially at rest.  This
is a classical problem and a solution can be obtained by describing
the beam behaviour in terms of free vibration modes (1). The solution
involves some mathematical complexity and this, coupled with the
difficulty of measuring transient phenomena, has hindered progress.
The advent of computers and improved instrumentation have facilitated
the investigation, and it is hoped that an understanding of this
relatively simple dynamic problem will shed some light on some of the
more complex dynamic problems which may occur in practice. The Cement
and Concrete Association has completed a two year investigation of
this problem in which both pin-ended and simply supported beams were
tested.  The work is detailed in Ref. 2.  This paper describes some
of the work on pin-ended beams and, in particular, the use of
parameters to facilitate the analysis.  Similar parameters can be
used for impact tests in general for both planning test programmes
and analysing results.

## NOTATION

A   area of beam cross-section
E   strain modulus
$E_c$   strain modulus of concrete
$F$   impact force
$F_m$   maximum impact force
$F_\infty$   maximum impact force when beam is massive
$I$   second moment of area
$I_c$   second moment of area of a cracked concrete beam
$K$   deformation constant of the impact zone
L   beam length
$M_y$   beam yield moment
N   number of vibration modes excited
$T_1$   period of free vibration of the first mode
$V_y^1$   first mode elastic energy

260

a      deformation of the impact zone
i      mode number
$m_b$      mass of beam
$m_s$      mass of striker
t      time
$t_m$      time of maximum impact force
$v_o$      impact velocity of striker
$w_i$      frequency of free vibration of the i'th mode
x      beam co-ordinate
y      beam displacement
$y_y$      first mode elastic displacement
$y_{max}$      beam maximum displacement
$y_{res}$      beam residual displacement
$y_\infty$      first mode displacement for force $F_\infty$

$\alpha$      mass ratio = $m_b / m_s$
$\beta$      pulse ratio = $\tau_\infty / T_1$
$\rho$      beam density
$\tau$      pulse duration
$\tau_\infty$      pulse duration when the beam is massive
$\chi$      coefficient

THEORY

Figure 1  Midspan impact of a pin-ended beam.

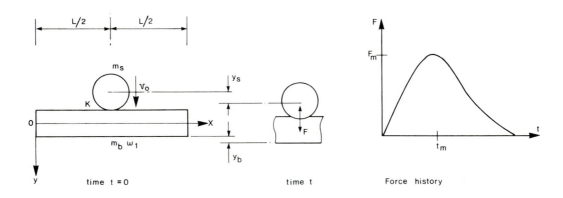

Fig. 1 shows the impact of a rigid spherical striker of mass $m_s$ with impact velocity $v_o$ on the midpsan of a pin-ended beam of mass $m_b$ and fundamental frequency of free vibration $w_1$. The Hertz contact law $F=Ka^{3/2}$ can be used to relate the impact force, F, to the deformation, a, of the impact zone.  The deformation constant K is a measure of the stiffness of the impact zone.  The impact equation is given by

$$a = y_s - y_b$$

where $y_s$ is the striker displacement and $y_b$ the beam midspan displacement.  This equation can be solved by expressing each term as a function of the impact force F.  The beam behaviour can be modelled in terms of its force vibration modes which are solutions of the 'simple' beam equation

$$EI \frac{\partial^4 y}{\partial x^4} + \rho A \frac{\partial^2 y}{\partial t^2} = 0$$

This equation describes elastic behaviour, and allows for strain energy of bending and transverse inertia.  The impact equation becomes

$$(F/K)^{2/3} = \left[v_o t - (1/m_s) \int_o^t\!\!\int F dt^2\right] + \left[(2/m_b) \sum_{i=1,3}^{\infty} (1/w_i) \int_o^t F(\overline{t}) \sin\{w_i(t-\overline{t})\} d\overline{t}\right]$$

with $w_1 = (\Pi^2/L^2)\sqrt{EI/\rho A}$ and $w_i = i^2 w_1$

This is an integral equation for F which is not soluble in closed form and requires numerical integration. It is found unnecessary to use all beam modes $i = 1,3\ldots\infty$ in the solution, and an accurate solution may be obtained using a finite number of modes N, with $i = 1,3\ldots$N in the summation. N will be referred to as the number of modes excited. The force history is thus obtained, and may then be used in the calculation of other quantities of interest e.g. beam moment and shear, energy distributions etc. Thus, the beam impact problem may be solved for quantities $m_s$, $v_o$,K,$m_b$ and $w_1$. This type of elastic analysis is classical and found in textbooks. (1)

Parameters (2) can be used which relate this beam impact problem to the limiting case when the beam is massive with movement limited to deformation at the impact zone. This limiting case is soluble for quantities $m_s$, $v_o$ and K to give an approximately sinusoidal pulse as shown in Fig. 2, with maximum impact force $F_\infty$ and pulse duration $\tau_\infty$ given by

$$F_\infty = (5m_s v_o^2 K^{2/3}/4)^{3/5}$$

$$\tau_\infty = 2.94 (5m_s/4Kv_o^{1/2})^{2/5}$$

This case, in essence, defines the striker, the impact velocity and impact zone.

Figure 2    Limiting case of the beam impact problem.

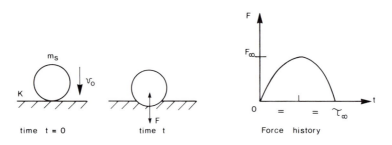

If the beam impact problem is dimensionally analysed it is found that 'output' parameters such as force/$F_\infty$ , time/$\tau_\infty$ , energy/$\frac{1}{2}m_s v_o^2$ are functions of only two 'input' parameters :

$\alpha = m_b/m_s$, the mass ratio

and    $\beta = \tau_\infty/T_1$, the pulse ratio ($T_1 = 2\Pi/w_1$)

The physical significance of the mass ratio is apparent. The pulse ratio measures the stiffness of the impact and mobility of the beam. Thus, the results of computations using the method described can be generalised to give parametric curves as shown in Fig. 3. The curves relate to the midspan impact of pin-ended beams for $\alpha = 1.43$ and varying $\beta$, and relate to most of the tests referred to in this paper. The variation with $\beta$ of the number of modes excited, N, the maximum impact force, $F_m$, and peak time, $t_m$, are shown. Also, the variation with time of the force, F, and beam midspan displacement, y, are shown for three values of $\beta$ . It was found convenient to use the parameter $y/y_\infty$ for the displacement, with $y_\infty$ the midspan displacement of the first mode for a static midspan load of $F_\infty$.

Figure 3  Parametric curves for the midspan impacts of pin-ended
         beams with   $\alpha = 1.43$.

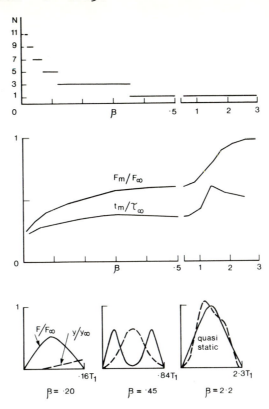

It is seen that for soft impacts (large $\beta$) the response is quasi-
static, while for stiff impacts more modes are excited and the beam
displacements during impact are relatively small.   The excitation of
higher modes gives the beam greater mobility to 'get away from' the
striker and consequently $F_m/F_\infty$ decreases as the impact becomes
stiffer.

The tests described later in this report were of the stiffer variety
( $\beta < .45$ ) with the maximum beam displacement occurring subsequent to
the first impact.  The beams remained elastic during the first impact
but, in most cases, yielded at midspan during the subsequent motion.
(Secondary impacts also occurred during this period).  Moreover, the
impact zones were not completely elastic and showed some residual
deformation.  Both the beam maximum and residual displacements were
measured.  It can be shown (2) that the beam displacements, subsequent to the first impact, approximately correspond to energy
$\chi\, m_s v_o^2 /2$ in the first mode, with the coefficient $\chi$ in the range

$$1/(1 + \alpha/2) \;<\; \chi \;<\; 1$$

$\chi$ is seen to be a function of the mass ratio alone.  The lower bound
corresponds to a completely plastic impact zone with the 'lost' energy
being deformation energy of the impact zone and higher mode energy.
The upper bound corresponds to an elastic impact zone and no lost
energy.  Thus, for a beam with first mode elastic displacement $y_y$ and
elastic energy $V_y$ (corresponding to beam yield moment $M_y$), the
recovery $y_{max} - y_{res}$ is ideally equal to $y_y$, and the residual displacement for an under-reinforced beam is

$$y_{res} = (\chi\, m_s v_o^2 /2 - V_y)\,(L/4\, M_y)$$

TESTS

80 pin-ended beams were tested by dropping a striker onto the beams
at midspan.  The beams remained elastic during the first impact, but

in most cases yielded subsequently at midspan. The impact force history was recorded using a load cell incorporated in the striker together with a high speed recorder. The beam maximum displacement was measured using a nail in a fixed tube of plasticene, and the beam residual displacement was measured using the floor as reference. Fig. 4 summarises the tests and results. The details may be found in Reference (2).

Figure 4 Summary of tests and results

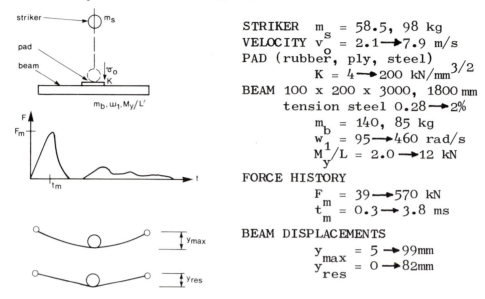

STRIKER $m_s$ = 58.5, 98 kg
VELOCITY $v_o$ = 2.1 ➛ 7.9 m/s
PAD (rubber, ply, steel)
$\qquad$ K = 4 ➛ 200 kN/mm$^{3/2}$
BEAM 100 x 200 x 3000, 1800 mm
$\qquad$ tension steel 0.28 ➛ 2%
$\qquad$ $m_b$ = 140, 85 kg
$\qquad$ $w_1$ = 95 ➛ 460 rad/s
$\qquad$ $M_y/L$ = 2.0 ➛ 12 kN

FORCE HISTORY
$\qquad$ $F_m$ = 39 ➛ 570 kN
$\qquad$ $t_m$ = 0.3 ➛ 3.8 ms

BEAM DISPLACEMENTS
$\qquad$ $y_{max}$ = 5 ➛ 99mm
$\qquad$ $y_{res}$ = 0 ➛ 82mm

In a typical test, a striker with mass ($m_s$) 98 kg and an impact velocity ($v_o$) of 7.3 m/s struck at midspan a 3 m long pin-ended beam with .85% tension reinforcement. The beam mass ($m_b$) was 140 kg and the fundamental frequency ($w_1$) was 166 rad/s. A 12mm thick ply pad was in position at the impact zone so that the deformation constant (K) was 24 kN/mm$^{3/2}$. Fig. 5 shows the recorded force history, and the theoretical first pulse computed as discussed previously. Three modes i = 1,3,5 were required for the theoretical solution. The crack pattern is the total effect of the modes excited. The flexural cracks on top of the beam at x/L ≃ 1/6, 5/6 and the diagonal shear cracks at x/L ≃ 1/3, 2/3 are not inconsistent with the presence of the third mode.

Figure 5 Recorded and theoretical force histories ; beam subsequent to test, and the theoretical modes excited.

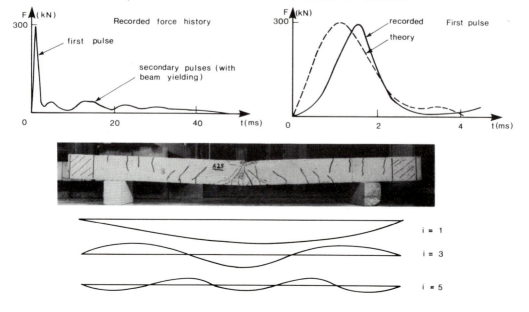

264

In parametric terms, the tests covered the ranges :

$$\alpha = .86 \text{ and } .23 < \beta < .65$$
$$\alpha = 1.43 \qquad .03 < \beta < .32$$
$$\ddot{\alpha} = 2.4 \qquad .02 < \beta < .17$$

Figure 6   Test results for maximum impact force $F_m$ and peak time $t_m$ with $\alpha = 1.43$.

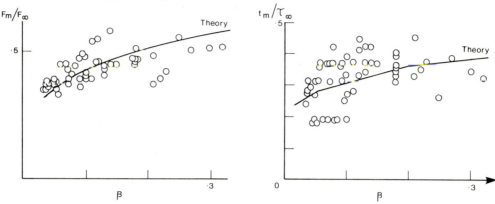

Figure 7   Test results for beam recovery

Figure 8   Test results for $\chi$ based on measured $y_{res}$

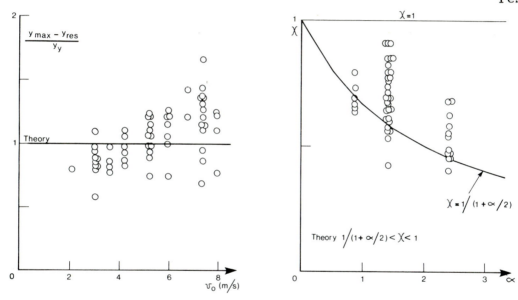

Figs. 6 and 7 show the test results in general. It is interesting that Fig. 7 shows an apparent increase in recovery with increasing impact velocity. This could be a consequence of an increased yield stress due to strain rate effects. Numerous investigators have noted the phenomenon (3). The test results for $\chi$ are not inconsistent with the theoretical bounds. However, it was noted that the results tended to the lower (plastic impact zone) bound with increasing impact velocity, while the recorded force histories did not show plasticity to the same degree. This again could be a consequence of an increased yield stress due to strain rate effects. In general, the test results were not inconsistent with the theory, although it would appear that some correction is required for strain rate effects.

CONCLUSIONS

This type of analysis is not without reason called the 'simple' beam analysis. Amongst other effects it neglects shear and rotary inertia, and is not a 'wave travel' type solution. In reality, the beam does not 'know' its end conditions until a wave has travelled from the impact point to the beam supports, and so, for very short duration impacts, the beam response is initially independent of the end conditions. The bounds of applicability can really only be found by experimentation.

It has been shown that the impact of a rigid striker on an elastic beam at midspan can be completely described in terms of two non-dimensional parameters : the mass ratio, $\alpha$, and the pulse ratio, $\beta$ . All other parameters can be derived as a function of these. Even though the case discussed is a relatively simple dynamic problem, the mathematics involved in the numerical solution of the impact equation are somewhat wearisome. Other problems, e.g. continuous beams, slabs and cylinders, would undoubtedly involve more complexity. However, if the relevant parameters were obtained by dimensional analysis, then the parametric curves could be defined by experimentation rather than theory. The experimentation could be of either prototype or model specimens ; in practice, the experimentation would probably involve models because impacts of interest generally involve objects of large magnitudes, e.g. ships, aeroplanes, cars, etc. Parameters clearly provide a framework within which design information can be generated from model tests.

An effectively rigid striker causes a 'hard' impact. If the striker undergoes deformation which is significant compared with its overall dimensions, then the impact may be described as 'soft', e.g. an aeroplane crumpling up on the roof of a nuclear reactor building. The rigid and soft striker problems are not dissimilar, the impact zone in the former is characterised by a deformation constant, K, while in the soft case it would be characterised by some yielding force or such-like. A dimensional analysis would give the parameters.

It is concluded that the use of parameters such as mass ratio and pulse ratio can greatly facilitate both theoretical and experimental investigations of impact problems.

ACKNOWLEDGEMENTS

The work described in this paper was carried out in the Design Research Department of the Cement and Concrete Association. The author wishes to thank his colleagues for their assistance ; in particular, Dr A W Beeby for many valuable suggestions, Mr D M Speirs and Dr R Sym for their contributions to the practical and computing aspects respectively, and also Dr G Somerville, Director of Research and Development, for his permission to publish this paper.

REFERENCES

(1) GOLDSMITH, W, Impact, The Theory and Physical Behaviour of Colliding Solids, Edward Arnold (Publishers) Ltd., London (1960).

(2) HUGHES, G and SPEIRS, D M, 'Investigation of the Beam Impact Problem', Cement and Concrete Association, London (To be published).

(3) 21-IL RILEM COMMITTEE, 'The Effect of Impact Loading on Buildings', RILEM Bulletin, Vol. 8, No. 44, (1975).

# Contribution to discussion of Paper No. 27

Norman Jones
*Department of Mechanical Engineering, The University of Liverpool, UK*

The writer was quite interested to learn that higher mode responses had been detected in the dynamic plastic range which gave rise to the failure pattern illustrated in Figure 5. Our work on higher modal responses of metal beams (Int. J. Mech. Sci., Vol. 20, 135-147, 1978) has shown that transverse shear effects become more important for the higher modes, as expected. Thus, could the discrepancy, which is attributed to material strain rate sensitivity, be partly due to transverse shear effects which are not retained in the theoretical work. Does the experimental work support this conjecture?

# Contribution to discussion of Paper No. 27

G Hughes
*Research & Development Division, Cement & Concrete Association, UK*

<u>Authors Reply to Professor N. Jones</u>

In all the tests, the measured force histories and observed failure patterns were consistent with an analysis based on an elastic beam. The author believes that a plastic analysis is not appropriate for impacts such as those described in the paper.

The elastic analysis used neglects both shear and rotary inertia; and, of course, these become more important for smaller pulse durations when more modes are excited. Thus, the discrepancies in measured beam recovery and residual deformation may be partly attributable to the neglect of these effects. A plot of the beam recovery against the pulse ratio, $\beta$, (which determines the number of modes excited) showed no particular trend, but a plot against the impact velocity (which is more or less proportional to strain rate) did indicate a trend. This suggested that the effect of strain rate was far more important than the effect of both shear and rotary inertia.

# Dynamic testing of a circular pretensioned cable roof model

## 28

H A Buchholdt,
*Professor, Polytechnic of Central London, UK*

SUMMARY

This paper describes resonance testing of a 6m diameter
cable roof structure. The work was carried out in order to
study the effect of cladding on natural frequencies and
modal damping in the lower modes.
The structure was tested without cladding, with roof cladding only and with roof and wall cladding. In each mode of
vibration mass was added to the roof to reduce the natural
frequencies of vibration.
The results given show how mass and cladding effect the
modal damping as well as the natural frequencies.
By plotting the ratio of the logarithmic decrement of
damping $\delta$ and the frequency f against $f^{-2}$ it is shown that
for the structure tested Raleigh damping may be assumed for
vibration at resonance, but not for vibration in general.
Finally an approximate method is used in an attempt to
estimate the level of aerodynamic damping.

INTRODUCTION

The author has in a previous paper (1) proposed a theory
for the calculation of the nonlinear dynamic response of
cable structures. Numerical experimentation using the
theory has shown that the degree of assumed damping can
significantly effect the predicted response. References to
published work on damping are given in (1). The total
amount of information available is, however, limited and
more work to study the degree and effect of damping is
desirable.
The test summarised in this paper describes one of several
dynamic tests on tension structures carried out or due to
be carried out at the PCL in order to study the dynamic
response of this class of structures.

SPECIFICATION AND DESCRIPTION OF STRUCTURAL MODEL

A diagram of the structure tested is shown in Fig. 1.
The main specifications of the roof are given below.

| | |
|---|---|
| Diameter of 7in x 1.5in channel ring beam | 6000mm |
| Maximum depth of central parabolic girder | 600mm |
| EA value of 7 x 7 stainless steel strand | 223730 N |
| EA value of 3/8in diameter aluminium struts | 4985820 N |
| Weight of cables | 0.16677 N/m |
| Weight of aluminium struts | 1.83859 N/m |
| Weight of plastic roof cladding | 5.25 N/m$^2$ |

The ring beam itself was supported by eight columns made
from scaffolding tubes  The wall cladding consisted of
corrugated plastic sheets.

## INSTRUMENTATION

The exciting equipment consisted of a forced draught cooled vibrator with a maximum thrust of 222 N and a 300 watt solid state amplifier with a built in oscillator. The recording equipment consisted of twelve displacement transducers, two pressure transducers, a nine channel modular carrier amplifier system, a six channel pen recorder, a two channel storage oscilloscope, two selector switch boxes and a six channel filter and amplifier unit which was used together with the pressure guages.
A diagram of the instrumentation is shown in Fig. 2.

## SUMMARY OF DYNAMIC TESTING

The dynamic testing was divided into two parts. The first part consisted of measuring the frequencies and damping in as many modes as it was possible to excite, with loading increasing from 10 to 90 N/joint for each mode shape. This was first done with both roof and wall cladding fixed, then with roof cladding only and finally with no cladding. The variations in the measured frequencies and damping are shown in Figs. 3 to 8. Using the order in which the mode shapes appeared when the structure was fully clad as reference it can be seen that the gradual removal of the cladding changed the order in which the different modes appeared.
The second part of the experiment consisted in measuring the pressure fluctuations in the different modes. The measurements showed that the air vibrated with the roof and that the maximum pressures lagged the response amplitudes by a phase angle $\phi$ which was nearly independent of the amplitude, but which varied with the frequency and mode shape as well as with the degree of cladding. At positions where the response was lagging the exciting force by $90^{\circ}$ the pressure curves were sinusoidal. Where this was not the case the sinusoidal curves were deformed, Fig. 9.
For a given mode shape and frequency the pressures measured varied linearly with the response amplitudes. In Fig.10 is shown how the pressures in Mode 4 varied above and below the roof with variations in amplitude and in Fig. 11 how the pressure gradients in the same mode varied with the applied load. The variations in the phase lag $\phi$ in Mode 4 is shown in Fig. 12.

## STRUCTURAL AND AERODYNAMIC DAMPING

In dynamic analysis of linear structures it is convenient to assume damping matrices which are orthogonal with respect to the mode shape vectors. One assumption leads to Raleigh damping which may be written as:

$$C = a_o M + a_1 K \tag{1}$$

where M and K are the mass and stiffness matrices respectively and $a_o$ and $a_1$ are constants of proportionality. Pre and post multiplication of Eq.(1) by the r-th mode shape vector $X_r$ yields after simplification

$$2 \xi_r \omega_r = a_o + a_1 \omega_r^2 \tag{2}$$

where $\xi_r$ and $\omega_r$ are the damping ratio and natural angular frequency respectively. Writing Eq.(2) in terms of the logarithmic decrement $\delta_r$ and frequency $f_r$ yields:

$$\delta_r/f = \tfrac{1}{2}a_o/f^2 + 2\Pi^2 a_1 \qquad (3)$$

In Figs. 13, 14 and 15 $\delta/f$ is plotted against $1/f^2$. The resulting graphs show that Raleigh damping applies to vibration in the individual modes, but not, since the values of $a_o$ and $a_1$ vary, to vibration in general.

When the pressures and displacements vary sinusoidaly and the pressure gradients and phase angles $\phi$ are constant both above and below the roof cladding, the expressions for the logarithmic decrement of aerodynamic damping and the mass of oscillating air per unit area may be written as:

$$\delta_{air} = \Pi(\tan A.\sin\phi_A + \tan B.\sin\phi_B)/\omega^2 m \qquad (4)$$

$$m_{air} = (\tan A.\cos\phi_A + \tan B.\cos\phi_B)/\omega^2 \qquad (5)$$

where m is the mass of roof per unit area.

Eq.(4) has been used to make an approximate estimate of the aerodynamic damping in modes 1, 2, 3 and 4. The calculated values are given in Table 1 below. For loads greater than 30N/joint the results agree reasonably well with the curves in Figs. 6, 7 and 8. For loads less than 30 N/joint, however, the values calculated are too large. This was due to the fact that the pressure gradients did reduce significantly towards the edges of the roof as the frequencies increased and when it was difficult to excite the roof in pure modes.

Table 1

| Load N/J | Modes 1 and 2 | | Mode 3 | | Mode 4 | |
|---|---|---|---|---|---|---|
| | Wall cladding | No wall cladding | Wall cladding | No wall cladding | Wall cladding | No wall cladding |
| 10 | 11.40% | 11.67% | 11.78% | 10.97% | 10.10% | 43.59% |
| 30 | 3.04% | 3.96% | 4.22% | 2.19% | 2.50% | 7.47% |
| 50 | 1.56% | 1.94% | 2.22% | 0.78% | 1.15% | 2.94% |
| 70 | 0.96% | 1.04% | 1.22% | 0.36% | 0.49% | 1.45% |
| 90 | 0.76% | 0.66% | 0.73% | 0.25% | 0.18% | 1.04% |

CONCLUSIONS

The tests showed that for the roof in question the natural frequencies and damping decreased with increasing load and that Raleigh damping only applied at resonance.
The tests also showed that at resonance the surrounding air vibrated together with the roof and increased the overall damping, which was greatest in the absence of wall cladding. It was also found that the degree of cladding affected the order in which the different modes appeared.

REFERENCES

1. Buchholdt, H. A.: Dynamic analysis of cable structures. International Journal of Structures, Roorkee, India. (To be published)

2. Jensen, J. J.: Das dynamische Verhalten eines vorgespannten Kabelnetzes. Division of Structural Mechanics, The University of Trondheim, Norway, Report 7, 1971

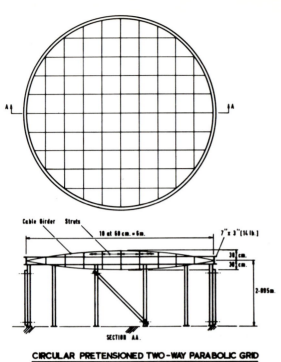

CIRCULAR PRETENSIONED TWO-WAY PARABOLIC GRID

Fig. 1

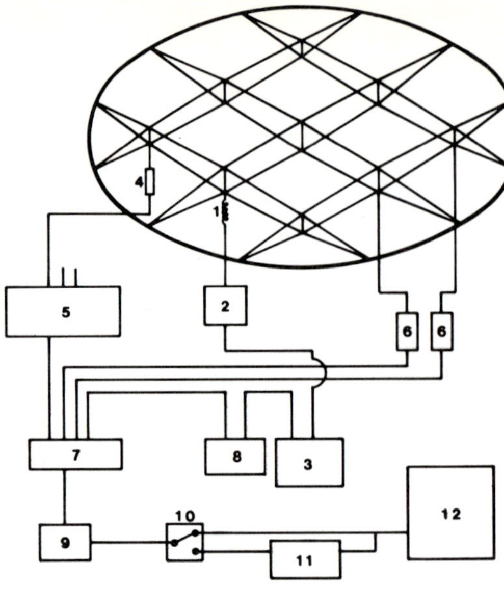

GENERAL LAYOUT OF EXCITING AND RECORDING SYSTEM
1. SOFT SPRING
2. VIBRATOR
3. POWER AMPLIFIER
4. LINEAR DISPLACEMENT TRANSDUCERS
5. SELECTOR SWITCH BOX
6. PRESSURE TRANSDUCERS
7. AC MODULAR CARRIER AMPLIFIER
8. STORAGE OSCILLOSCOPE
9. SELECTOR SWITCH BOX
10. SWITCH BOX
11. FILTER CIRCUIT AMPLIFIER
12. PEN RECORDER

Fig. 2

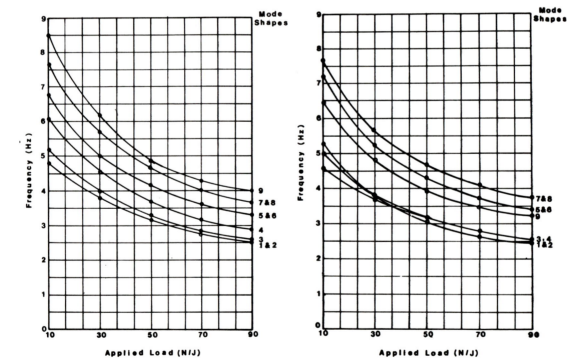

STRUCTURE + ROOF + WALL CLADDING

STRUCTURE + ROOF CLADDING

Fig. 3

Fig. 4

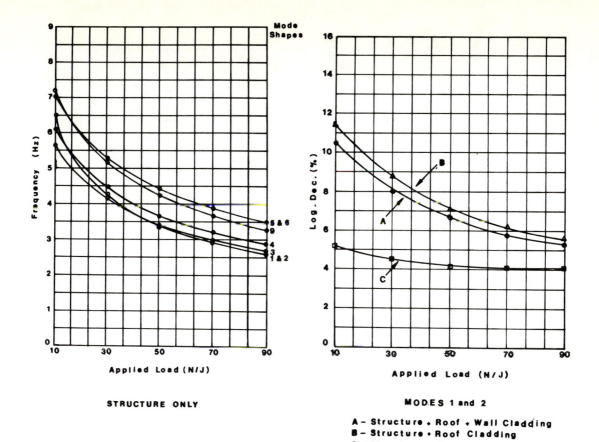

STRUCTURE ONLY

Fig. 5

MODES 1 and 2

A – Structure + Roof + Wall Cladding
B – Structure + Roof Cladding
C – Structure only

Fig. 6

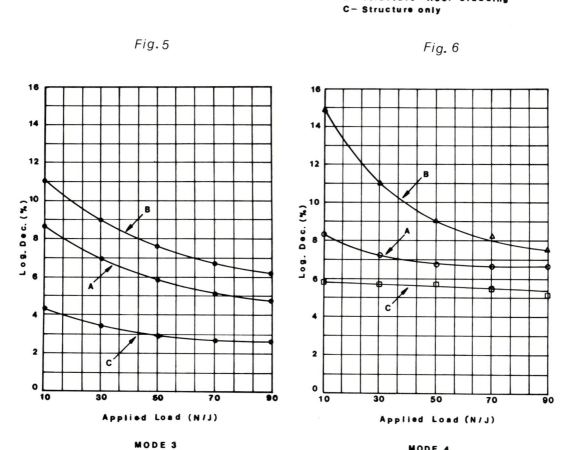

MODE 3

A – Structure + Roof + Wall Cladding
B – Structure + Roof Cladding
C – Structure only

Fig. 7

MODE 4

A – Structure + Roof + Wall Cladding
B – Structure + Roof Cladding
C – Structure only

Fig. 8

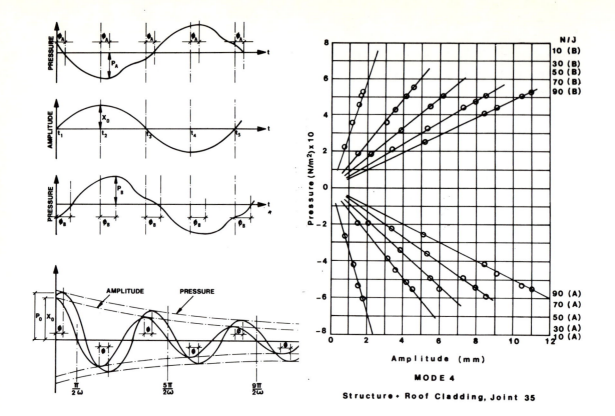

Fig.9                                              Fig.10

MODE 4

Structure + Roof Cladding, Joint 35

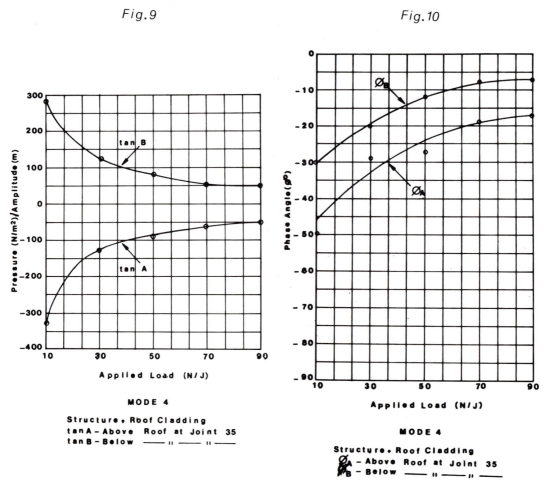

MODE 4

Structure + Roof Cladding
tan A – Above Roof at Joint 35
tan B – Below ——— " ——— "

MODE 4

Structure + Roof Cladding
$\phi_A$ – Above Roof at Joint 35
$\phi_B$ – Below ——— " ——— "

Fig.11                                              Fig.12

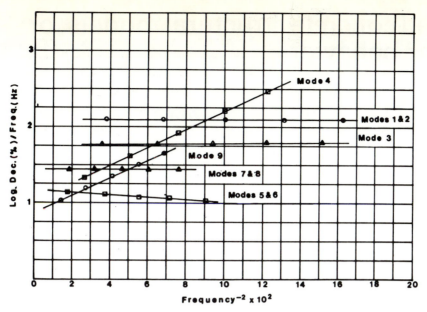

Fig. 13

STRUCTURE + ROOF + WALL CLADDING

STRUCTURE + ROOF + WALL CLADDING

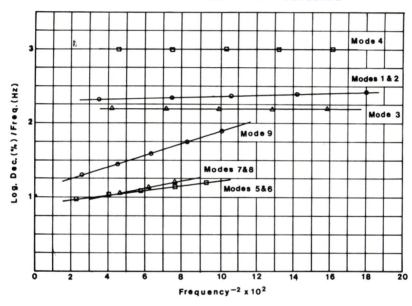

Fig. 14

STRUCTURE + ROOF CLADDING

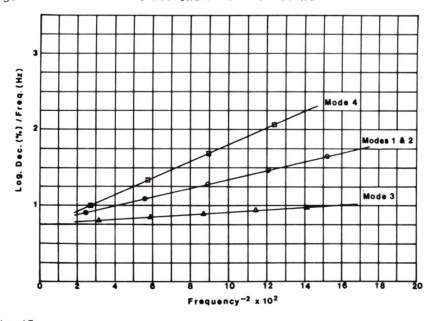

Fig. 15

STRUCTURE ONLY

# Reinforced concrete structural models for non-linear behavior

29

Theodor Krauthammer
*Applied Research Associates, Inc, Albuquerque, NM, USA*

## SUMMARY

Models for the analysis of reinforced concrete structural elements under severe loading conditions are presented and discussed. These models were successfully employed for the analysis of such systems as beams, beam-columns, and slabs both in the static and dynamic domains of behavior, and consider the influence of flexure, shear, and axial forces on the structural performance.

## INTRODUCTION

The purpose of these studies (4, 5) was to enhance the understanding of the behavior of reinforced concrete (RC) protective structures under high intensity pressure loading, such as that occurring in high explosive (HE) or nuclear detonations. The analytical procedures resulting from these studies are applicable to the analysis of slender and deep RC beams and RC slabs as based on synthesis of existing experimental data and interpretation thereof, and considers combined effects of flexure, axial forces, and shear on the behavior of such systems. The uncertainty involved in the design of protective structures, as a direct result of the nature of the expected force systems, may require the structure to provide its full resistance if necessary, and any structural element failing to perform as expected may reduce the probability of survival for the entire structure. The high intensity of the loads and the necessity of maximum performance require accurate and reliable analytical methods by which the performance of the structure can be evaluated.

## ANALYTICAL APPROACH - BEAMS AND BEAM-COLUMNS

At present, the design procedures that are employed in the U.S.A. and Europe require the provision of shear reinforcement for controlling the influence of shear on the performance of reinforced concrete members. However, these methods do not explicitly relate ultimate moment capacity to shear influence. The artificial separation between shear and moment capacity may prevent the achievement of an optimum design, and may result in premature failures. Therefore, it is reasonable to expect that if methods could be developed for predicting the moment capacity, and at the same time include the shear influence a design could assure full moment capacity. Furthermore, the incorporation of the method into numerical procedures for evaluating the moment and rotation capacities of structural members, as for example, the procedure proposed in Ref. (4), can result in an improved method for evaluating the performance of such members. The method which results from the present study is described next.

## Analysis For Flexure

The stress-strain relationship for confined concrete in compression was presented in the literature previously (4) and is described by the following formulation.

Part 1: For the strain range $0 \leq \epsilon \leq \epsilon_0$ (where $\epsilon$ is the concrete strain, and $\epsilon_0$ is the strain which corresponds to maximum stress) the stress-strain relationship is parabolic, as presented in Eq. (1).

$$f_c = f_c'[(E_c \epsilon_0/f_c')(\epsilon/\epsilon_0) - K(\epsilon/\epsilon_0)^2]/1 + [E_c \epsilon_0/(Kf_c') - 2](\epsilon/\epsilon_0) \qquad (1)$$

Part 2: For the strain range $\epsilon_0 < \epsilon \leq \epsilon_{0.3K}$ (where $\epsilon_{0.3K}$ is the strain at which the concrete stress is reduced to 0.3 of the previously reached maximum) the stress-strain relationship is linear as presented in Eq. (2).

$$f_c = f_c' K [1 - 0.8 Z \epsilon_0 (\epsilon/\epsilon_0 - 1)] \qquad (2)$$

Part 3: For the strain range $\epsilon > \epsilon_{0.3K}$ the stress-strain relationship is presented in Eq. (3).

$$f_c = 0.3 K f_c' \qquad (3)$$

where:

$$\epsilon_0 = 0.0024 + 0.005 (1 - 0.734 S/h'') \rho'' f_y''/(f_c')^{1/2} \qquad (4)$$

$$K = 1 + 0.0091 (1 - 0.245 S/h'') [\rho'' + (d''/D) \rho] f_y''/(f_c')^{1/2} \qquad (5)$$

$$Z = 0.5/[3/4 \rho''(h''/S)^{1/2} + (3+0.002 f_c')/(f_c'-1000) - 0.002] \qquad (6)$$

Equations (4), (5), and (6) were originally proposed in Ref. (14) for describing the stress-strain relationship of confined concrete in axially loaded columns, as a result various parameters are redefined herein for adapting the previous stress-strain relationship also for the analysis of beams, as follows: $\rho''$ is the ratio of the volume of transverse reinforcement to the volume of confined concrete, both for the compressive zone of the beam cross section; $\rho$ is the ratio of the cross-section area of longitudinal bars to the total concrete area, both in the compressive zone; $h''$ is the average dimension of the rectangular compressive zone is defined by the following formulation: $h'' = 0.5 (h_1 + h_2)$, where: $h_1$ and $h_2$ are the dimensions of the two sides that describe the rectangular compressive zone in a beam, and measured to the outside of the steel hoops; $E_c$ is the tangent modulus of concrete and is obtained from the following expression as recommended in Ref. (14), $E_c = 316.9 (f_c')^{1/2}$; $S$ is the spacing of the transverse reinforcement; $f_y''$ is the static yield stress of the transverse reinforcement; $f_c'$ is the unconfined compressive strength of the concrete; $d''$ is the nominal diameter of the transverse reinforcing bars; K is a constant that defines maximum concrete strength thus $K = (f_c/f_c')_{max}$; Z is a constant that indicates the amount of confinement, as obtained from Eq. (6).

Several assumptions are made here for providing a consistent description of the behavior of the concrete cover, and the contribution of the concrete cover to the behavior of a flexural member as follows.

a. The concrete cover is assumed to crush at a strain of 0.004.

b. The concrete cover may continue to resist compressive stresses even after crushing, as long as the cover has not spalled; however, spalling is assumed to occur only when the compressive bars reach a strain of 0.004.

c. It is assumed that at strains smaller than 0.004 on first loading the concrete cover behaves as the confined concrete, and the same stress-strain relationships describe the behavior. At strains larger than 0.004 the concrete cover exhibits a behavior different from that of the confined concrete, as noted in (d) next.

d. Concrete cover located at an elevation higher than the bottom of the compressive reinforcement spalls off at strains which exceed 0.004, as described previously in (b). However, concrete cover located between the bottom of the compressive reinforcement and the neutral axis may remain on the member even at strains larger than 0.004, and contribute to the load resistance.

e. The concrete cover remaining on the member at strains which exceed 0.004 may resist compressive stresses up to $0.85f_c'$.

The stress-strain curve for steel reinforcing bars employed herein is composed of the following three parts: the linear elastic part, the yield plateau and the nonlinear strain hardening region. The nonlinear part of the curve can be described by a polynomial formulation which was proposed by Burns and Siess (1). Another model for the polynomial was proposed by Park and Paulay (13), and which was employed for the present study, as follows:

$$f_s = f_y \, [m(\epsilon_s - \epsilon_{sh}) + 2]/[60(\epsilon_s - \epsilon_{sh}) + 2] + (\epsilon_s - \epsilon_{sh})(60 - m)/[2(30r + 1)^2] \quad (7)$$

where

$$m = [(f_u/f_y)(30r + 1)^2 - 60r - 1]/(15r^2) \quad (8)$$

$$r = \epsilon_{su} - \epsilon_{sh} \quad (9)$$

$f_s$ is the stress in the steel bars which corresponds to the strain $\epsilon_s$; $\epsilon_{sh}$ is the strain at which strain hardening begins, and $\epsilon_{su}$ is the ultimate strain for a given bar, $f_y$ and $f_u$ are the yield and ultimate steel stresses respectively. These parameters can be obtained experimentally.

Analysis For Shear

The analysis for evaluating the influence of shear and shear reinforcement on the flexural moment capacity is based on the results which were obtained by Leonhardt and Walther (9, 10), and Kani (3) on singly reinforced beams without shear reinforcement. The beams, and beam-columns can be classified as either "deep", or "slender" based on the a/d ratio (shear span to effective depth ratio) as follows. Define the value a/d = P2 (where 2 < P2 < 3), if the a/d ratio for a given beam is larger than P2 it is a "slender beam", and if a/d is less than P2 it is a "deep beam". Results (3, 9, 10) show that for slender beams if a/d > P3 (where P3 is about 7) there is practically no shear influence on the flexural moment capacity. Similarly, for deep beams if a/d < P1 (where 1 < P1 < 1.5) no shear influence on the flexural moment capacity was noticed.

The following equations and parameters have been employed herein for describing the influence of shear on the flexural moment capacity as based on the a/d ratio definitions provided above. The present models consider a general case including compressive, and shear reinforcements.

Slender beams without web reinforcement:

$$(M_u/M_{f\ell}) = 1.0 + [M_u/M_{f\ell})_m-1.0] \ (a/d-P3)/(P2-P3) \leq 1.0 \qquad (10)$$

Deep beams without web reinforcement:

$$(M_u/M_{f\ell}) = 1.0 + [(M_u/M_{f\ell})_m-1.0](a/d-P1)/(P2-P1) \leq 1.0 \qquad (11)$$

Where: $M_u$ is the ultimate moment capacity with shear influence, $M_{f\ell}$ is the ultimate flexural moment capacity without shear influence, and $(M_u/M_{f\ell})_m$ is determined from the following formulation, as related to the longitudinal reinforcement ratio:

$$\rho \leq 0.0065 \qquad (M_u/M_{f\ell})_m = 1.0$$

$$0.0065 < \rho \leq 0.0188 \qquad (M_u/M_{f\ell})_m = 1 - 0.366(\rho - 0.0065) \qquad (12)$$

$$0.0188 < \rho \leq 0.028 \qquad (M_u/M_{f\ell})_m = 0.6$$

When shear reinforcement is present the ratio $(M_u/M_{f\ell})_m$ is replaced by $(M_u/M_{f\ell})_m$, as discussed in Ref. (4):

$$(M_u/M_{f\ell})'_m = (M_u/M_{f\ell})_m + [1 - (M_u/M_{f\ell})_m] \ \cot \alpha \leq 1.0 \qquad (13)$$

And $\alpha$ the inclination of cracks to cause failure (angle between crack and the horizontal axis) can be computed from the following equations which relate the geometry and shear strength of the beams with the inclination of measured cracks, as derived in Ref. (4).

Define: $\rho_2'' = \rho'' \ f_y''/(f_c')^{1/2}$ \qquad (14)

The variables $\rho''$, $f_y''$, and $f_c'$ have been defined earlier, and $\alpha$ is the crack inclination to the horizontal at failure, in degrees.

For slender beams (data from Ref. (1)):

$$\alpha = - 3.68 \ \rho_2'' \ (a/d) + 107.46 \qquad (15)$$

And for deep beams (data from Ref. (2)):

$$\log \alpha = 0.042 \ \log [\rho_2'' \ (a/d)] + 1.903 \qquad (16)$$

## ANALYTICAL APPROACH — SLABS

The resistnace of reinforced concrete slabs, as reported in the literature (5, 7, 11, 12, 15) can be described as follows. Initially, the slab resistance is provided by the conventional one-way, or two-way slab mechanism. However, event at relatively small central deflections the resistance is provided by a compression membrane mechanism, as a result of restraining the outward movement of the slab edges. The peak resistance could be between 2 to 8 times higher than the resistance predicted by the yield line theory (15). The central deflection associated with the peak resistance is in the range between 0.5d to 0.5t (where d and t are the slab effective and total depths, respectively). After the peak has been reached, a gradual decrease in resistance can be noticed, and eventually the membrane forces in the slab central region are transformed from compression to tension. At that stage the central deflection is in the range of 2d to 2t (d and t have been defined earlier), and the slab boundaries provide restraint against inward movement of the edges. Extensive tensile cracking is associated with the transition into a tensile membrane, and further deflection of the central zone is associated with an increased resistance which is provided by the steel reinforcement. Failure may be controlled either by shear along the edges (punch-through), or by the fracture of the steel bars in tension. The slab resistance in the compression membrane mode can be computed from the formulation proposed by Park (11). However, the analysis of hardened box-type structures has to consider the presence of externally applied thrust forces, and their influence on the slab resistance. Therefore, the formulation proposed by Park (11) was modified herein to include the effects of thrust, as follows.

$$w L_x^2 (3L_y/L_x - 1)/24 = k_1 k_3 f_c' t^2 [(L_y/L_x)(0.188 - 0.281 \, k_2)$$

$$+ (0.479 - 0.490 \, k_2)] + 2[(L_y/L_x)T_x(d_{1x} - d_{2x})$$

$$+ T_y(d_{1y} - d_{2y})] + [(L_y/L_x)N_x + N_y] \tag{17}$$

where:

| | |
|---|---|
| $w$ | = Uniformly distributed load on the slab (force per unit area). |
| $L_x, L_y$ | = Short and long dimensions of a rectangular slab, respectively. |
| $k_1, k_2, k_3$ | = Parameters for describing the compressive stress distribution in concrete, and related to the concrete compressive strength. |
| $f_c'$ | = Uniaxial compressive strength of concrete. |
| $t$ | = Slab total thickness. |
| $T_x, T_y$ | = Tensile forces in steel bars along the x and y directions, respectively. |
| $d_{1x}, d_{1y}$ | = Slab effective depth along the x and y directions, respectively. |
| $d_{2x}, d_{2y}$ | = Distance from slab compressive face to the compressive reinforcement along the x and y directions, respectively. |
| $N_x, N_y$ | = Thrust in the x and y directions, respectively. |

It is clear from Eq. (17) that increasing the externally applied thrust has a positive effect on the load carrying capacity of the slab under consideration. However, the lateral stability of the slab, and the compressive strength of reinforced concrete may limit that contribution.

The resistance provided by the tensile membrane can be computed from another expression proposed by Park (12). Nevertheless, a generalized formulation is presented here, as based on the mechanism that was proposed in Ref. (5). The tensile membrane resistance is formulated as follows:

$$w = 8 (M_n + T_\Delta)/L^2 \tag{18}$$

where:

w     = Uniformly distributed load on membrane (force per unit length).
$M_n$  = Bending moment in the slab central region.
T     = Tensile force in reinforcement.
Δ     = Central deflection.
L     = Membrane span length.

Equation (18) describes the load capacity of the slab immediately before, during, and after the formation of a hinge in the central region of the slab. After the hinge is formed Eq. (18) will transform into the solution provided by Park (11). When a square slab is considered one may employ Eq. (18) after introducing an appropriate expression for $M_n$.

## ANALYTICAL RESULTS AND EXPERIMENTAL DATA

The methods described previously in this paper were employed for the analysis of RC beams, beam columns, and slabs, as described in Refs. (4, 5, 6, 7, 8). Here, only the results of these studies will be briefly discussed, and for further information the reader is referred to the above mentioned publications.

Twenty slender RC beams (a/d > 3.0) were analyzed by employing the method proposed in this paper, and the results were compared to experimental data that has been reported in Ref. (1). It was found that the analytical results were in the range from 4 percent above to 5 percent below the experimental data, and the average of the ratio of computed to measured results for twenty beams was 0.987.

Ten deep RC beams (1.6 < a/d < 3.0) were analyzed based on the proposed method, and the results were compared to experimental values as reported in Ref. (2). The analytical results were in the range from 5 percent above to 5 percent below the experimental data, and the average ratio of computed to measured values for ten beams was 1.005.

Four beam-columns were analyzed by employing the proposed methods and the results were compared to experimental data, as described in Ref. (4). The accuracy of analytical results was similar to that obtained for slender and deep beams.

The method for the analysis of slabs was employed to study the behavior of hardened structure roofs (5, 7), both under static and dynamic conditions. The results obtained may provide a different explanation as to the mechanisms that influence the behavior of such systems.

## REFERENCES

(1)     BURNS, N H, and SIESS, C P, Load Deformation Characteristics of Beam-Column Connections in Reinforced Concrete, University of Illinois Department of Civil Engineering, SRS Report No. 234, (January 1962).

(2)     CRIST, R A, Static and Dynamic Shear Behavior of Uniformly Loaded Reinforced Concrete Deep Beams, Technical Report No. AFWL-TR-71-74, Air Force Weapons Laboratory - Kirtland AFB, (November 1971).

(3)      KANI, G N J, 'Basic Facts Concerning Shear Failure', _Journal ACI_, Vol. 63, pp. 675-692, (June 1966).

(4)      KRAUTHAMMER, T, and HALL, W J, _Resistance of Reinforced Concrete Structures Under High Intensity Loads_, University of Illinois Department of Civil Engineering, SRS Report No. 463, (May 1979).

(5)      KRAUTHAMMER, T, 'Resistance of Hardened Shallow-Buried Structures', _Second ASCE/EMD Speciality Conference on Dynamic Response of structures_, Atlanta, (January 1981).

(6)      KRAUTHAMMER, T, 'Response of Reinforced Concrete Structures to High Intensity Loads,' invited paper to _AFWL-DNA-NMERI Workshop on Constitutive Relations for Concrete_, Albuquerque, New Mexico, (April 1981).

(7)      KRAUTHAMMER, T, 'Reevaluation of Test Results on Hardened Shallow-Buried Structures', _10th International Conference on Soil Mechanics and Foundation Engineering, Sweden (June 1981)_.

(8)      KRAUTHAMMER, T, 'Moment-Shear Interaction in Reinforced Concrete Beams', _Joint ASME/ASCE Mechanics Conference_, Boulder, Colorado (June 1981).

(9)      LEONHARDT, F, and WALTHER, R, 'Contribution to the Treatment of Shear Problems in Reinforced Concrete', (in German), _Beton und Stahlbetonbau_, Vol. 56, No. 12 (December 1961); Vol. 57, No. 2, (February 1962); No. 3 (March 1962); No. 6 (June 1962); No. 7 (July 1962); No. 8 (August 1962).

(10)     LEONHARDT, F, and WALTHER, R, 'Wandartiger Träger', _Deutscher Ausschuss für Stahlbeton_, Bulletin No. 178, Wilhelm Ernst und Sohn, Berlin, (1975).

(11)     PARK, R, 'Ultimate Strength of Rectangular Concrete Slabs Under Short-Term Uniform Loading with Edges Restrained Against Lateral Movement', _Proceedings, Institutions of Civil Engineers - London_, Vol. 28, pp. 125-150, (June 1964).

(12)     PARK, R, 'Tensile Membrane Behaviour of Uniformly Loaded Rectangular Reinforced Concrete Slabs with Fully Restrained Edges', _Magazine of Concrete Research_, Vol. 16, pp. 39-44, (March 1964).

(13)     PARK, R, and PAULAY, T, _Reinforced Concrete Structures_, Wiley, (1975).

(14)     VALLENAS, J, BERTERO, V V, and POPOV, E P, _Concrete Confined by Rectrangular Hoops and Subjected to Axial Loads_, Report No. UCB/EERC-77/13, University of California, Berkeley, (August 1977).

(15)     WOOD, R H, 'How Slab Design has Developed in the Past, and What the Indications are for Future Development', _ACI Publication SP-30_, (1971).

# Dynamic behaviour of cores of tall buildings

## 30

J E Gibson
*Department of Civil Engineering, The City University, London, UK*

SYNOPSIS

This research was concerned with the investigation of the dynamic
behaviour of the cores of tall buildings, in particular with the
determination of their natural frequencies.   A series of different
cross sections were constructed from perspex and then micro concrete
and subjected to various types of transverse forces and displacements.
The resulting dynamic behaviour of these cores was then investigated
by examining the output from foil strain gauges fixed at various
points on the models.   In this way the natural frequency of the
core could be readily determined experimentally.   Simultaneously,
a theoretical investigation was carried out using in the first
instance a vibrating beam and a Rayleigh method of attack to
determine the natural frequency and a comparison was made with
experimental values.

## DESCRIPTION AND CONSTRUCTION OF MODEL CORES

In the initial investigation the core sections were constructed
from Perspex sheets as these were relatively easy to machine.   The
prepared sheets were then clamped in position by various clamping
devices to form the required cross section and were then glued to
one another by special adhesives.   Some considerable time was spent
in the selection of the correct adhesive and the necessary length
of time for curing under given temperature conditions to produce a
joint of equal strength to that of the adjacent plates.
Eventually a technique was developed that produced satisfactory
joints.

Two initial model types were selected for the first series of tests,
one having a rectangular hollow section and the other a triangular
hollow section.   The rectangular section is probably the most
popular type of core used in practice whilst the triangular was
typical of some tall structures.

The cross section of the first rectangular core model was 75 mm by
37 mm outer dimensions and had a uniform wall thickness of some
6 mm.   The structure was 584 mm in height and was firmly fixed
at the base to a relatively thick perspex block in order to ensure
as far as possible an encastre condition.   Four holes in the base
allowed the model to be firmly fixed to a rigid metal base by
means of Allen screws.

Holes were drilled in the core itself at three levels to allow
metal blocks to be firmly attached to the core at these points
and thus allow the effect of using lumped masses in the
theoretical calculations to be investigated.   The model is shown
in Figure 1 and schematically in Figure 8.

The triangular hollow section model was formed in much the same way
and again strain gauges were attached at relevant positions to allow
the dynamic behaviour to be examined.

The remaining models in this series of tests were constructed using
micro concrete suitably reinforced with a rectangular steel mesh.
Again in the first model the cross section was rectangular and of
the same dimensions as the perspex model, i.e. 75 mm by 37 mm with
a uniform wall thickness of 6 mm.  This required the manufacture
of a very accurate mould and very careful casting of the micro
concrete.  The mix had to very carefully selected so that a
relatively high strength could be achieved after some two or three
days, to allow ease in stripping, and at the same time had to be
sufficiently fluid to fill the mould easily.

The mould was placed on a vibration table during the casting process
thus considerably reducing voids in the model section.  At the same
time small rectangular beams, to allow the tangent modulus to be
determined experimentally and cubes for strength evaluation, were
cast.   The normal process of humidity curing was carried out in all
cases.

A photograph of the rectangular and triangular micro concrete models
is shown in Figure 2.

All models were tested in a free undamped mode condition initially
and later where possible in a forced vibration mode.


PHYSICAL PROPERTIES OF THE MODEL CASE

In assessing the theoretical dynamic behaviour of the models it was
of course necessary to determine the elastic rigidity EI of the
cores.   Obviously in the case of the perspex models measurement
of the cross section by micro meter allowed the second moment of area
to be accurately determined.   As Young's modulus for perspex had
been accurately measured by the manufacturers then the rigidity EI
could then be readily obtained.

However, when considering the micro concrete cores, as these were
cast in one pour, only the end dimensions of the cross section could
be accurately determined.   Any variation of the wall thickness
internally along the length could not be measured readily.

The problem was finally resolved by conducting a preliminary
investigation into the load deflection characteristics of the
model itself in the following manner.   The model was erected in the
testing frame and transverse load was applied through a wire
attached to a specific point on the model.   The transverse
deflections at the loading point were measured by a dial gauge
graduated in .01 mm for a series of known applied loads thus
allowing a load deflection graph to be constructed.   The loading was
kept well within the linear elastic range for the micro concrete

so that the elementary deflection equation namely,

$$\Delta = WL^3/3EI \quad \text{giving} \quad EI = WL^3/3\Delta$$

could be employed to determine the required quantity EI.

As this method was so successful and relatively quick in the initial experiment with micro concrete, it was decided to use it throughout the range of tests to determine the rigidity constant EI. A typical load deflection curve for the micro concrete rectangle section is shown in Figure 3. It may be seen that the load/deflection relationship is quite linear and that EI may then be accurately determined.

## EXPERIMENTAL PROCEDURE

As the main subject of this investigation was to determine not only the natural frequency of the various core sections, but at a later date their dynamic behaviour under various disturbing forces, then these factors governed the design of the testing apparatus. The main frame as shown in Figure 4 was constructed from rectangular hollow sections and the models were bolted to the rigid plate with large Allen screws.

To determine the elastic rigidity EI of the cores the tip of the core was connected to a Bordon cable which passed over an adjustable pulley connected to the frame and carried a weight pan. Thus, any desired transverse static load could be applied to the top of the core and the resulting deflection was measured by an dial gauge calibrated in .01 mm.

The natural frequency of the cores was determined as follows. The top of the model was displaced by a known distance and then suddenly released and the resulting dynamic behaviour was recorded by two foil strain gauges located at the base of the core. The electrical responses from the gauges were amplified by standard d.c. amplifiers and then directly fed onto a U.V. recorder, the mini galvonometers then produced permanent traces on sensitised paper.

A typical set of wave forms from the U.V. recorder is shown in Figures 5 and 6 where the typical decaying free damped vibrations may be readily seen. The vertical streaks on the paper were set at intervals of 0.1 seconds for the purpose of these experiments and thus allowed the natural frequency to be readily measured. The gains on the amplifiers could be adjusted to give any initial desired amplitude and thus allow clearly defined traces to be produced.

In order to produce forces vibrations, pairs of large electro magnets could be located in any desired positions on opposite sides of the models. The magnets were mounted so that they could be adjusted to any desired distance from the face of the core. The magnets were driven by a large power amplifier and were so ganged that they could be driven in a push pull mode on opposite faces through any desired frequency. The varying magnetic field thus produced, attracted the steel mesh in the cores of the micro concrete models and thus caused the structures to vibrate. Again the dynamic behaviour of the cores was portrayed on the U.V. recorder. The complete apparatus is shown in Figure 4.

The results of this forced vibration investigation will be published in a later paper.

## METHODS OF ANALYSIS

In the initial investigation, since in principle the core may be considered as a cantilever beam, two elementary methods of analysis were used to determine the natural frequencies.

The first simply considered flexural vibrations of a uniform beam subject to the typical cantilever end condition, i.e. the complementary function of the standard dynamic partial differential equation

$$(1) \quad \frac{\partial^2}{\partial x^2}\left(-EI\,\frac{\partial^2 y}{\partial x^2}\right) = \rho A\,\frac{\partial^2 y}{\partial t^2}$$

If the beam is of uniform cross section then (1) reduces to

$$(2) \quad EI\,\frac{\partial^4 y}{\partial x^4} + \rho A\,\frac{\partial^2 y}{\partial t^2} = 0$$

For free vibrations y is a harmonic function of time and may be written

$$(3) \quad y = Y \sin(\omega t + \beta)$$

in which Y is purely a function of x.  Substituting (3) in (2) yields after reduction

$$(4) \quad \frac{d^4 Y}{dx^4} - \frac{\rho A \omega^2}{EI}\,Y = 0$$

an ordinary differential equation of the fourth order with the general solution.

$$(5) \quad Y = A_1 \sin \beta x + A_2 \cos \beta x + A_3 \sinh \beta x + A_4 \cosh \beta x$$

in which

$$(6) \quad \beta^4 = \rho\, A\omega^2/EI$$

The arbitrary constants $A_1$ to $A_4$ are determined by the boundary conditions of the problem.  Thus, in the case of the cantilever at x = 0 then y = 0 and dy/dx = 0 whilst at x = L, $d^2y/dx^2 = 0$ and $d^3y/dx^3 = 0$ and these four conditions eventually yield the frequency equation

$$(7) \quad \cos \beta L \cosh \beta L + 1 = 0$$

which in turn yields the natural frequency mode

$$(8) \quad \beta_1 L = 1.875 \qquad \text{and hence finally the natural frequency}$$

$$(9) \quad f_1 = \frac{0.5595}{L^2}\left[\frac{EI}{\rho A}\right]^{\frac{1}{2}}$$

286

While the equation (9) is sufficiently adequate for determining the
natural frequency of cantilevers with uniform cross sections it does
not provide a solution for a uniform cantilever with masses located
at various points throughout its length.  For this purpose it is
convenient to use the Rayleigh method of analysis which will now be
examined.

The well known Rayleigh method for determining natural frequencies
based on the conservation of energy yields

$$(10) \quad \omega_1^2 = \frac{\int_o^L EI \left[\frac{\partial^2 Y}{\partial x^2}\right]^2 dx}{\int_o^L \rho A Y^2 dx + \Sigma m_r \{Y(x_r)\}^2} \qquad \text{and } f_1 = \omega_1/2\Pi$$

in which $\rho$ is the mass density of the core, Y is the deflected form,
A is the cross sectional area and $m_r$ are the individual masses
located at specific distances $x_r$ from the cantilever root.

Equation (9) was used to determine the natural frequency of the
various cantilever cases, while in the case of those models with
lumped masses at specific points equation (10) was used.  In point
of fact, equation (10) only could have been used as with $m_r$ set to
zero throughout, the resulting equation yields values identical
to those given by (9).

In further experiments, however, where a more detailed exploration
will be made a standard dynamic finite element program will be used
to determine the dynamic characteristics of the cores.

Both of these elementary methods are of course ideally suited for
programming on a mini computer and in point of fact this was
carried out as it did tend to speed up computation.

## INVESTIGATIONS WITH PERSPEX MODEL CORES

The rectangular perspex core shown in Figure 8 was designed so that
lump masses could be located at any of the three levels (1),(2) and
(3) as shown.  These lumped masses each consisted of five weights,
A,A,B,C and C.  The sets A and C were all of equal weight 0.796 lb
whilst B was equal to 0.615 lb.  This allowed a thorough examination
of natural frequency under various lumped mass conditions to be made.
Only two of these cases will be considered in detail, the first
being that in which (A+B+C) are located at the tip of the core as
shown in Figure 8.  With the weights firmly bolted in position, the
model was slightly displaced at the tip and released, the strain
gauge at the base recorded the trace as shown in Figure 5 on the
U.V. recorder.  The core, after release, assumes a typical freely
damped vibration mode, the frequency of which is extremely close to
its natural frequency.  From this trace, as the verticals are at
intervals of 0.1 seconds, the natural frequency is 15.8 Hz as against
a computed value of 15.18 Hz.  A repeat of this investigation with
(2A+B+2C) equal to 4.02lb gave an experimental value of 12.5 Hz as
against a computed value of 12.0  Hz, as shown in Figure 6.

## INVESTIGATION OF MICRO CONCRETE MODEL CORE

The rectangular micro concrete model was mounted in the testing
frame and a load deflection experiment yielded the plot shown in

Figure 3 from which EI was determined. Again the trace is characteristic of free damped vibration and the natural frequency was determined as 38 Hz. The computed value was 38.5 Hz using the experimental value for EI.

To check the operation of the forced vibration apparatus shown in Figure 4 the model was then located between the top pole pieces of the electo magnets and subjected to forced vibrations. When the generator emitted a steady frequency of 38 Hz, the model responded by exhibiting resonance as shown by the U.V. recorder trace. It would thus seem that the method of driving the reinforced micro concrete models should prove satisfactory.

## CONCLUSIONS

It would seem that with the reasonable correlation between theory and experiment that exists, the elementary methods of analysis for natural frequencies for cores is sufficiently accurate. However, final conclusions must await the experiments both with lumped masses and different cross sections.

## ACKNOWLEDGEMENT

The author wishes to thank the S.R.C. for providing a research grant for part of this investigation and to the various technicians at City University for their help in construction of apparatus and to Mrs. Prange for typing the manuscript.

Figure 1  Perspex Model

Figure 2  Micro Concrete Models

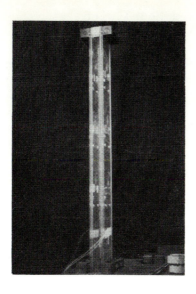

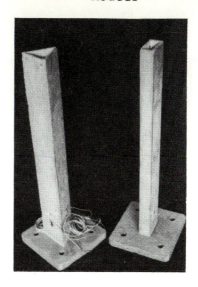

Figure 3  Load Deflection Curve for Micro Concrete Model

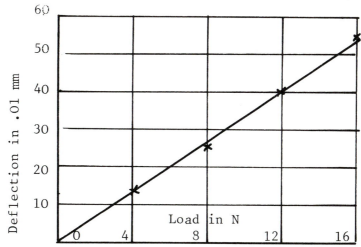

Figure 4  Vibration Frame

Figure 8  Perspex Model with Lumped Masses

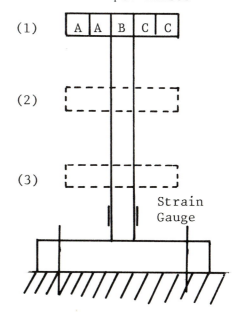

Figure 5  U.V. Recorder Trace for Perspex Model with
Mass (A + B + C) at Level (1)

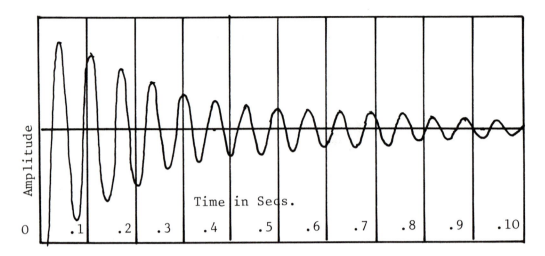

Figure 6  U.V. Recorder Trace for Perspex Model with
Mass (2A + B + 2C) at Level (1)

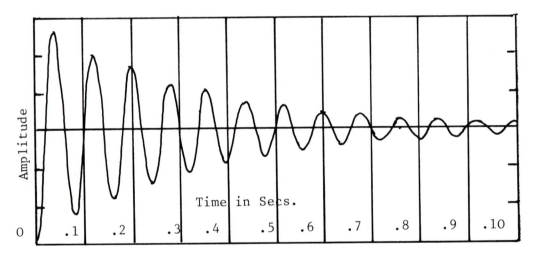

Figure 7  U.V. Recorder Trace for Micro Concrete Model

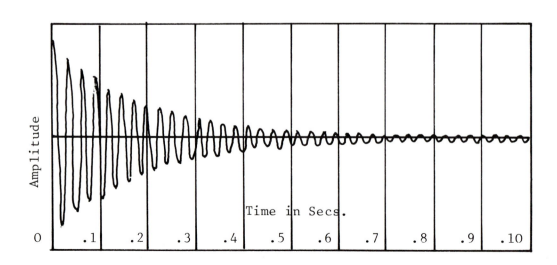

# Modelling of damped structures in resonance

## 31

V Askegaard and H Schmidt
*Department of Structural Engineering, T.U. of Denmark*
F Bligaard Pedersen
*R & H Consultant Engineers, Copenhagen, Denmark*

SUMMARY

An attempt has been made to determine whether damping at resonance frequencies in structures that are artificially damped with Neoprene can be predicted on the basis of model tests. The results of tests on damped steel beams in the scale 1:4:12 are described.

INTRODUCTION

Common construction materials such as steel and concrete have a low damping capacity, which means that structures made from these materials and loaded dynamically, especially at resonance, can produce vibrations which are unacceptable to the human organism or which can build up such big strains that fatigue failure can occur.

These vibrations can be damped by combining the construction material with high-damping materials, Bligaard Pedersen (1), (2). For more complicated structures, the numerical method of analysis proposed in (2) can be used in principle, but is likely to present great problems in practice.

The alternative, experimental method of solution, viz. modelling has been used on several occasions for determining resonance frequencies that are rather independent of whether the damping properties are correctly modelled, whereas only a few references, for example Borges and Pereira (3) mention model tests for determining the anticipated damping properties of structures.

MODEL CONDITIONS

Modelling phenomena in which the loading is dynamic presents fundamental problems. Modelling is thus hardly a viable method in the case of steel structures, where air damping, friction and damping in cracks largely determine the maximum strains caused by resonance.

Some of the materials combined with, for example, a steel structure in order to achieve increased damping can be characterized with good approximation as linear visco-elastic, isotropic and homogeneous; thus, when the load is harmonic, such materials can be characterized by a complex modulus of elasticity $E^*$ and a complex Poisson's ratio $\nu^*$.

The properties of a few of the materials, for example Neoprene, are approximately independent of frequency in a certain frequency range. This opens the way for satisfying the model conditions for a composite structure when the loading is dynamic, provided only that the frequency spectrum of the load lies within the above-mentioned frequency range. This applies particularly to resonance, where the strain field can be assumed to vary harmonically.

An additional condition is naturally that by far the greater part of the total damping occur in the high-damping material.

If we assume that the low-damping construction material is linearly elastic, isotropic and homogeneous, the model conditions for the composite structure become as follows, where the effect of the mass forces is neglected:

$$K_\sigma = K_e \, K_E = K_e \, K_{|E*|}, \quad K_t = (\frac{K_\rho}{K_E})^{\frac{1}{2}} K_l,$$

$$K_\nu = K_{\nu*} = K_\delta = 1$$

Here, $K$ denotes the ratio between corresponding quantities at corresponding points at corresponding times in the structure $A$ and the model $B$. Thus, $K_\sigma = \sigma^A/\sigma^B$ is the ratio between corresponding stresses in $A$ and $B$. The following designations are used: $\sigma$: stress, $l$: length, $\rho$: density, $e$: strain, $t$: time, $E$: modulus of elasticity, $\nu$: Poisson's ratio, $E* = |E*| \cdot$epx. $i\delta$: complex modulus of elasticity, $\nu*$: complex Poisson's ratio, $\delta$: loss factor.

The same materials can be used in $A$ and $B$ as long as it is ensured that the frequency range in which $|E*|$ and $\delta$ are constant contains the eigenfrequencies in both $A$ and $B$ which are assumed to be of interest. In this case, $K_E = K_{|E*|} = K_\rho = 1$.

MATERIAL TESTS

Tests on Neoprene were carried out with the test arrangement shown in fig. 1, in which the damping material was subjected to pure shear, and where the material parameters $|G*|$ and $\delta$ were determined. The relationship between the complex shear modulus $G*$ and $E*$ is given by: $E* = 2 G*(1 + \nu*)$. The complex Poisson's ratio $\nu*$ for a rubber material like Neoprene is assumed to be real, with the value 0.5, independent of the frequency.

The results from tests using a 3 mm thich Neoprene plate are shown in fig. 2. $|G*|$ and $\delta$ are approximately constant in the frequency range 5 - 100 Hz, which means that it should be possible to use the model technique in connection with the design of structures with a low 1st eigenfrequency.

Similar material tests were carried out on 10 and 40 mm thick Neoprene plates. In principle, the curves have the same appearance as in fig. 2, but a difference was found in $|G*|$ for the different plate thicknesses, the biggest measured value being twice the smallest measured value at the same frequency.

The material parameters do not appear to be particularly sensitive to temperature changes in the range 0 - 20°C.

Figure 1 Test set up with electrooptical extensometer for measuring |G*| and δ.

Figure 2 |G*| and δ for 3 mm Neoprene vs. frequency.

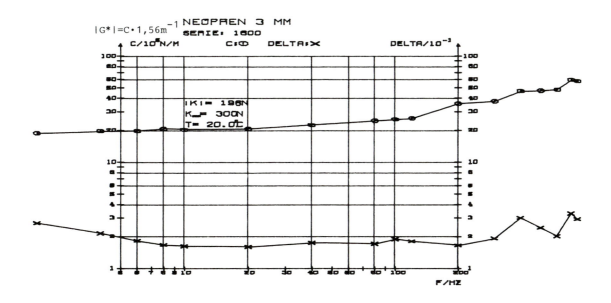

## BEAM TESTS

### Test arrangement

Tests were carried out on three I-steelbeams made as geometrically similar as possible. The lengths were 6 m, 1.5 m and 0.45 m respectively.

Each beam was suspended at its centre of gravity so that they performed a free-free vibration in their symmetry plane. The suspension, and a cross-section of a beam, undamped and damped, are shown in fig. 3.

### Measuring technique

The inertance defined as the ratio between the spectra for an acceleration and a force, was used as a measure of the frequency response of the beam. The acceleration was measured at one end of the beam, since all the interesting natural frequencies gave a measurable amplitude at this point. The measurements were carried out with a two-channel FFT-analyser, Hewlett Packard 358 A, which can directly calculate the desired functions and which also gives the possibility of using "ZOOM-FFT", i.e. the possibility of using a special sampling technique that improves the resolution around the eigen frequencies.

Figure 3  Suspension of 1.5 m and 6 m beams.

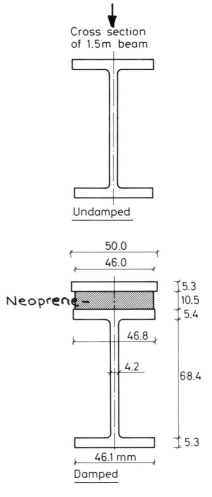

294

## Test results

The results of the tests on both undamped and damped beams are shown in table 1. Eigen frequencies and system loss factors were measured. The figures in parantheses indicate that the material parameters for the damping material were not present at the frequency in question.

It will be seen that, with good approximation, the frequency scale is, as anticipated, inversely proportional to the length scale. This applies to both damped and undamped beams. For the undamped beams, it will be seen that the damping depended on the type of steel used, which differed in the three beams. The damping seems to depend on the frequency for the 6 m and 1.5 m beams, the change from 1st to 2nd eigen frequency corresponding to a doubling and a quadrupling.

It will be seen that the type of construction chosen was not ideal, since the first eigen frequency for the three beams lay in the range 57 - 861 Hz and the second eigen frequency in the range 253 - 4060 Hz, i.e. in a frequency range in which $|G^*|$ and $\delta$ were not constant. $|G^*|$ was thus approximately twice as big at 745 Hz as at 57 Hz. A not insignificant scale error can therefore be anticipated in the model test.

In table 2 the system loss factors measured are compared with system loss factors calculated in accordance with (2), using the values for the material parameters determined in the material tests. Accordance between the measured and the calculated values appears to be reasonably good.

Table 1.

| beam | length scale ratio | undamped beam | | | | | |
|---|---|---|---|---|---|---|---|
| | | 1' eigenfrequency | | | 2' eigenfrequency | | |
| | | eigen-frequency Hz | scale ratio for eigen-frequency | system loss factor $\delta$ | eigen-frequency Hz | scale ratio for eigen-frequency | system loss factor $\delta$ |
| 6 m | 1 | 63 | 1 | $0,3 \cdot 10^{-3}$ | 299 | 1 | $0,6 \cdot 10^{-3}$ |
| 1,5 m | 0.25 | 252 | $\frac{1}{0.25}$ | $0,6 \cdot 10^{-3}$ | 1195 | $\frac{1}{0.25}$ | $2,4 \cdot 10^{-3}$ |
| 0,45 m | 0.075 | 861 | $\frac{1}{0.073}$ | $0,4 \cdot 10^{-3}$ | 4060 | $\frac{1}{0.074}$ | $0,4 \cdot 10^{-3}$ |

| beam | length scale ratio | damped beam | | | | | |
|---|---|---|---|---|---|---|---|
| | | 1' eigenfrequency | | | 2' eigenfrequency | | |
| | | eigen-frequency Hz | scale ratio for eigen-frequency | system loss factor $\delta$ | eigen-frequency Hz | scale ratio for eigen-frequency | system loss factor $\delta$ |
| 6 m | 1 | 57 | 1 | $1,7 \cdot 10^{-2}$ | 253 | 1 | $1,2 \cdot 10^{-2}$ |
| 1,5 m | 0.25 | 223 | $\frac{1}{0.26}$ | $1,9 \cdot 10^{-2}$ | 999 | $\frac{1}{0.25}$ | $(3,2 \cdot 10^{-2})$ |
| 0,45 m | 0.075 | 745 | $\frac{1}{0.077}$ | $1,7 \cdot 10^{-2}$ | 3310 | $\frac{1}{0.076}$ | $(3,0 \cdot 10^{-2})$ |

Table 2.

| beam | System loss factor $\delta$ 1' eigenfrequency | | System loss factor $\delta$ 2' eigenfrequency | |
|---|---|---|---|---|
| | measured | calculated | measured | calculated |
| 6 m | $1,7 \cdot 10^{-2}$ | $9,7 \cdot 10^{-3}$ | $1,2 \cdot 10^{-2}$ | $7,4 \cdot 10^{-3}$ |
| 1,5 m | $1,9 \cdot 10^{-2}$ | $1,6 \cdot 10^{-2}$ | $(3,2 \cdot 10^{-2})$ | $1,3 \cdot 10^{-2}$ |
| 0,45 m | $1,7 \cdot 10^{-2}$ | $2,2 \cdot 10^{-2}$ | $(3,0 \cdot 10^{-2})$ | |

If the deviation is attributed only to experimental error, it
will be seen that if the measured system loss factor for 1st eigen
frequency for the 0.45 m beam is used for predicting for a 12
times bigger beam, i.e. the 6 m beam, using the scale $K_\delta = 1$,
then the value thus determined will be approximately 70% bigger
than the calculated value. The reasons for the deviation have not
been clarified, but it is believed that the deviation on the
material properties within the same Neoprene plate may constitute
a considerable source of error.

Strains were not measured, but since $\delta$ expresses the ratio
between energies and is thus proportional with the ratio between
strains to the second power, the corresponding error on a strain
measurement can be expected to be about 35%.

The following conclusions can be drawn from the measurements:

1) The system loss factor in the damped beam is 20 - 50 times
   greater than in the undamped beam, for which reason damping
   in the steel in the damped beam can be neglected.

2) The dependence on frequency of the material properties of
   Neoprene is small in the frequency range 5 - 100 Hz, which
   means that it should be possible to use model tests for
   determining the condition in structures with low eigen frequencies when the length scale is not too big.

3) The eigen frequencies in the frequency range in question must
   lie so far from each other that small geometrical inaccuracies etc. in connection with the modelling do not give rise
   to superimposed eigen frequencies. This can be achieved by
   only considering the lowest eigen frequencies, at which the
   energy content is also often greatest.

4) The type of construction used is not ideal because Neoprene
   does not have constant material properties in the frequency
   range bounded by the eigen frequencies of interest.

5) If the test on the 0.45 m beam is used for predicting the
   system loss factor for the 6 m beam at 1st eigen frequency,
   the value thus determined will be about 70% higher than the
   calculated value. For strains, the deviation is expected
   to be about 35%.

## ACKNOWLEDGEMENT

Statens teknisk-videnskabelige forskningsråd has granted financial support for the work described in this paper.

## REFERENCES

(1)  BLIGAARD PEDERSEN, F, Measurement of the complex modulus
     of viscoelastic materials, Struct. Res. Lab., T.U. of
     Denmark, report R 66, (1975).

(2)  BLIGAARD PEDERSEN, F, Vibration analysis of viscoelastically
     damped sandwich structures, Struct. Res. Lab., T.U. of
     Denmark, report R 88, (1978).

(3)  FERRY BORGES, J, PEREIRA, J, Dynamic model studies for
     designing concrete structures, Models for concrete Structures,
     ACI publication No. SP 24, (1970).

# Contribution to discussion of Paper No. 31

J M Wilson
*Department of Engineering, University of Durham, UK*

The analysis of structural vibration is generally undertaken with damping assumed to be viscous and proportional (1). For a structure under design, it is customary to estimate the amount of damping to be adopted for the analysis from the known behaviour of other similar structures. One of the pitfalls associated with this practice is that designers are wont to base estimates of damping solely on the logarithmic decrements obtained for similar structures without taking into account either their natural frequencies or those of the new structure. Such a practice is inconsistent with the nature of viscous proportional damping as will be shown. A rational method of estimating damping is proposed, in which the natural frequencies of vibration as well as the associated logarithmic decrements are taken into account, and linear visco-elastic material behaviour is assumed. It is shown that viscous proportional damping, for which the damping matrix is assumed to be proportional to the stiffness matrix, can be regarded as a special case of this type of material behaviour.

To illustrate this method, let us assume that the amount of damping in the structure is slight, that this damping is due solely to material behaviour which is of a linear viscoelastic nature and tha Poisson's ratio is constant. Subject to these restrictions it can be shown that the damped structure vibrates freely in the natural mode shapes (2). For small damping, the amplitude of free vibration decays exponentially with time and this motion can be characterised by the logarithmic decrement which for the $r^{th}$ mode of vibration is given by (3)

$$\delta_r = \pi\, G_2(\omega_r)\, G_1(\omega_r) = \pi\, \tan\, \psi(\omega_r) = \pi\, \lambda(\omega_r) \quad \cdots (1)$$

where $\omega_r$ is the    natural frequency

$G_1$    and $G_2$ are real and imaginary components (the storage and loss modulus respectively) of the complex modulus $G(\omega_r)$

$\psi$    is the loss angle, and

$\lambda$    is the loss tangent.

The frequency dependent qualities defined in equation (1) can be determined in principle from dynamic tests on the viscoelastic material. Thus the in-phase storage modulus, $G_1(\Omega)$, the out-of-phase loss modulus, $G_2(\Omega)$, and hence the loss tangent, $\lambda(\Omega)$ can be found by applying a steadystate constant amplitude sinusoidal stress of frequency $\Omega$ to the test material and measuring the strain response. A typical form of loss characteristic (a plot showing the variation of loss tangent with forcing frequency) for a linear viscoelastic material is shown in Figure 1a. If the natural frequencies of a structure made of this material are known, or can be estimated, then an estimate of the damping, in terms of the logarithmic decrement for each mode of vibration, can be made using equation (1) and the loss characteristic.

Unfortunately the determination of the loss characteristic of common structural materials is difficult to measure in practice because the loss angle is so small in the frequency range of interest. However, by using the corollary of equation (1) the loss characteritic can be deduced from tests on freely vibrating structures composed of a single material. If as a result of such tests the logarithmic decrement is found to depend only on frequency and it can be shown that the damping arises predominantly the material behaviour, then the structural material can be regarded as linear viscoelastic. The loss characteristic can be deduced from a plot of the logarithmic decrement against natural frequency (Figure 1b) because from equation (1)

$$\lambda_r = \frac{\delta_r}{2\pi} \qquad \ldots (2)$$

This plot can be used to predict the damping in each mode of any structure made of the same material provided that the natural frequencies are taken into account and that the underlying assumptions hold.

The behaviour of a linear viscoelastic material can be represented by a rheological model consisting of linear elastic springs and linear viscous dashpots as shown in Figure 2 (4). Each spring or dashpot obeys a constitutive law of the form

$$\sigma = E\varepsilon \qquad \ldots (3a)$$

or

$$\sigma = C\varepsilon \qquad \ldots (3b)$$

where E is the spring constant and C is the dashpot constant.

Viscous damping for which the damping matrix is solely proportional to the stiffness matrix, can be shown to be equivalent to damping produced by a linear viscoelastic model consisting only the spring $E_0$ and the dashpot $C_0$. Such a model is known as a Kelvin model and its loss tangent is proportional to frequency as shown in Figure 3a. Thus

$$\lambda (\Omega) = \frac{C_O}{E_O} \qquad \Omega = \alpha \Omega \qquad \ldots (4)$$

where $\alpha$ is the constant of proportionality relating the damping and stiffness matrices. It follows that the logarithmic decrement of a structure composed of a material exhibiting this behaviour will be proportional to the assocated natural frequency as shown in Figure 3b. Thus

$$\delta_r = \pi \lambda (\omega_r) \qquad \pi \alpha \omega_r \qquad \ldots (5)$$

A consequence of the above theory is that, for different structures composed of the same linear viscoelastic material and having a common natural frequency (or for a given structure having two or more coincident natural frequencies), the damping will be the same for each mode shape associated with that frequency.

References

1.	Hurty, W.C., and Rubenstein M.F., Dynamics of Structures, Prentice Hall, New Jersey, (1964).

2.	Valanis, K.C., Exact and Variational Solutions to a General Viscoelastic-Kinetic Problem, J.App. Mech., Trans,Am.Soc., Mech.Engs., 33, 888-892, (1966).

3.	Struik, L.C.E. Free Damped Vibration of Linear Viscoelastic Materials, Rheologica Acta, 6, (2), 199-129, (1967).

4.	Gross, B. Mathematical Structure of the Theories of Visco-elasticity, Hermann, Paris, (1963).

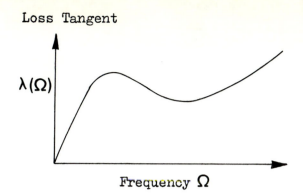

Figure 1a

Typical loss characteristic
for viscoelastic material

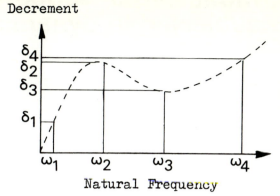

Figure 1b

Logarithmic decrement obtained
from structural tests

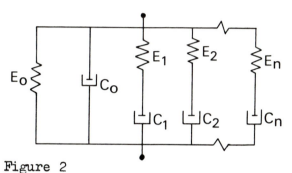

Figure 2

Rheological model representative of linear
viscoelastic behaviour

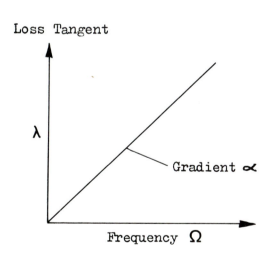

Figure 3a

Loss characteristic for
proportional viscous damping

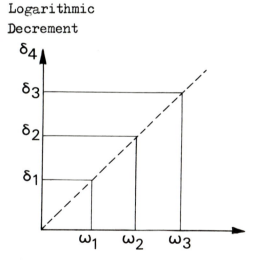

Figure 3b

Logarithmic decrements for
proportional viscous damping

# Modelling of supporting cores subjected to dynamic loading

32

B Goschy
*Geotechnical Institution, Budapest, Hungary*

## 1. INTRODUCTION

The objective of this study is to present a simplified
method of dynamic modelling of reinforced concrete
supporting cores. In modern high-rise buildings
supporting cores are often employed to resist lateral
forces of static and/or dynamic nature.

The structural system under investigation is represented
in Fig.1. This consists of a thin walled shear core in
partially closed box section surrounded symmetrically
by beam-column members with hinged connections.

## 2. SELECTION OF THE DYNAMIC MODEL FOR THE STUDY

The inter-system connectivity between environmental and
structural systems is formulated in structural dynamics
by equations of motion:

$$k\, u^{IV} - m\, \ddot{u} + c\, \dot{u} = f(t) \qquad (1/a)$$

for continuous models, and

$$\underline{M}\, \ddot{U} + \underline{C}\, \dot{U} + \underline{K}\, U = F(t) \qquad (1/b)$$

for discrete models.
The left hand side describes properties of the structural
system in motion /$\underline{M}$,m = mass; $\underline{C}$,c = damping; $\underline{K}$,k =
= stiffness/, while the right hand side informs about
the force - time history termed as forcing function.
Equalities (1/a) and (1/b) also imply a balance in
modelling procedure, in other words a compatibility in
objectivity of both sides. If, for example, the

302

excitation /force, displacement, acceleration etc./ can be determined on stochastic basis, using statistical techniques, structural properties should be evaluated with the help of probability considerations. On the other hand the structural response to artificial /constructed, deterministic/ dynamic events may be analysed with codified /standard/ values.

## 3. CHARACTERISTICS OF THE STRUCTURAL MODEL

### 3.1. Mass

Structural mass /uniformly distributed/ to be considered can be directly deduced from actual gravity loads /DL = dead load; LL = semi-permanent, live load with design values/ and their error bounds such as

$$m = \left[ DL\left(1,0 \pm 0,10\right) + LL \left( \frac{1,2 \pm 0,20}{2} \right) \right] 10^{-1} \quad (2)$$

### 3.2. Damping

The system damping experiences variations, which are response range and time dependent. Accordingly, the damping ratio of R/C structures in elastic range has the value of $\eta = 1,2 - 2,0\%$, while in plastic range $\eta = 4,0 - 5,0\%$ of the critical, considering deviation bands.

Non-linear variance in time may be estimated from [1]

$$\eta(t) = \frac{\eta}{1 - \exp\left(-2 \eta \omega_n t\right)} \quad (3)$$

where $\omega_n$ is the modal circular frequency.

### 3.3. Stiffness

### 3.31. Elastic systems

Performing flexural vibrations, the bending stiffness of the uniform cantilever /Fig.1./ may be described as

$$K = EJ_{ee} \frac{R_N}{R_r} \quad (4)$$

where

E is the Joung modulus of the concrete,

$J_{ee}$ is the equivalent moment of inertia,

303

$R_N$, $R_r$ are action factors.

Recommended values for E or G = 0,4 E to be considered in dynamic calculations of R/C structures are:

- E = $(1,25 - 1,40)E_o$ for impact loadings,
- E = $(0,9 - 1,1)E_o$ for short term excitations,
- E = $(0,4 - 0,5)E_o$ for long term excitations,

where $E_o$ is the quality dependent codified /specified/ value.

The gross moment of inertia of the investigated shear core in combined bending and shear takes the form [3], [4]

$$J_e = \sum_{i=1}^{2} \bar{J}_{yi} + \frac{J_s}{1 + D} \tag{5}$$

with

$$J_s = \frac{A_1 A_2}{A_1 + A_2} L^2; \quad D = \frac{h\ell_c^3}{24 \, \bar{J}_c} \frac{J_s}{H^2 L^2} \left(\frac{\pi}{2}\right)^2;$$

$$0 \le \bar{J}_c = \frac{J_c}{1 + 3\left(\dfrac{h_c}{\ell_c}\right)^2} \le \infty \quad \text{/gross moment of inertia of coupling beams/}$$

$$\sum_{i=1}^{2} \bar{J}_{yi} = \frac{\displaystyle\sum_{i=1}^{2} J_{yi}}{1 + \left(\dfrac{\pi}{2H}\right)^2 \dfrac{E}{G} \dfrac{\displaystyle\sum_{1}^{2} J_{yi}}{\displaystyle\sum A_{yi}}};$$

$$A_{yi} = \rho_i A_i \text{ /equivalent cross sectional area in pure shear/.}$$

It is recommended to consider the influence of shear in case of squat cantilever walls with an aspect ratio of H/b ≤ 4,0 [2].

Due to formation of cracks and combined actions gross values are to be replaced by the equivalent second moment of area /$J_{ee}$/.

At first approximation we may accept a reduction of 60% of the gross values, based on uncracked concrete area:

$$J_{ee} = 0,6 \, J_e \tag{6}$$

Some more detailed estimates to improve the values of $J_e$ are given in [2].

Considering, however that the cracking process is loading-time dependent, it is resonable to separate cross-sectional characteristics in to two main cathegories.

Accordingly:

- for short term, impact like loadings, when we may have cross sections up to - or beyond the first crack, with

$$J_{ee} = 0,8 \ J_e; \qquad A_{ee} = A_e \qquad (7)$$

- for long term excitations, when crack are fully developed due to previous cycles, with

$$J_{ee} = 0,4 \ J_e; \qquad A_{ee} = 0,5 \ A_e \qquad (8)$$

The presence of axial loading /N = 10 $m_{co}$ H; $m_{co}$ = mass of the core/ is noted with:

$$R_N = 1 - \frac{N}{N_E} = 1 - 1,258 \ \frac{H^2}{EJ_{ee}} \ N \qquad (9)$$

A theoretically produced estimate of soil structure interaction is given by [3]

$$R_r = 1 + 9,0 \ r + 21,0 \ r^2 \quad \text{with} \quad r = \frac{EJ_{ee}}{C_r \ BH} \qquad (10)$$

where

$C_r = \left(500 - 1500\right) \ \dfrac{f_s}{B} \ \left(kNm^{-3}\right)$ is the subgrade modulus

of soils in tilting /rocking/,

$f_s$ is the limit strength of the soil.

The error bounds in accessing $C_r$ are over 50%.

## 3.32. Inelastic systems

Structures responding plastically dissipate energy by post elastic deformations. Consequently a rational criterion to qualify the ability of a structure to resist at a limit a given forcing excitation may be formulated on energy basis by establishing the balance of energies written as

$$T = V_{max} \qquad (11)$$

where: T is the induced energy,

$V_{max}$ is the potential energy stored at maximum displacement.

The local /cross sectional/ energy dissipation ability
is quantified by the ductility factor:

$$\mu = \frac{u_{max}}{u_E} \qquad (12)$$

while the energy absorbing potencial of the structural
system is generally expressed in terms of energy
dissipation ratio:

$$\mu_O = \frac{V_{max}}{V_e} \qquad (13)$$

where

$u_E$; $V_E$ is the elastic displacement; elastic strain
    energy,

$u_{max}$; $V_{max}$ is the maximum displacement; maximum strain
    energy.

The ratio of ther energy absorbing parameters is about
$\mu/\mu_O = 2-4$.

Flexural members may have a ductility factor ranging as:

$4 \leq \mu \leq 12$ in pure bending /e.g. lintels/

$1,5 \leq \mu \leq 5$ in combined bending and compression
    /e.g.walls/.

Values for member ductility factors are to be derived
from stable or unstable hysteritic characteristics of
moment,- curvature relationships. As a primary model,
however a bi- or three-linear moment-rotation diagram
is acceptable. Degrading stiffness response models are
applied when the duration of input motions or forces is
long enough to produce a low or high cycle fatique.
It is also accepted to derive the maximum potential
energy of the total structural system from static
collaps mechanisms.
It is recommended to perform a mechanism in which
plastification process begins in coupling beams while
cantilever walls behave elastically /Fig.2./.
Reasons:
- higher ductility of lintels in pure bending,
- possibility of regulation ductility demands by proper
  detailing,
- prevention of a progressive collapse by keeping walls
  in the elastic range.

The induced energy in [11] is given by

$$T = \frac{1}{2} \int_{O}^{H} m\, v^2\, dz \qquad (14)$$

or more detailed

$$T = \frac{1}{2} \int_{O}^{H} m \left(\frac{t_n}{2\pi}\right)^2 a_s^2\, dz \quad /e.g.\ earthquakes/$$

$$T = \frac{1}{2} \int \frac{I^2(z,t)}{m}\, dz \quad /e.g.\ blast\ when\ t_d \le \frac{t_n}{3}\ /$$

where

$t_n$ is the modal period of the core,
$t_d$ is the duration of the loading pulse,
$I(z,t)$ is the impulse,
$a_s$ is the maximum spectral acceleration.

In case of impulsive loads it is of primary concern to determine the time $/t_y/$ at which a first excursion in plastic range begins

$$t_y = f\left(t_n,\ R_y,\ P_d\right)$$

where:

$R_y$ is the structural resistence at first yield,
$P_d$ is the impulsive load /peak value/.

A true-to-life inelastic analysis is based on a step-by-step integration procedure, which takes into account generated changes of system states in assessing the energy content of the structure /Fig.2./.

Many types of moderate geometric and/or physical non-linearities can be dealt with successfully by using equivalent linear analysis, adjusting stiffness and damping values to be compatible with plastic deformations /Newmark's, Hall's model/.
The linearization technique, used mostly in earthquake engineering is based on the application of reduction factors /R/ to the peak accelerations, within the bounds

$$R_1 = \frac{1}{\mu} \le R_2 = \frac{1}{\sqrt{2\mu - 1}} \le R_3 = 1 \qquad (16)$$

for different frequency bands of the response spectrum /e.g. $f_1 < 2,0\ s^{-1}$; $2,0 \le f_2 < 10\ s^{-1}$; $f_3 \ge 10\ s^{-1}$/

# 4. CONCLUSIONS

This short survey shows a large variety in dynamic modelling of R/C shear core structures acted on by various environmental events of dynamic nature. It is apparent, that the quality of dynamic modelling is depending upon:
- response state of the structural system /elastic, plastic/,
- time history of forcing motions or forces /short,-long term/,
- importance and life time of the building structure /damage evaluation, probability of failure/, and
- purpose of analysis /predesign, control/.

A proper detailing seems however to be mach more important in keeping dynamic stability, than a sophisticated dynamic model simulation.

## References

1. Parkus,H.: Random excitation of structures by earthquakes and atmospheric turbulences. Springer Wien-NY 1977 /CISM/

2. Paulay,T.: The design of reinforced concrete ductile shear walls for earthquake resistance. Univ.of Canterbury,Christchurch, 1981.-1.

3. Goschy,B.: Statics and dynamics of shear wall structures /in hungarian/ Müszaki Könyvkiadó, Budapest, 1981.

4. Coull,A; Mukherjee P.R.: Natural vibrations of shear wall buildings on flexible support. Earthquake Eng. and Structural Dyn. 1978.Vol.6.

# ELEVATION

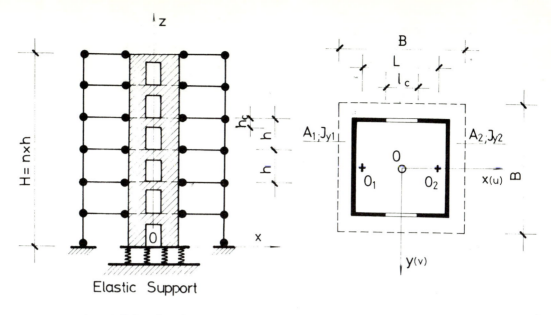

# SHEAR CORE

# CROSS SECTION

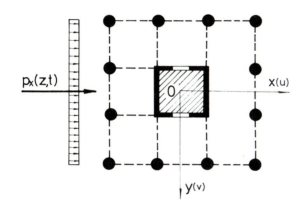

Figure 1.

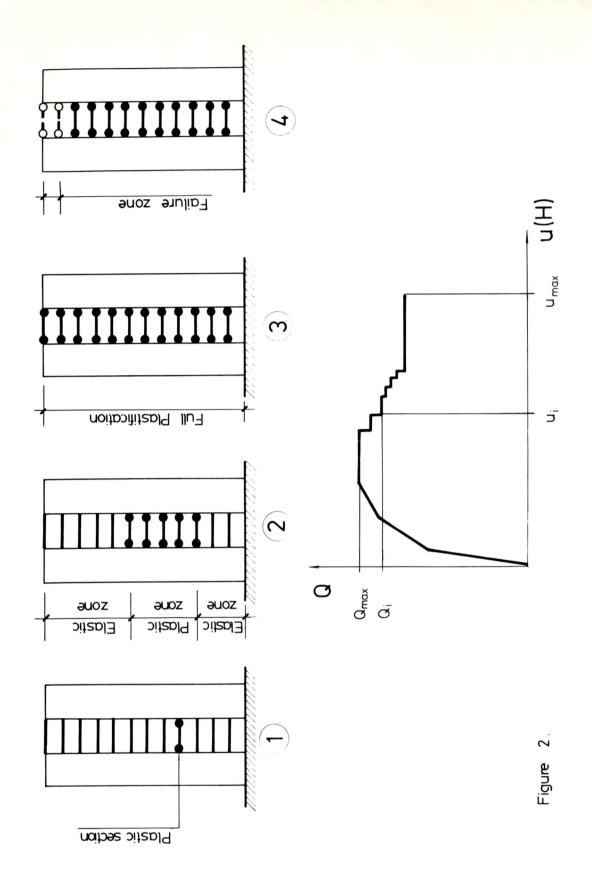

Figure 2.

310

## 33

# The rigid-plastic model for structures loaded dynamically

Norman Jones
*Department of Mechanical Engineering, The University of Liverpool, UK*

SUMMARY

This article focuses on the response of structural members subjected to dynamic loads, or sudden displacements, which are
sufficiently severe to cause plastic flow of the material and
permanent displacements or structural damage. The rigid-plastic
idealisation, which has proved useful for static problems, has been
developed in recent years to successfully predict the response of
structures loaded dynamically. A broad summary of this idealisation
is presented herein.

INTRODUCTION

Simple rigid-plastic methods of analysis have been used extensively
in civil engineering design and in other branches of engineering.
They have predicted in a straightforward fashion acceptable
estimates of the maximum load carrying capacities of beams, frames,
plates, shells and shell intersections subjected to a variety of
static loads as well as the behaviour of other bodies which are of
less interest here.

The rigid-plastic procedure has been extended in order to study the
response of structures subjected to large dynamic loads which cause
inelastic material behaviour and give rise to finite permanent deformations or damage. They have been used to solve many practical
problems but this article will concentrate on the structural
behaviour due to large dynamic or impulsive loads which act only
once in contradistinction to repeated loads which might raise
concerns with fatigue and shakedown. Furthermore, emphasis is
placed on the late-time response, or structural behaviour, while
the early-time response, which is governed by wave propagation
effects, is not mentioned again.

The idealisation of a rigid-plastic material and the various con-
comitant simplifications are discussed in the next section, while
the following sections briefly discuss the rigid-plastic theoretical
procedure and its application to the dynamic plastic behaviour of
beams and plates.

MATERIAL IDEALISATION

The uniaxial tensile behaviour of a typical ductile material is
compared in Fig. 1 with several material idealisations. In
Reference (1), Hodge has examined the validity of a rigid, perfectly
plastic idealisation of a real material for static loading
situations.

It is possible to solve some dynamic plastic structural problems
with the aid of wholly numerical methods which use the actual

material properties rather than any of the idealised ones illustrated in Fig. 1. However, these numerical schemes are expensive and can be time-consuming and sometimes of dubious value because of uncertainties in the dynamic loading and material properties as well as convergence and uniqueness difficulties in some cases.

Simple methods of analysis are invaluable for preliminary design purposes when feasibility decisions are required. If a design is taken further and more accurate information is needed then the parameters of the preliminary design can be used as input to a numerical scheme. In this sense, both simple design methods and numerical schemes are necessary and complementary. This article focuses on simple methods of analysis, but an interested reader may consult References (2) and (3) for brief literature reviews on numerical schemes.

## Elasticity

It is well known that the static plastic collapse load of a body is independent of material elasticity according to the limit theorems of plasticity. Similarly, for dynamic loads, one would anticipate that elastic effects are not important when the external dynamic energy is much larger than the maximum amount of strain energy a structure could absorb in a wholly elastic manner. In fact, good agreement between theoretical and experimental results has been found for energy ratios as small as 2.01 (4). However, for a rigid-plastic model to predict acceptable results the duration of dynamic loading must be short compared with the period of natural elastic vibration (5).

## Strain Hardening

The neglect of material strain hardening in structural plasticity problems involving static loading is usually acceptable because good, albeit conservative estimates of the static collapse load are predicted. A similar situation appears to prevail for dynamic loads which produce an inelastic response.

However, strain hardening effects should be retained in the analysis of structures made from materials which exhibit strong strain hardening, such as 304 stainless steel which has widespread use in the nuclear industry. In this circumstance the mathematical formulae which govern material strain hardening are the same as those which have been developed for the static behaviour of structures.

## Strain Rate Sensitivity

The tensile stress-strain diagram in Fig. 1 was obtained from a 'static' uniaxial test on a ductile material. However, it is well known that the curve depends on the magnitude of the uniaxial strain rate. Generally speaking, the yield stress increases with increase in strain rate, while the ultimate stress increases but less markedly and is reached at smaller strains. Thus, strain rate sensitive materials have larger plastic flow stresses at high strain rates and rupture at smaller strains than the corresponding static behaviour.

The uniaxial yield stress of a hot rolled mild steel, for example, is double the corresponding 'static' value at a strain rate of $40 \text{ sec}^{-1}$, approximately. Aluminium 6061 T6 on the other hand is essentially strain rate insensitive at the strain rates of interest.

Thus, even though the strain rate may be large only at isolated locations in a structure, it is important to take this phenomenon into account because it could have important consequences on the energy absorbing capacity of plastic hinges and therefore on the response of structures made from strain rate sensitive materials. There are may constitutive equations which describe the strain rate sensitivity of a material, but the one due to Cowper and Synonds (5) (see equation (1) in Reference (6)) has been almost universally used in theoretical and numerical studies.

## Yield Criteria

The previous discussion has been largely confined to a discussion of the uniaxial stress-strain curve. The generalisation of these results to a multiaxial situation follows an identical procedure to that undertaken for the static behaviour with the exception of the extra considerations required to account for the phenomenon of material strain rate sensitivity. Thus, generalised stresses, such as bending moments and membrane forces, are used to construct the yield criterion for structural members (beams, frames, plates and shells). These yield criteria are complicated and nonlinear, particularly for shells, so that many authors have sought linearisations.

It may be shown that the yield condition in stress or generalised stress space must be convex, while the normality requirement, or flow rule of plasticity, governs the increments of generalised strain. Generalised strains, it should be recalled, are defined such that the product of generalised stress and associated generalised strain gives the internal work of that particular component. Thus, if a yield criterion is plotted in generalised stress space, then for plastic flow the increment of generalised strain associated with a point on the yield surface must be perpendicular (normality requirement) to the yield surface when the axes of generalised stresses and associated generalised strains are made coincident.

## Transverse Shear

The yield criteria for structural members referred to in the previous section invariably disregard the influence of transverse shear forces which are nevertheless retained in the equilibrium equations and are therefore known as reactions because transverse shear deformations are neglected. However, it appears that transverse shear effects exercise a more important influence on the response of structures loaded dynamically than on similar structures loaded statically (7).

This phenomenon can possibly be ignored for most structural problems except those which involve transverse shear failures (e.g., at supports of impulsively loaded beams (8), higher modal responses (4), or highly anisotropic structures (3)). Generally speaking, transverse shear is more important for structural members with open or non-compact cross-sections, but Reference (9) could be consulted for guidance in doubtful situations on beams and Reference (3) contains additional citations for beams and plates.

## THEORETICAL RIGID-PLASTIC METHODS

It is not possible to present here a comprehensive overview of all the theoretical work which has been published in this field. However, an interested reader may find more detailed literature reviews in References (2), (3) and (6).

A theoretical solution is 'statically' admissible if it satisfies the equilibrium equations and yield condition for a rigid, perfectly plastic material as well as the boundary conditions for generalised stresses. A theoretical solution is kinematically admissible if the displacement and velocity profiles satisfy the normality requirements of plasticity, any corresponding boundary conditions and are everywhere continuous except at stationary or travelling plastic hinges where certain kinematic conditions must be satisfied. A theoretical rigid-plastic solution is exact when it is simultaneously statically and kinematically admissible.

Exact theoretical rigid-plastic solutions have been reported for the behaviour of many beam, plate and shell structures which undergo a stable response when loaded dynamically. However, most of these analyses have been developed for infinitesimal displacements using equilibrium equations derived in the undeformed configuration. This assumption might provide reasonable results for some practical problems despite the apparent contradiction between the neglect of finite displacements and the requirement of a sufficiently large energy ratio. Nevertheless, it is necessary to retain the influence of geometry changes or finite transverse displacements in the governing equations for many practical beam and plate problems. The initial predominantly bending only response gradually changes into a string or membrane response governed by in-plane forces with zero bending moments at sufficiently large transverse displacements. Actually, this phenomenon is similar to that encountered for static loads and is important for maximum transverse displacements of axially restrained beams, plates and some shells as small as one-half or so of the corresponding structural thickness. Thus, simple theoretical bending only analyses are of limited value in practical problems which entail these particular members, but as is evident from the following section relatively simple methods are available which may be used for this class of problems.

DYNAMIC RESPONSE OF BEAMS AND PLATES

It was remarked in the previous section that it was necessary to retain the influence of finite transverse displacements, or geometry changes, in the governing equations for the dynamic plastic response of some structures. The influence of this phenomenon on the behaviour of beams and plates loaded with transverse dynamic loads is discussed further in this section because of the importance of these structural members in practical design. It is not possible to obtain simple exact analytical solutions for this class of problems except in exceptional circumstances. However, approximate and relatively simple, yet surprisingly accurate, theoretical methods have been developed in Reference (10) for the response of beams and arbitrarily shaped plates and for shells in Reference (11).

The theoretical procedure in Reference (10) for the particular case of a fully clamped strain rate insensitive rigid, perfectly plastic rectangular plate subjected to an impulsive velocity $V_0$ predicts maximum permanent transverse displacements which agree favourably with the corresponding experimental results as shown in Fig. 2 of Reference (6).

It is evident from Fig. 2 here that the theoretical predictions for a plate with a vanishingly small aspect ratio (i.e., beam) provide reasonable agreement with the maximum permanent transverse displacements recorded on strain rate insensitive (aluminium 6061 T6) beams loaded impulsively. The same theoretical rigid-plastic procedure has been used to generate simple formulae for fully clamped rectangular plates subjected to rectangular and triangular shaped pressure-time histories, which arise in a number of practical situations. In addition, theoretical results have been reported for the simply supported case, but the method in Reference (10) is sufficiently simple that many more beam and arbitrarily shaped plate problems are amenable to analysis.

Symonds and Jones (12) examined the simultaneous influence of geometry changes and material strain rate sensitivity on the dynamic plastic response of fully clamped beams loaded impulsively. It transpired that surprisingly accurate predictions for the influence of strain rate sensitivity in this particular finite deflection problem were obtained by using a modified yield stress $n\sigma_y$ as indicated in Fig. 2 for fully clamped beams made from hot rolled mild steel which is a strain rate sensitive material (see also Reference (6)).

The permanent transverse displacements of the impulsively loaded, fully clamped beams presented in Fig. 2 indicate the relative importance of material strain rate sensitivity and geometry changes. It is clear that a classical infinitesimal rigid-plastic theory is inadequate for this particular problem, when the deflections are larger than half of the beam thickness approximately. The theoretical predictions with $\sigma_y$ replaced by $0.618 \, \sigma_y$ for an inscribing yield criterion, provide a simple way of bounding a more exact theory developed in Reference (10) and give acceptable bounds for the experimental results recorded on the strain rate insensitive beams (aluminium 6061 T6). The theoretical results which retain the simultaneous influence of geometry changes and material strain rate sensitivity, agree quite well with the corresponding predictions according to the numerical scheme of Witmer et al. (13) and the experimental results on the strongly strain rate sensitive mild steel beams. It is evident from Fig. 2 that geometry changes exercise a dominant effect on the structural response, while material strain rate sensitivity causes a further significant reduction of the permanent transverse displacements of the mild steel beams.

DISCUSSION

A small part of the literature available on the dynamic plastic response of structures, which is reviewed more thoroughly in References (2), (3) and (6), has been touched upon in this article. The objective is to sketch the main features of the rigid-plastic idealisation because it can be used for the dynamic modelling of structures which is the concern of this conference. The agreement between the theoretical predictions and the corresponding experimental results shown in Fig. 2 is typical of what can be achieved with the aid of simple rigid-plastic calculations when the limitations of the method referred to earlier are respected.

Main emphasis in this article has been placed on the influence of dynamic loads which give rise to large permanent ductile deformations. Clearly, for sufficiently severe external dynamic loadings failure may occur due to exhaustion of the material ductility (i.e., tensile bearing). This mode of failure has not been examined for any structures except the impulsively loaded beam problem in Reference (8). It is also possible that structural members with open cross-sections, or other members subjected to severe dynamic loadings, may fail due to the development of large transverse shear forces as discussed earlier.

The foregoing work has only examined the behaviour of structures subjected to dynamic loads which produce a stable response. However, experimental and theoretical studies have been reported on the dynamic plastic buckling of columns, beams, plates, rings and shells as recounted in Reference (6).

A number of investigations have been undertaken on the impact resistance of prestressed and reinforced concrete members (14, etc.) some of which have considered the design of buildings.

REFERENCES

(1)    HODGE, P G, 'On real and ideal materials', Experimental Mechanics, Vol. 11, No. 1, 1-7 (January 1981).

(2)    JONES, N, "A literature review of the dynamic plastic response of structures', The Shock and Vibration Digest, Vol. 7, No. 8, 89-105 (August 1975).

(3)    JONES, N, 'Recent progress in the dynamic plastic behaviour of structures', The Shock and Vibration Digest, Part 1, Vol. 10, No. 9, 21-33 (September 1978); Part 2, Vol. 10, No. 10, 13-19 (October 1978); Part 3, In Press and Univ. of Liverpool, Dept. of Mech. Eng. Rep. ES/03/81 (April 1981).

(4)    JONES, N, and GUEDES SOARES, C, 'Higher modal dynamic plastic behaviour of beams loaded impulsively', Int. J. of Mech. Sci., Vol. 20, 135-147 (1978).

(5)    SYMONDS, P S, 'Survey of methods of analysis for plastic deformation of structures under dynamic loadings', Brown Univ. Rep. BU/NSRDC/1-67 (1967).

(6)    JONES, N, 'Response of structures to dynamic loading', Inst. Phys. Conf. Ser. 47, 254-276 (1979).

(7)    GOMES de OLIVEIRA, J, and JONES, N, 'Some remarks on the influence of transverse shear on the plastic yielding of structures', Int. J. Mech. Sci., Vol. 20, No. 11, 759-765 (1978).

(8)    JONES, N, 'Plastic failure of ductile beams loaded dynamically', J. Eng. Industry, Trans. ASME, Vol. 98, No. 1, 131-136 (February 1976).

(9)    JONES, N, and GOMES de OLIVEIRA, J, "The influence of rotatory inertia and transverse shear on the dynamic plastic behaviour of beams', J. Appl. Mechs., Vol. 46, No. 2, 303-310 (June 1979).

(10)   JONES, N, 'A theoretical study of the dynamic plastic behaviour of beams and plates with finite-deflections', Int. J. Solids and Structures, Vol. 7, 1007-1029 (1971).

(11)   WALTERS, R M, and JONES, N, 'An approximate theoretical study of the dynamic plastic behaviour of shells', Int. J. Nonlinear Mechs., Vol. 7, 255-273 (1972).

(12)   SYMONDS, P S, and JONES, N, 'Impulsive loading of fully clamped beams with finite plastic deflections and strain rate sensitivity', Int. J. Mech. Sci., Vol. 14, 49-69 (1972).

(13)   WITMER, E T, et al. 'Large dynamic deformations of beams, rings, plates and shells', AIAA Journal, Vol. 1, No. 8, 1848-1857 (1963).

(14)   BATE, S C C, 'The strength of concrete members under dynamic loading', proc. Symp. on the Strength of Concrete Structures, pub. by Cement and Concrete Assoc., 487-556 (1956).

Figure 1   Uniaxial tensile stress ($\sigma$) - strain ($\varepsilon$) curve for a typical ductile material with elastic, perfectly plastic and rigid, perfectly plastic idealisations.

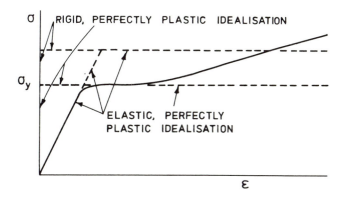

Figure 2   Maximum permament transverse displacements ($W_m$) of a fully clamped  beam loaded impulsively with a velocity $V_0$. □, △, ○, ◇ experimental results on aluminium 6061 T6 beams (10). ■, ▲, ▼, ◆ experimental results on mild steel beams (12). ● a-e, numerical finite-difference results of Witmer et al. (13) (from (12)).  1, Infinitesimal analysis (bending only); 2, circumscribing yield criterion (10); 3, inscribing yield criterion with $\sigma_y$ replaced by $0.618\sigma_y$; 4, theoretical solution of (10) for 'exact' yield criterion; 5, strain rate correction according to References (6) and (12) and H = 0.1 in (0.254 cm) (circumscribing yield criterion); 6, as 5, but with $\sigma_y$ replaced by $0.618\sigma_y$ (inscribing yield criterion).

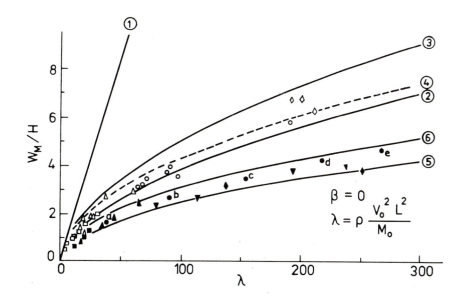

# Chairman's summing up

J E Gibson
*Department of Civil Engineering, City University, London, UK*

In this session the authors have devoted themselves to an investigation of the effect of impact, nuclear explosion and vibration on a variety of structural forms.

In the opening paper Mr. Davies of Taylor Woodrow considered the effect on reinforced concrete structures of short duration dynamic loads such as nuclear blast. Having outlined the theoretical aspect of dynamic modelling he then gave an account of two types of model tests. One on which the effect of nuclear blast against a model shelter complex was studied eventually yielded useful design data. In the second, in which long thin walled cylindrical missiles were fired to impact end on against reinforced concrete targets, also yielded results which were predictable.

The second paper by a group of authors, Barr, Howe, Nielsen and Carter from the U.K. and Nachtsheim, Riech and Rudiger from Germany presented an interesting collaboration between the two countries on the effect of missile impacts on reinforced concrete structures. The scale of the missiles and targets at Meppen were four times those at Winfrith and from the records from instrumentation, the effects of scaling could be measured.

Dr. Hobbs dealt with the strain rate sensitivity in impulsively loaded structures and derived a simple analytical theory. This theory was then applied to the case of simply supported beams subjected to uniformly distributed impulsive pressure loading and theoretical values showed reasonable agreement with the experimental work of Florence and Firth.

Watson and Tang described how simply supported reinforced micro concrete beams were subjected to static and impact loading, the aim being to determine the load deflection and cracking response and also the residual strength of the beam after impact damage. The instrumentation used in these tests was worthy of note.

In this second paper on impact, Perry described experiments in which a long steel cylinder was allowed to fall onto a prestressed concrete slab. The impact force was measured by a load cell fixed to the head of the cylinder. In one of the illustrations it was to be noted that the slab was completely penetrated by the cylinder causing total fracture of the prestressing wires.

Continuing the theme of impact Dr. Hughes described tests in which a rigid striker was dropped onto the midspan of pin ended beams. The impact force and beam displacements were measured and it was shown that simple beam vibration theory allowing for strain energy of bending and transverse inertia was able to explain the experimental phenomena.

On a different theme Professor Buchholdt in his paper described the resonance testing of a circular pretensioned cable roof model of 6m diameter. The structure was tested with and without cladding and the results showed how mass and cladding effect the natural frequencies and damping. The effect of air damping proved of great interest.

Dr. Krauthammer again on the subject of the response of reinforced concrete strucures to nuclear detonations, emphasised the non linear approach to analysis. Models developed for predicting the moment capacity and shear influence at the same time, gave reasonable agreement with tests.

Professor Gibson described the dynamic behaviour of cores of tall buildings. It was shown that the Rayleigh method for determining natural frequencies gave reasonable agreement with perspex and micro concrete model cores tested in the laboratory.

The final paper of the session by Askegaard, Schmidt and Pedersen was concerned with the prediction of the damping effect by model tests. Experiments on three steel beams undamped and damped were reported and the results proved of interest.

All authors are to be congratulated on the excellence of their papers and their standard of presentation.

# Vibration tests on open spherical shells    34

V C M de Souza
*Univeridade Federal Fluminense, Niterói, Brazil*
J G A Croll
*University College London, UK*

SUMMARY

Fabrication and testing of vibrations in small scale, electro-plated
nickel, open spherical shells are briefly described. Comparisons
with theoretical predictions show that with moderate levels of
excitation energy classical thin shell theory provides a close
description of observed resonance in both axisymmetric and non-
axisymmetric vibration modes. For larger excitation energies results
indicate a possible need to include the effects of geometrically non
-linear, modal interactions.

INTRODUCTION

Large orthotropically stiffened metal and concrete spherical shells
form important components of many pressure vessels, including most
recently those constituting the bouyancy chambers for certain semi-
submersible, floating and tethered offshore drilling and production
platforms. Despite the known importance of dynamic loading on these
components there is a notable absence of relevant experimental and
theoretical information on their forced dynamic response. This paper
describes a recent test programme, carried out at University College
London in collaboration with the Building Research Station, to
observe the vibratory characteristics of a representative range of
small scale, elastic, isotropic, open spherical shells.

The major advances in understanding the dynamics of open and closed
spherical shells have been concentrated into two distinct periods.
In the first, occurring in the late 19th century, the impetus derived
from attempts to understand the relationships between the geometry
of bells and their acoustic behaviour. The second phase had to wait
the combined demands of the nuclear and aerospace industries of the
1950's and 60's. The result has been considerable advances in
theoretical understanding of the free vibrations of isotropic
spherical shells of which the major aspects have been reviewed in
references (1,2). Nevertheless, the problem of predicting the
dynamic response of orthotropically stiffened spheres, which is
more characteristic of current marine applications, has only recently
been explicitly addressed. In an analysis of the seperate contributions from each of the membrane and bending strain energy components
of an isotropic shell (1) it has been possible to better understand
how changes in these energies, through the use of orthotropic
stiffening, effect the vibration characteristics (2).

But despite these theoretical advances there is surprisingly little empirical evidence on the nature of vibrations in spheres. Even that reported (3) seems to have resulted in controversy (4) regarding the role of membrane energy, not unlike that occurring in the late 1880's. It appears that the most complete and reliable tests on the vibrations of spherical shells have been those of Glockner and Tawadros (5) but even these were limited to the study of symmetric modes. There is an evident need to examine the dynamic response for all possible vibration modes, including those exhibiting non-axisymmetry, and to validate linear classical shell thoery as a reliable analysis procedure. Quite apart from its other importance, the prediction of the dynamic response of these and other shells could be crucial in the estimation of fatigue lives of such components.

The following reports a recent test programme to observe the vibratory characteristics of a representative range of nickel-plated shell models fabricated using an electro-plating technique and excited by means of an audio vibratory source. The response spectra for geometries covering the range of shallow and moderately thin to deep and very thin are described and correlated against the theoretical predictions of classical thin shell theory (1). It is shown how for moderate levels of excitation, in which the displacements are small, linear shell theory provides reliable descriptions of the observed behaviour for both symmetric and non-symmetric vibration modes. Where excitation energies are large, the tests suggest that it may be necessary to take into account the possibility of geometrically non-linear modal interactions.

## DESIGN, FABRICATION AND USE OF TEST FACILITIES

To allow satisfaction of the objectives of the present experimental study, the test facilities had to fulfill a number of design require-ments. Firstly, the experimental set-up had to ensure that the supporting frame had a negligible influence on the vibration characterisitics of the shells. Secondly, the source of vibration had to be capable of exciting a great number of natural modes and the instrumentation able to record vibrations of very small amplitudes. Thirdly, the fabrication procedure adopted had to allow very thin small scale shells to be manufactured with strict tolerance limits on geometric imperfections and thickness variation.

Previous experimental investigations of the dynamics of spherical shells were handicapped by the problems of fabricating accurate models. Electro-plating (5) seems to have provided a more reliable and versatile means of fabrication than either vacuum pressing of hot PVC sheet (7), or a metal spinning process (3); it has the disadvantages of being time consuming to initially implement and subsequent production tends to be slow and expensive.

For reasons described elsewhere (8), the present models were constructed from nickel, plated on a non-collapsible metallic former. The formers, shown in Fig.1, were made from duraluminium by a profile turning technique. With duraluminium having a coefficient of expansion considerably higher than that of the nickel, the plated models could be removed by freezing. Fig.2, shows the schematic arrangment for the electroplating process, in which a "Ni-speed"(nickel sulphamate) bath solution was used.

The experimental rig, shown in Fig.3, was constructed in two sections. The first consisted of the supports for the model, the edge clamping facilities and transducer probe scanning device. The second provided support for the vibration source (the Loudspeaker)and was dynamically

Figure 1. Duraluminium former positioned for electroplating

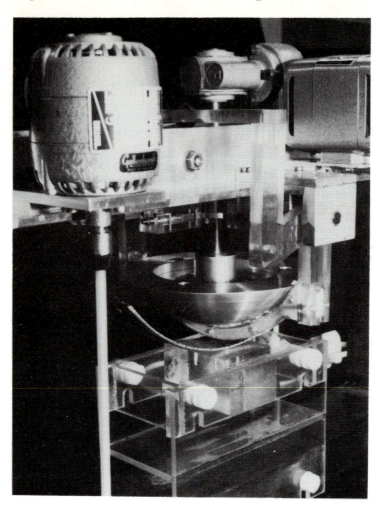

Figure 2 Schematic arrangement for electroplating

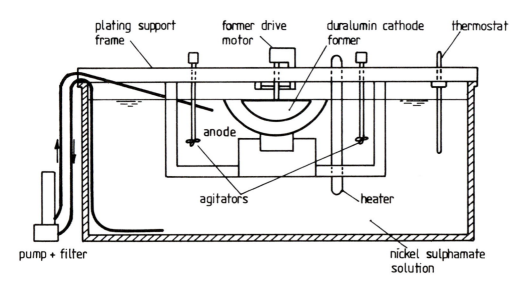

isolated from the shell's support frame by a rubber seal, which
ensured that the least possible acoustic energy would be lost.
The vibratory energy was provided by a D50/153 Dalesford 6½"Bass/
mid range loudspeaker, with nominal frequency response range 30Hz
to 8kHz, and a cone resonant frequency of 32Hz, well below the
expected range of natural frquencies of the models. A Gould L.F.
Signal Generator J3A, with a frequency range 10Hz to  100 kHz, was
used in conjunction with a Feedback FM 610 Digital Frequency Meter
to generate and record accurate signals. The power of the loudspeaker was provided by a purpose constructed amplifier having 25W RMS
into 8Ω and frequency range 10Hz to 45kHz. Displacements were
recorded using either a capacitive displacement transducer TE
100 MK 11 DIMEQ, with a MC1 type probe, (Transducer), for reading
in frequencies up to 2 kHz, or for higher frequencies a non-contact
displacement measuring Distec Model 915 signal conditioning unit
having a 250-X probe, non-calibrated. An Advanced Type OS250
oscilloscope was used for readout.

The general experimental set-up is illustrated in Fig.3, and 4.
Frequency spectra of the shells were obtained as follows: (a) the
probe was placed so that the midpoint of the linear displacement
range of the transducer was coincident with the surface of the model
and the reading instrument was set equal to zero; (b) the excitation
circuit was switched on at a frequency that had been previously
chosen to be somewhat below the first theoretical natural frequency
of the model; (c) the frequency of the excitation circuit was changed
until a reasonance condition occurred, charactised by a relative
maximum in the displacement (peak-to-peak) frequency curve;(d)the
process was repeated until the upper limit of the linear field
of the transducer was reached. This process was repeated for at
least three to four different positions of the probe, its position
varying in the circumferential and meridional directions, to ensure
that all resonant modes could be observed. Because no filter was
available it was difficult to observe precisely some of the resonant
frequencies, and over certain frequency intervals the energy input
had to be frequently changed to allow the exact resonant frequencies
to be observed.

When mapping mode-shapes the following somewhat different procedure
was adopted: (a) the probe was place at a certain position
(initially at the apex), and the instrument was zeroed as before;
(b) the excitation circuit was switched on at a frequency coincident
with the lowest resonant frequency of the shell, and the peak-to-
peak displacements measured; (c) the frequency of the excitation
circuit was moved to the second resonant frequency of the shell and
so on, and the displacements measured, until the last resonant
condition of the selected range was reached;(d) the probe was moved
to another position and the process repeated.

Since the models were reasonably thin, having maximum thickness to
radius ratio, h/R, of around 1/1000, any asymmetry in the mounting
of either the model or that of the loudspeaker caused the excitation
of the symmetric modes, as in the work of Glockner and Tawadros(5),
as well as those of the nonsymmetric modes.The ease of exciting some
of the nonsymmetric modes depended on the degree of asymmetry in the
mounting of the loudspeaker relative to the model. For this reason
some mode-shapes were easier to identify than others, since at
certain resonant frequencies the modal participations of some non-
resonant modes were higher than that in the resonant modes.

Figure 3  General set-up for supporting and testing shell
vibrating

Figure 4 Schematic arrangment for tests

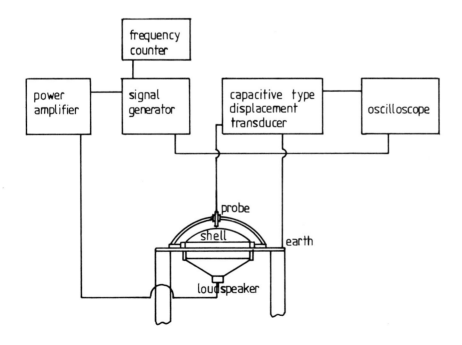

## EXPERIMENTS,RESULTS AND COMPARISONS

Five different shell models were tested, and care taken in the manufacture, mounting and instrumentation of the models has allowed: (a) comparisons with the results obtained from theoretical predictions obtained from classical shell theory;(b) an investigation of the nonsymmetric modes of vibration of spherical shells;(c) careful assessment of the dynamic characteristics of modes in which resistance is dominated by membrane strain energy.

Three of the models were specially constructed for this work.

Figure 5    Theoretical (———) and experimental fit (---) of recorded (+ ) mode-shapes, shell VC2

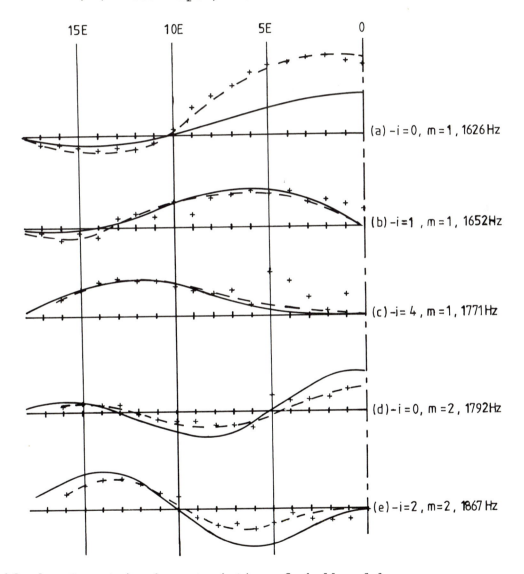

Table 1    Geometric characteristics of shell models

| Shell | Half-opening angle ($\phi$N) | Thickness (h) | Radius (R) | h/R |
|---|---|---|---|---|
| VC1 | 10° | 0.127mm | 475.4mm | 1/3700 |
| VC2 | 9.26° | 0.264mm | 475.4mm | 1/1800 |
| VC3 | 43° | 0.102mm | 116.75mm | 1/1140 |

For these, the Young's modulus was $1.8 \times 10^2$ kN/mm$^2$, Poisson Ratio 0.3, specific gravity 8910 kg/m$^3$ and the geometric characteristics are listed in Table 1. Of these models shells VC2 and VC3 presented no significant variations of thickness. Shell VC1 showed a thickness variation from 0.135mm near the support to 0.124mm near the pole. The average value of 0.127mm was taken to compute the theoretical natural frequencies and associated mode shapes.

The behaviours of these shells are exemplified by the results of Table 2, where the observed resonant frequencies are compared with the theoretical predictions. Representative mode-shapes, obtained for shell VC2, are depicted Fig.5.

The other two shells tested were available from earlier buckling studies carried out by Tillman(6,7). They were both very shallow ($\phi_N$= 2.81° and R$\cong$ 1525mm; $\phi_N$= 3.13° and R$\cong$1397mm), and were used in the form of feasibility-development tests in order to perfect the experimental set-up. Both shells were significantly imperfect. One shell,VT1, was badly mounted in its support ring, and showed deviations from a true spherical surface due to improper handling of the model in its storage. It had significant variations in thickness throughout its surface, ranging from 0.09mm at the apex to 0.13mm near the support implying that h/R varied from 1/17000 to 1/11700 $^*$. The other shell, VT2, showed the same imperfections as shell VT1. For this shell, thickness variations were much more extreme, ranging from 0.10mm to 0.188mm, with a central region near the pole much thinner than the rest of the shell. Furthermore, this reduced thickness region near the pole showed considerable flattening.

Table 2  Experimental resonant frequencies and theoretical predictions
R=(fexp-ftheo)/fexp

| SHELL MODE (i,m) | VC1 | | | VC2 | | | VC3 | | |
|---|---|---|---|---|---|---|---|---|---|
| | ftheo (Hz) | fexp (Hz) | R (%) | ftheo (Hz) | fexp (Hz) | R (%) | ftheo (Hz) | fexp (Hz) | R (%) |
| 0,1 | 1566 | 1529 | -2.42 | 1619 | 1626 | 0.43 | 6091 | 6126 | 0.57 |
| 0,2 | 1609 | 1595 | -0.88 | 1799 | 1792 | -0.31 | 6381 | 6410 | 0.45 |
| 0,3 | 1704 | 1696 | -0.47 | - | - | - | - | - | - |
| 0,4 | 1872 | 1848 | -1.30 | - | - | - | - | - | - |
| 1,1 | 1576 | 1550 | -1.68 | 1647 | 1652 | 0.30 | 6222 | 6281 | 0.94 |
| 1,2 | 1635 | 1627 | -0.49 | 1802 | 1806 | 0.21 | 6414 | 6475 | 0.94 |
| 1,3 | 1736 | 1724 | -0.70 | 2049 | 2036 | -0.73 | - | - | - |
| 1,4 | 1853 | 1824 | -1.59 | - | - | - | - | - | - |
| 2,1 | 1587 | 1561 | -1.67 | 1682 | 1685 | 0.13 | 6329 | 6357 | 0.44 |
| 2,2 | 1662 | 1645 | -1.03 | 1874 | 1867 | -0.37 | 6442 | 6515 | 1.12 |
| 2,3 | 1773 | 1755 | -1.03 | - | - | - | - | - | - |
| 2,4 | 1920 | 1911 | -0.47 | - | - | - | - | - | - |
| 3,1 | 1601 | 1576 | -1.59 | 1726 | 1718 | -0.41 | 6366 | 6392 | 0.41 |
| 3,2 | 1690 | 1669 | -1.26 | 1999 | 1996 | -0.20 | 6453 | (6546) | 1.42 |
| 3,3 | 1817 | 1792 | -1.40 | - | - | - | - | - | - |
| 4,1 | 1619 | 1612 | -0.43 | 1788 | 1771 | 0.95 | 6389 | 6450 | 0.95 |
| 4,2 | 1723 | 1712 | -0.64 | - | - | - | 6467 | (6564) | 1.48 |
| 4,3 | 1882 | 1867 | -0.80 | - | - | - | - | - | - |
| 5,1 | 1641 | 1631 | -0.61 | 1880 | 1885 | 0.27 | 6407 | 6465 | 0.90 |
| 5,2 | 1766 | 1744 | -1.26 | - | - | - | - | - | - |
| 6,1 | 1667 | 1654 | -0.79 | 2006 | 2016 | 0.49 | 6421 | 6491 | 1.08 |
| 6,2 | 1822 | 1803 | -1.05 | - | - | - | - | - | - |
| 7,1 | 1701 | 1685 | -0.95 | - | - | - | 6435 | 6502 | 1.03 |
| 7,2 | 1904 | 1876 | -1.49 | - | - | - | - | - | - |
| 8,1 | 1744 | 1731 | -0.75 | - | - | - | 6451 | (6525) | 1.13 |
| 9,1 | 1797 | 1784 | -0.73 | - | - | - | 6467 | (6573) | 1.61 |
| 10,1 | 1862 | 1849 | -0.70 | - | - | - | - | - | - |
| 11,1 | 1940 | 1926 | -0.73 | - | - | - | - | - | - |

The behaviour of these thin shells displayed characteristics which
for the relatively thicker shells, and the limited power of the
vibratory source, could not be duplicated. This consisted of a
potentially serious, non-linear, interaction between two or more
modes in which the resistance was probably dominated by membrane
energy. Fig.6 shows curves of displacement against energy input for
the shell VT1, at the resonant frequencies of 460.4Hz and 533.8Hz.
At 460.4 Hz a non-linear behaviour of the displacement response,
measured in Fig.6 with the probe located at the apex, is accompanied
by a "beating" with a much lower frequency. It was not possible to
stabilize the oscilloscope to accurately record this lower frequency,
but it appeared to be in the region of 150 to 155Hz. At 460.4Hz.
there is a very short region I, where the out-of-plane displacement
at the apex has a linear response at the same frequency as the input
of energy. This is followed by region III, where the appearance of
the second frequency in the response results in increasingly non-
linear behaviour in displacement as the input energy is increased.
For the 533.8Hz response the picture is similar. In this case
however, the linearity of displacement in region I persists even
when the response curve shows a coupling with the lower frequency.
Eventually in region III an apparent loss of stiffness in the
resonant mode occurs as the input energy is increased. For both
resonant conditions, once the displacement had reached the value of
around 100 to 110μm(the same order of the shell's thickness) it
became impossible to observe any kind of periodicity in the response.
It is possible that this was either due to significant contributions
of more than two modes in the overall response or due to a condition
of dynamic instability of the structure. This form of non-linearity
was also observed for shell VT2 and at other frequencies for shell
VT1.

Figure 6 Displacement against energy input, at resonant frequencies
         460.4Hz and 533.8Hz for shell VT1

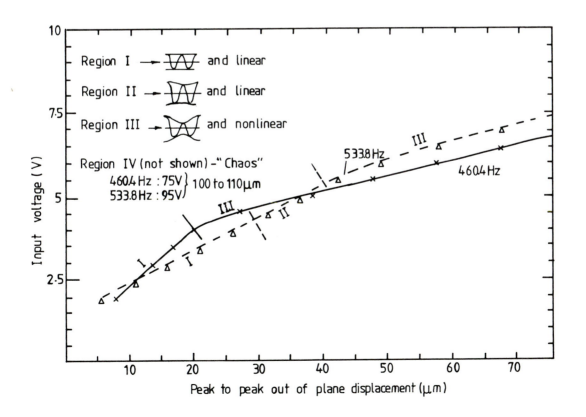

## CONCLUSIONS

The tests have shown that use of accurate models and an appropriate choice of instrumentation can lead to experimental results in close conformity with predictions from classical thin shell theory. This agreement has been shown to extend to the non-axisymmetric modes, first detected but not analysed by Eikrem and Doige (9).

The models tested have allowed observation of very shallow shells ($\phi_N = 2.81^o$ and $\phi_N = 3.13^o$) in which membrane action dominated (shells VT1 and VT2); shallow shells ($\phi_N = 9.26^o$ and $\phi_N = 10^o$),where bending provided a significant contribution to the overall resistance of the shell; and a deep shell ($\phi_N = 43^o$) which, for the modes observed had almost all of its strain energy arising from membrane contributions. It was observed that of the imperfect shells (VT1,VT2 , and VC1) it was shell VC1, which was the one exhibiting a significant parcel of bending strain energy, that was relatively insensitive to imperfections.Furthermore, it was shown how at a resonant condition for the very shallow membrane dominated shells the periodicity can be lost on account of a non-linear coupling between modes.The results suggest that modes that derive a large part of their strain energy from membrane action are more susceptible to potential destabilisation arising from non-linear mode interactions than are the bending dominated modes. Limitations in the power input of the excitation system made it impossible to confirm this for all the shell models.

Although the present tests have increased confidence in the use of linear classical thin shell theory for very small vibrations,further investigations appear to be necessary to elucidate some of the observed non-linear dynamics.

## ACKNOWLEDGEMENTS

V.C.M. de Souza would like to thank the National Council for Scientific and Technological Development,Brazil (CNPq) for the grant that made this work possible (Proc.30.0689/81).

1, DE SOUZA,V C M ,CROLL,J G A , 'An energy analysis of the free vibrations of isotropic spherical shells', Journal of Sound and Vibration, Vol.73, No.3,(1980),pp 379-404.

2, DE SOUZA,V C M ,CROLL,J G A , 'Free vibrations of orthotropic Spherical shells', Engineering Structures, Vol.3,(April,(1981), pp 71-84.

3, HWANG,C , 'Some experiments on the vibrations of a hemispherical shell', Journal of Applied Mechanics, Vol.33,No.4 (1966).

4, KALNINS,A, 'A discussion of some experiments on the vibration of a hemispherical shell', Journal of Applied Mechanics,Vol.34 No.3(1967).

5, GLOCKNER,P G ,TAWADROS,K Z ,'Experiments on free vibration of shells of revolution',Experimental Mechanics,Vol.13, No.10(1973).

6, TILLMAN,S C , 'The manufacture of accurate thin shells by the electro-deposition of nickel', University College London,1969.

7, TILLMAN,S C ,'On the buckling behaviour of shallow caps under a uniform pressure load', International Journal of Solids & Struct. Vol.6,(1970).

8, DE SOUZA,V C M , 'Vibrations of Orthotropic Spherical Shells',PhD, Thesis, University of London,(1980 ).

9, EIKREM,A K A ,DOIGE,A G ,'Natural Frequencies of a hemispherical shell', Experimental Mechanics,Vol.12,No.12,(1972).

# Simplified analysis and design for mechanical transient behaviour of structures and fluids

## 35

B Saravanos and K F Allbeson
*National Nuclear Corporation Ltd, UK*

Summary

A simplified, approximate analytical model is proposed for the prediction of the transient mechanical response of fluids and structures, accounting where appropriate for interaction effects. A commensurately simple experimental programme for validation is described. The analysis is based on the use of unimodal response with Lagrangian equations of motion. The results of a parameter survey are presented as related to the behaviour of a fast reactor primary containment subjected to a hypothetical core-disruptive accident.

Some experimental work is described on an annular plate subjected to a transient uniform pressure, from which it is shown that the plastic mode shape is virtually time-independent.

Design implications of analytical methods are discussed.

## INTRODUCTION

The use of large-scale computer programs for the analysis of the response behaviour of structures, and also fluids, is common industrial practice, especially for technologically advanced projects. Finite-element codes typify this approach; and at an appropriate stage in design development the suitability of a proposed design is explored experimentally by means of commensurately elaborate physical models.

There are disadvantages to these practices. The computational codes tend to embody levels of analytical complication which are inconsistent with the limited possibilities of accurate description of physical behaviour. These limitations are imposed largely by the inadequacies of the experimental information on materials behaviour required for mathematical and computational modelling. An example of such difficulties occurs for multi-dimensional elasto-plastic behaviour of metal continua when subjected to force systems which induce internal stress and strain components with time histories which differ for each set of components along, say, orthogonal axes. It is not practicable previously to store experimental information on multi-dimensional plastic behaviour with arbitrary component time histories for subsequent use in a given analytical case. Moreover, in the general case, available constitutive relationships are inadequate; and the post-yield behavioural models based on the theories of von Mises, Hencky, Tresca, et al, are defective also in the present context, and their use in elaborate computational codes with large potential capacity for accurate and detailed predictions is consequently inappropriate. A further disadvantage of large-scale codes concerns their unwieldiness and cost of operation when parameter surveys are required and when frequent changes in early design need to be analysed.

It would follow that simplified, approximate analysis and experiments are appropriate to meet needs for day-to-day design analysis. One such simplified approach is described in this paper, the application of which is related to the design of the primary containment of a fast-breeder reactor when exposed to the transient mechanical effects of a hypothetical core-disruptive accident.

## GENERAL DESCRIPTION OF THE SIMPLIFIED ANALYTICAL METHOD

The structural components and fluids are assumed to undergo transient displacements u, which are expressed in terms of generalised co-ordinates $u_k$ and, where required, mode shapes $\eta_k$, so that:

$$u = \eta_k u_k \qquad (1)$$

It is possible in the event that $\eta_k$ is time-dependent. Kinetic, potential (fluids), and elasto-plastic strain (structures) energies are formulated in terms of $\eta_k$ and $u_k$, from which equations of motion are formulated from Lagrangian equations of the form

$$\frac{d}{dt}\left(\frac{\partial T}{\partial \dot{u}_k}\right) + \frac{\partial U}{\partial u_k} = Fu_k \qquad (2)$$

Because of the use of mode-shapes $\eta_k$, the number of equations arising for solution are advantageously limited, and because of the use of Lagrangian equations finite displacements can be evaluated without numerical problems of instability and inaccuracy. A typical expression for kinetic energy is:

$$T = \frac{1}{2} m (\dot{u}_k)^2 \qquad (3)$$

Typical expressions for potential energy of the fluids, assumed to behave adiabatically, are:

$$U = -\int_{V_o}^{V} p \, dV \qquad (4)$$

where volumes V are expressed in terms of the generalised co-ordinates and account, where appropriate, for fluids-structures interaction.

$$
\left.
\begin{array}{l}
p(\text{LIQUID}) = - K.LN(V/V_o) + p_o \\
p(\text{GAS}) = p_o(V/V_o)^{-\gamma}
\end{array}
\right\} \qquad (5)
$$

The strain energies of the structural components are expressed as:

$$
\left.
\begin{array}{l}
U(\text{ELASTIC}) = K_k U_k^2 \\
U(\text{PLASTIC}) = K_k U_k^{n+1}
\end{array}
\right\} \qquad (6)
$$

where the stiffnesses $K_k$ are derived, for a given structural component, using the following "true" stress-strain relationships:

$$
\left.
\begin{array}{l}
\sigma(\text{ELASTIC}) = E\varepsilon \\
\sigma(\text{PLASTIC}) = A\varepsilon^n
\end{array}
\right\} \qquad (7)
$$

and using the von Mises post-yield behavioural model.

A switching procedure is introduced for transformation from the elastic to the plastic condition, and vice versa. Further details of the simplified method are given in References 1 and 2.

## NUMERICAL ANALYSIS AND PARAMETRIC SURVEY

An idealisation of the axisymmetrical primary containment of a pool-type fast breeder reactor is shown in Fig. 1. The primary vessel is assumed to be rigid and to transmit downward-acting bubble forces to the roof. The ferritic steel roof and shield behave elasto-plastically in biaxial bending in two design configurations: with pinned attachment of roof to shield ("roof-shield"); with this attachment providing a radial bending moment path ("circular plate"). The fluids are taken to be compressible and to undergo one-dimensional vertical displacements. The mechanical energy available from the expanding core bubble is taken at two levels: 1 GJ and 1/2 GJ, both of which are considered to be pessimistic for design purposes.

### Results of the Calculations

The calculated results are summarised in Figs. 2 and 3. The corresponding roof and shield peak interaction pressures are given as follows, with a rise time in each case of approximately 50 ms:

| Bubble Energy | Model | Peak Pressure (MPa) |
|---|---|---|
| 1 GJ | (Roof-shield | 11.7 |
| | (circular plate | 7.7 |
| 1/2 GJ | (Roof-shield | 4.0 |
| | (circular plate | 3.3 |

The curves of Figs. 2 and 3 indicate that deflections reach an asymptote with roof depth H, so that the design can be optimised.

## THE EXPERIMENTAL PROGRAMME

The purpose of the experimental approach is to provide, by means of small, very simple models, a physical verification of the theoretical analysis, in the following stages:

(i)   elasto-plastic response behaviour of plates, annuli, cylinders, and hemispheres, when subjected to forces with known levels and time histories

(ii)  elasto-plastic structural response behaviour with compressible fluids-structures interaction for: rigid vessel, flexible roof: rigid roof, flexible vessel; flexible roof and vessel. All experiments are to be accompanied by carefully controlled tests to provide stress-strain data on the materials actually used for the models.

### Some Results of the Experimental Programme and Corroborative Finite-Element Analysis

The simplified method described previously has been confirmed for annular plates subjected to time-dependent pressures by finite-element calculations of post-yield behaviour.

Experimental work is at an early stage of development and currently includes provision of uniaxial stress-strain data for aluminium, and for austenitic and ferritic steels. Some preliminary tests have been carried out on a steel annulus with simply-supported outer edges and on an aluminium circular plate with encastré edges.

The experimental results on the steel annulus have confirmed the virtually linear, time-independent mode shape predicted theoretically, as indicated in the photographs of Fig. 5. A C.R.T. recording of the pressure-time history is reproduced in Fig. 4; the scale for these pressures was unfortunately unresolved.

Certain experimental difficulties have been highlighted: short pressure rise-times have been difficult to achieve; sealing problems for simply-supported edges induce pressure fade; plastic response is highly sensitive to the shape of the stress-strain curve for true strains >0.05, so that a high level of accuracy will be needed to correlate with analysis.

## DESIGN IMPLICATIONS OF ANALYTICAL MODELS

The primary containment for the fast reactor, consisting of a primary vessel 35 mm thick x 19.2 m dia and a steel roof plate and web structure 3 m deep x 22.4 m dia, is designed to meet a wide range of loadings arising from both normal and abnormal operating conditions. The resulting design is also assessed against the transient mechanical loading arising from the hypothetical core disruptive accident, the HCDA. Alternative forms of design would be of reinforced and/or prestressed concrete: an essential pre-requisite here is the effect of fluids-structures interaction, which is not known to have been investigated as yet. For this assessment two general approaches have been tried. The first uses the large computer programme approach involving a         programme modelling the transient behaviour of a 2D compressible fluid in response to a given sodium fuel vapour bubble expansion, with boundary conditions imposed to represent the containment vessel and roof. The resulting loadings are subsequently applied to a 3D structural analysis programme modelling the response of the roof.

This analytical approach was supported by an experimental programme, which started with a simple representation of the containment boundary and proceeded by a series of elaborations to a fairly detailed model of the then current design.

In addition to the disadvantages noted in the introduction this approach cannot meet the designers' requirement for a ready means of assessing the effects of changes in parameters on the containment HCDA capability as the reactor system develops generally in related areas.

In recognition of this need NNC have also pursued a second course, the simplified 1D fluid-structure interaction model discussed in this paper, in which the containment boundary is represented by a flexible roof and a rigid primary vessel.

This approach, albeit not as yet completely experimentally validated has been considered successful to the point where a first degree of roof optimisation has been based upon the results of a parametric survey carried through with this programme, referred to as "FLUSTR" (see Figs. 2 and 3). It is acknowledged that some degree of pessimism must be allowed for in the 1D treatment and in the modelling of a rigid primary vessel boundary. In this respect developments of FLUSTR to model a flexible vessel are in hand. There is also reason to believe that the 1D treatment is not a serious drawback to the use of the programme as a design tool, enabling the containment capability to be reassessed and modified as the basic design against operational loading proceeds.

Consistent with this simplified approach the experimental validation of FLUSTR has also been considered as a step-by-step modelling process in which the response of simple structural forms employed in the containment design have and are being subject to transient mechanical loads (e.g. Fig. 5). This process, if carried forward to models of increasing complexity, will achieve two objectives:

(a)    in the early phase, demonstrate  the validity of the FLUSTR predictions and also builds up confidence in the large 3D structural programme and

(b)    in the later, more detail modelling phase, validate  the increasingly complex 3D analytical model as it approximates more closely to the actual design.

Because of inherent limitations to this approach the validation of the final design is seen to be through the agency of a dynamic test on a detailed model of the final design.

DISCUSSION

Modelling for the kind of problem here discussed is unlikely to enable accurate prediction and assessment to be made of response behaviour of fluids and structures. Experimentally, scale effects present formidable difficulties, especially for concrete designs. Experimental test facilities can be expensive, for example, the provision of means for achieving short rise times for the applied forces. The provision of experimental information on the behaviour of steel and concrete materials and structures is unlikely to be adequate for accurate response predictions.

For these reasons a simplified experimental and analytical approach may well be appropriate which gives an approximate indication of global response and smears detailed behaviour. A pessimistic estimate of response behaviour can be made acceptable for design purposes for the minimum level of complication in design idealisation.

REFERENCES

1.    SARAVANOS, B, Simplified methods of analysis for design of primary containment for a commercial fast reactor against a core disruptive accident ANS/ENS Conference, Seattle, (August 1979)

2.    SARAVANOS, B, and DANIELS, M, Transient fluids-structures inter- action effects on the roof of a fast reactor under a core- disruptive accident  SMIRT 6, Paris, (August 1981)

ACKNOWLEDGEMENTS

Thanks are due to Dr. J. Smart and Messrs. M.J. Harris, K.S. Crentsil and J. Abrefah of Simon Engineering Laboratories, Manchester University, for kind permission to reproduce some of their experimental work.

The views expressed in this paper are those of the authors.

Figure 1 Axisymmetrical Primary Containment Idealization

Figure 2 Parameter Survey

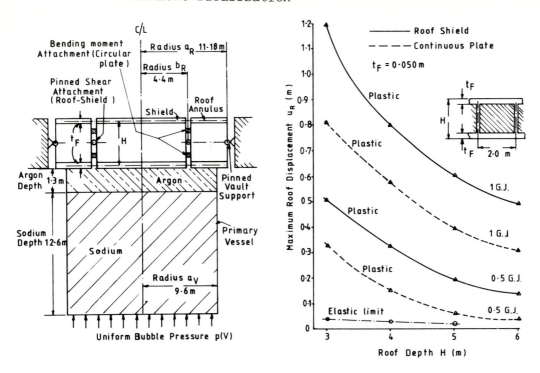

Figure 3 Parameter Survey

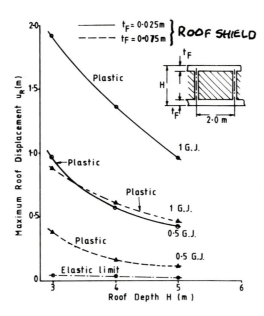

Figure 4  C.R.T. Trace of Pressure/Time History for
          Annular Plate (Manchester University)

## Scales:- For time 't' 1 div = 0·1 sec

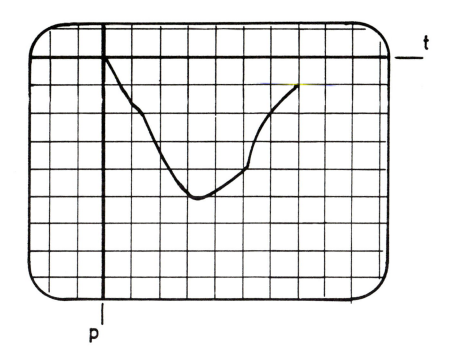

Figure 5  Steel Annular Plate After Plastic
          Deformation (Manchester University)

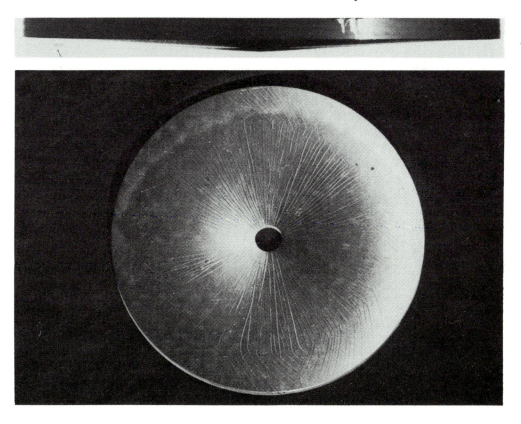

# Contribution to discussion of Paper No. 35

Professor Norman Jones
*Department of Mechanical Engineering, The University of Liverpool, UK*

The writer wishes to make some observations about the results reported in Figures 4 and 5 for an annular plate loaded dynamically.

It appears that the rise time for the pressure pulse is slightly over 0.3 sec while the total duration is 0.7 sec, approximately. The writer wonders if the response is therefore quasi-static which would give rise to the time-independent deformation profiles observed in the experiments.

Lance and Onat (J. Mech. Phys. Solids, Vol. 10, 301-311, 1962) observed that the Lueder's bands on the surface of a circular plate, fully clamped around the outer boundary and loaded uniformly, were radial in a central zone with an outer radius equal to 0.73 of the plate radius and had the form of logrithmic spirals in the outer zone. Similarly, it may be shown that the rigid-plastic analysis of an annular plate which is loaded uniformly and simply supported around the outer boundary (L. W. Hu, proc. A.S.C.E., J. Eng. Mech. Div., Vol. 86, No. EM1, 91-115, 1960) predicts radially orientated Lueder's bands on the plate surface. Thus it would appear that the disturbance to the radial pattern on the annular plate surface in Figure 5 is due to some clamping which prevents the development of a simply supported edge. Indeed, it should be possible to estimate the degree of clamping from the width of the zone.

# Contribution to discussion of Paper No. 35

B Saravanos and K F Allbeson
*National Nuclear Corporation Ltd, UK*

AUTHORS REPLY

Professor Jones' comments are appreciated.

It is true that the rise time associated with the plate of Figures 4 and 5 is too long for a "transient" effect. As we pointed out in our paper, this is due to initial difficulties in achieving experimentally short rise times. Methods of overcoming them are currently being investigated. Nevertheless, the experimental results we presented are not without relevance to our nuclear reactor design problems. Thus, our assumption of a nearly linear plastic mode shape calculated statically is validated as a phenomenon of time-history or "memory" in the plastic regime. Furthermore, in studying the reactor excursion phenomenon relative to improved, and consequently less onerous, core-bubble behaviour we need to consider the contingency of a slowing-down of the bubble expansion which might be accompanied by lower pressure levels but also by roof displacement responses which are removed from regimes associated with transient resonance. A general indication of effects such as increased rise time is given in Figure A. There is in this context an associated problem of elastic modal excitation in those spatial domains of the roof to which plasticity has not yet reached, accompanied simultaneously by the plastic response of the remaining regions.

Currently, we have some theoretical corroboration of our simplified model from comparative studies with finite-element codes. Typical calculational results are presented in Figure B. It may be noted here that although correlation is good, all calculations were based on the von Mises-Hencky post-yield behavioural model, which may not be adequate for complicated loading systems, so that finally experimental validation is required.

Regarding the clamping effects referred to by Professor Jones we have reason to believe that some rotational restraint on the boundaries might have occurred from the presence of O-rings, but that this is probably slight. Figure C indicates that response behaviour is significantly affected by the effects of overhang, which may well have offered constraint leading to the Lüder bands.

Thanks are due to Dr. J. Smart of Simon Engineering Laboratories for the information in Figure C; and to Mr. V. Washby of UKAEA (Risley) for the EURDYN 3 curve on Figure B.

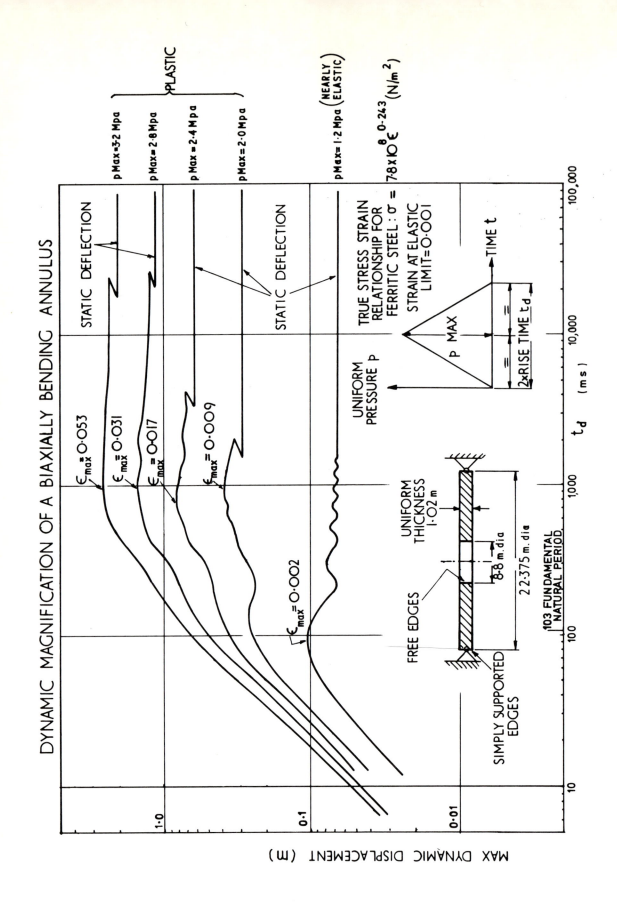

Figure A  —  Dynamic Magnification of a Biaxially Bending Annulus

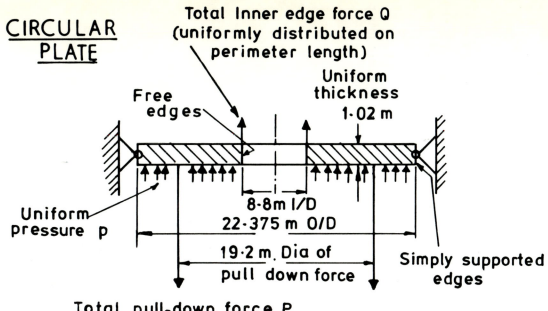

# CIRCULAR PLATE

Total Inner edge force Q
(uniformly distributed on
perimeter length)

Free edges

Uniform thickness 1·02 m

Uniform pressure p

8·8m I/D

22·375 m O/D

19·2 m. Dia of pull down force

Simply supported edges

Total pull-down force P

(uniformly distributed on
perimeter length)

## PLATE MATERIAL

FERRITE STEEL $\sigma = 7\cdot8 \times 10^{8} \, \epsilon^{0\cdot243}$ N/m$^2$

$E = 2\cdot01 \times 10^{11}$ N/m$^2$

STRAIN AT ELASTIC LIMIT = 0·001

Figure B — Comparative Calculations of Numerical Models

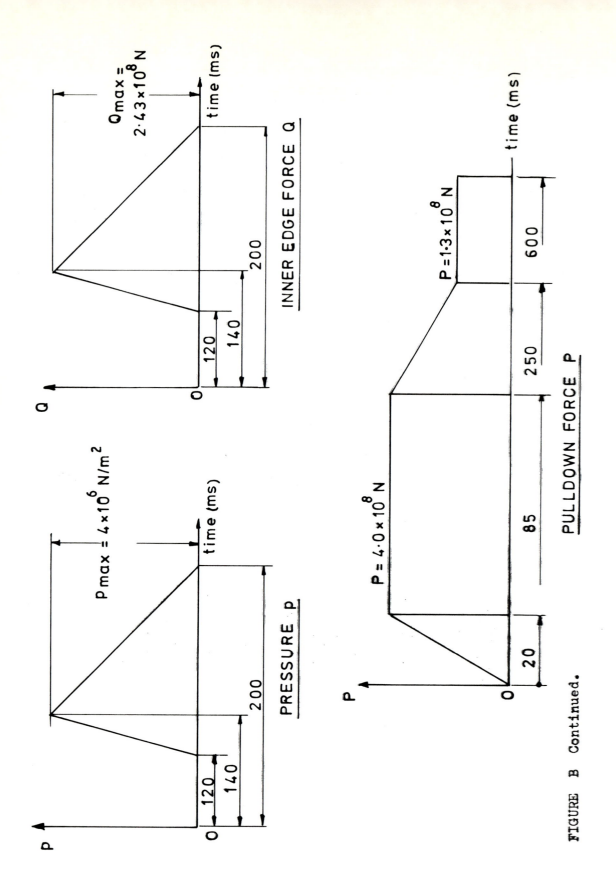

FIGURE B Continued.

340

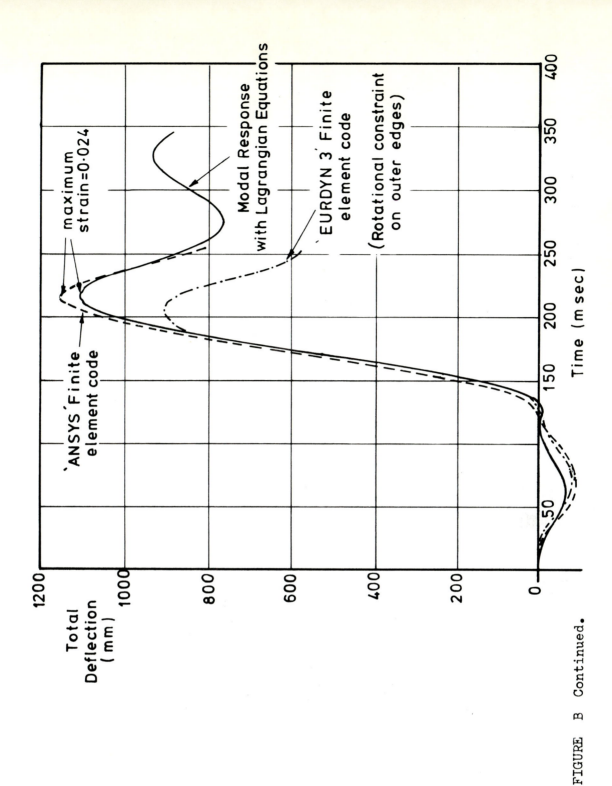

FIGURE B Continued.

341

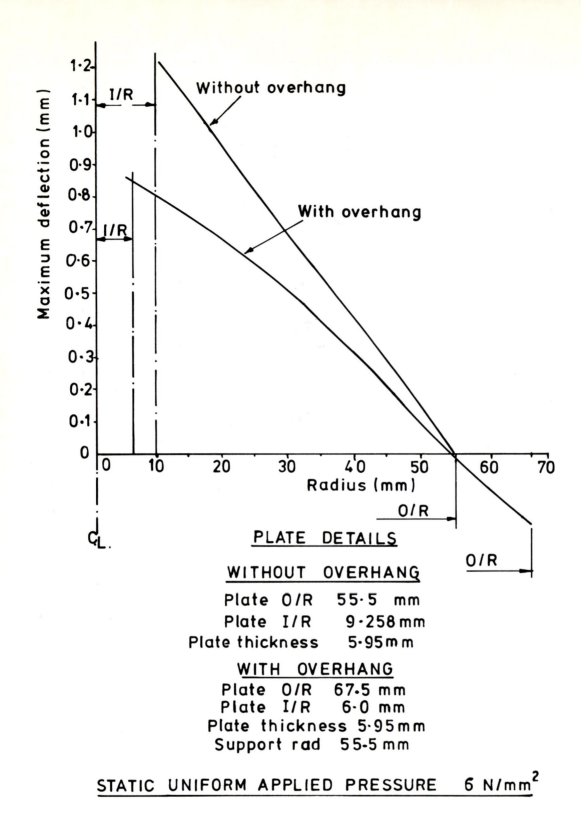

PLATE DETAILS

WITHOUT OVERHANG

Plate O/R    55·5  mm
Plate  I/R    9·258 mm
Plate thickness    5·95 mm

WITH  OVERHANG
Plate  O/R   67·5 mm
Plate  I/R    6·0 mm
Plate thickness 5·95 mm
Support rad   55·5 mm

STATIC UNIFORM APPLIED PRESSURE    6 N/mm$^2$

Figure C  –  Effects of Overhang on Static Response of an Annulus

342

# Load transfer and vibrational characteristics of steel joist-concrete slab floor systems

## 36

Francis M Thomas
*University of Kansas, USA*

## I    INTRODUCTION

Steel joist concrete floor systems are commonly used in construction because they allow creation of large uncluttered floor areas.  This type of construction is low in cost due to its lightweight joist members which require no special equipment for erection.  The light-weight characteristics possessed by steel joist-concrete floor systems are obtained by replacing heavy steel beams, rebar reinforcing and thick short spanned concrete floors used in conventional floors with lightweight economical joists, wire mesh reinforcing and thinner concrete slabs.  As a result of this reduced mass, span lengths can be increased accordingly giving way to the desired open space atmosphere.  The combination of long span design coupled with open area use of floor space has lead to floor systems which are structurally sound but have objectional vibration characteristics.

The movement of human occupants is the primary source of objectionable vibration for steel joist-concrete floor systems.  A person walking down a hallway or working at his desk can and often does excite the primary mode of vibration in the floor.  Because of its poor damping qualities, a vibration of long enough duration and high enough amplitude is produced to be annoying to others on the floor.  Since the source of the vibration cannot be eliminated or isolated, it is necessary that the floor system be designed so that movements by human occupants do not cause objectionable vibrations.

## II    PAST WORKS AND CURRENT INVESTIGATION

In order to determine if a proposed design of a steel joist-concrete slab floor system will exhibit objectionable vibration characteristics, the vibration displacement amplitude and frequency resulting from a human impact on the floor must be predicted.  Based on these quantities, the degree of human perception to the vibration may be predicted using the criteria presented in (3) and shown in Fig. 1.

The Steel Joist Institute commissioned Dr. Kenneth Lenzen to investigate the problem of vibration experienced in buildings with steel joist-concrete floors.  Through research done in various projects, Lenzen was able to develop theoretical floor responses for various dynamic impacts.  The amplitude of floor deflection, frequency of motion and the percentage of critical damping can be predicted. Also investigated was the measurement of dynamic impacts and their perception by humans at various distances from the impact force. Moderow (4) also presents a general "state of the art" analysis for joist-concrete floors.  The experimental work by both of these investigators is restricted to floors in which the ends of the joists are supported by rigid supports which yield no movement upon impact.

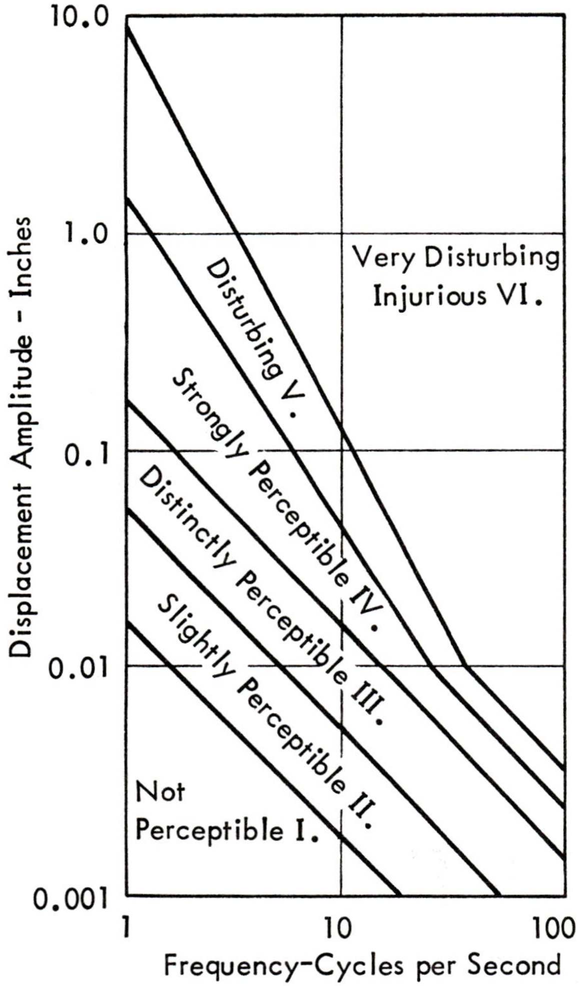

Figure 2   Test configuration of steel joist-concrete
floor system

Figure 1   Criteria for human perception
to transient vibration.

Such a support would exist when the joists rest on masonary walls.

This investigation is concerned with the vibrational behavior of a joist-concrete floor system in which the ends of the joists are supported by a flexible simply supported I beam rather than a fixed support as was done in the previous investigations. The intent of this configuration was to more closely model an actual floor as it might be found in a steel frame building. The ends of the joists were allowed to translate with the I beam and rotate. The previous work considered only the relationships between the applied loads and the resulting vibrations. This investigation has as one of its objectives the determination of the load transfer between the floor system and the supporting structure. This was accomplished by placing load cells between the end of each joist and the I beam supporting the joists as well as a load cell between the end of the I beam and the fixed base support. The other objective of this investigation was to determine the effect of the flexible supports on the frequency and amplitude of the response of the floor system and to compare the results with an untested relationship presented in (4) for floor systems with flexible supports.

III  TEST CONFIGURATION AND LOADING

A steel joist-concrete slab floor was constructed for the purpose of experimentally measuring its vibrational behavior. The floor was 20 feet x 20 feet and the slab was 3 inches thick (Fig. 2). Eleven 20 feet long 12H3 joists were used. The joist spacing was 24 inches. Standard corruform centering conforming to building code specifications was spot welded to the top of each joist and served as a base for the concrete.

The ends of the joists were supported by 20 foot simply supported W14x34 I beams. Load cells were positioned between the ends of the joists and the I beams and between the I beam and the support below the entire flooring system.

The floor was loaded with a human impact to simulate people walking on the floor. This loading procedure known as the "heel drop" is accomplished by rocking on the balls of the feet, lifting the heels approximately 2 to 2½ inches off the floor, then relaxing and striking the heels against the floor slab. The heel drop was performed at the floor's center and could be reproduced with some degree of consistancy.

Fig. 3. shows the actual force time curve created by the heel drop procedure along with a transformed triangular impact based on equivalent areas under the load time curve. This equivalent loading is used in the theoretical response calculations of the floor system. The actual load reached a peak of 905 lbs at 15 milliseconds. The loading based on equal area under the curve reaches a max load of 847 lbs. The duration of both impacts was 47 milliseconds.

Once the floor had been constructed the objective of the experimental investigation was to measure the vibration frequency and amplitude occurring throughout the flooring system as a result of the heel drop impact at the center of the floor. It was also desired to measure the force time relationship that exists between the end of each joist and the simply supported I beam and also the force-time relationship that exists at the base of the simply supported I beam. Once the measurements were made they were compared with the existing theory so that the effect of the flexible supports for the joists could be

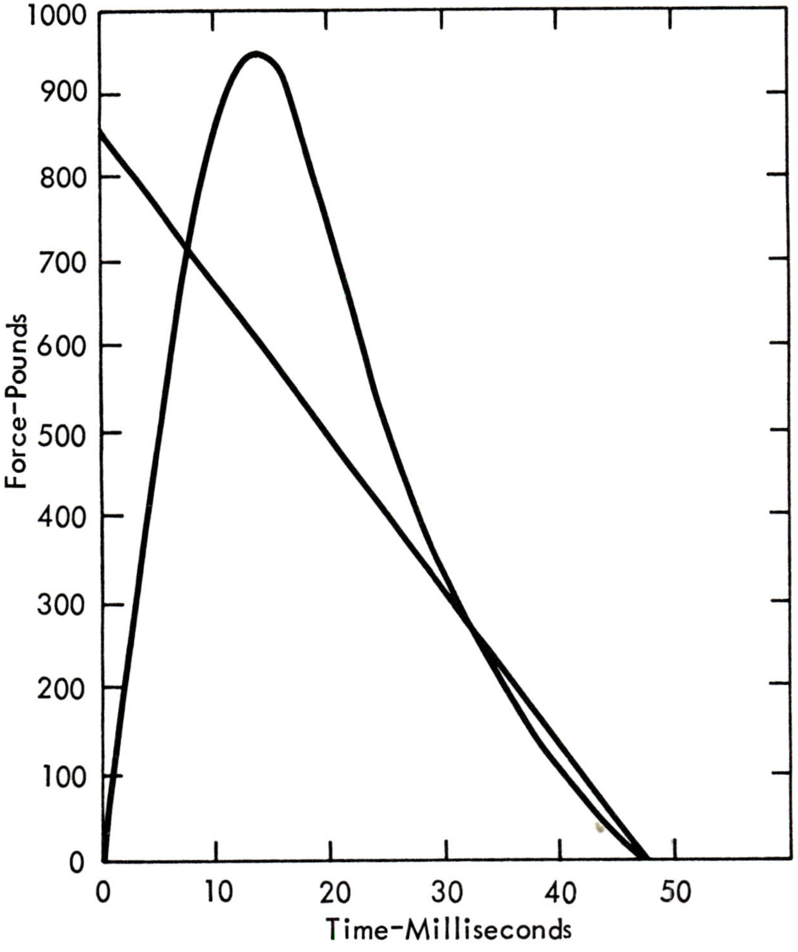

Figure 4  Typical floor response records.

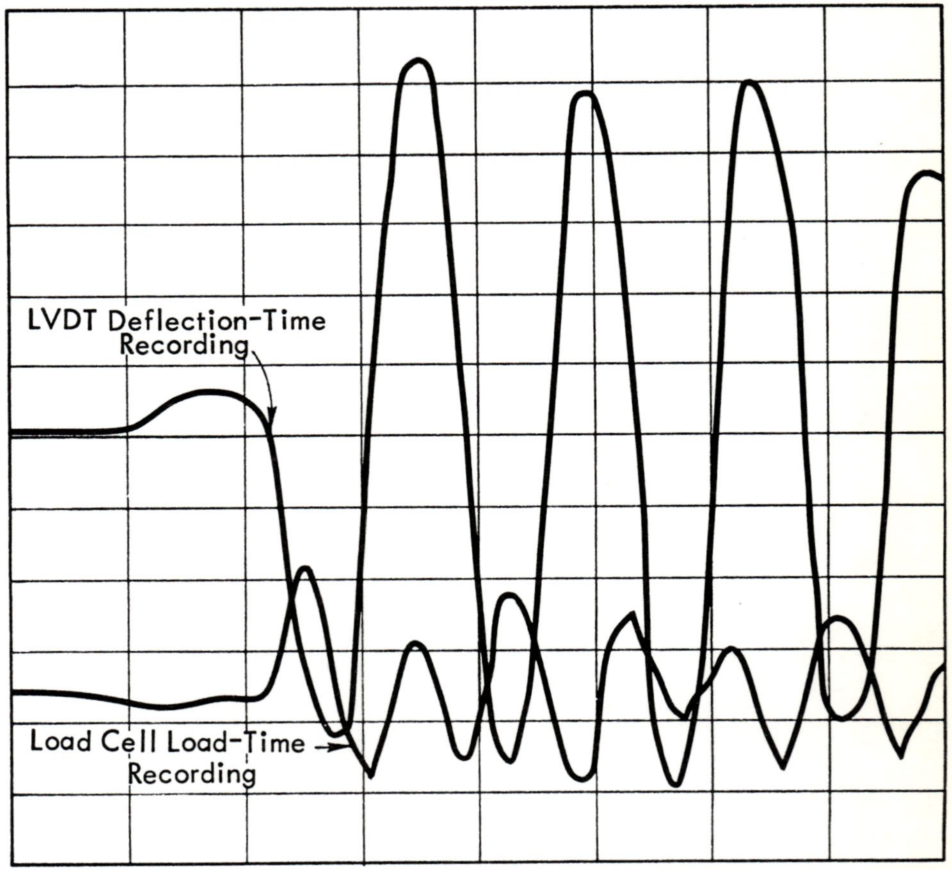

Figure 3  Actual and idealized force-time
curve for heel drop impact.

determined.

The loads were measured using the before mentioned load cells. The movements of the floor were measured using linear variable displacement transducers (LVDT).

## IV EXPERIMENTAL RESULTS AND COMPARISONS WITH EXISTING THEORY

As a result of a human heel drop loading on the constructed floor system, the following observations were made

1. The frequency of the load transferred from the floor joist to the supporting I beam was nearly equal to the frequency of vibration of the floor had it been supported on rigid supports. The measured values of frequency at the load cells between the joists and the I beam ranged from 10.2 to 10.8 cycles per second as compared to a value of 11.05 cycles/sec as predicted by theory in (4) for a floor system with fixed supports. The details of the frequency calculation are given in (5).

2. The total displacement of the center of the floor as measured by the LVDT produced a cyclic but not harmonic response since it was influenced by the steel joist floor that tried to vibrate at its frequency and also by the simply supported I beam which tried to vibrate at its frequency. However, based on 5 cycles of motion, the frequency of the combined beam and joist floor was measured by means of the LVDT to range from 7.7 to 8.3 cycles/sec. This compared to a calculated frequency for the simply supported beam of 32.1 cycles/sec (first mode) and a value of the floor of 11.05 cycles/sec (5). Based on an emperical approach used in (4) but here before not varified, the frequency of the combined system can be determined as

$$\frac{1}{f} = \frac{1}{f_{floor}} + \frac{1}{f_{beam}}$$

where f = frequency of combined floor and beam system (cycles/sec)
$f_{floor}$ = frequency of floor system as it rested on fixed supports (cycles/sec)
$f_{beam}$ = frequency of beam supporting the floor (cycles/sec).
which yields a calculated frequency equal to 8.22 cycles/sec for the floor system under consideration. A typical load versus time and displacement versus time recording is shown in Fig. 4.

3. The displacement of the center of the joists relative to the ends of the joists were measured with the LVDT. The maximum relative deflection of the joist directly below the impact was $1.52 \times 10^{-2}$ inch. The deflection decreased with distance from the impact until approximately the third joist from the impact. Beyond this point the joists did not carry any load resulting from the impact. This distance indicated the effective joist length for the floor system. The calculated maximum deflection of a joist floor resting on rigid supports which was subjected to an impulse load equal to that given the experimental floor was also $1.52 \times 10^{-2}$ inch. The calculated number of effective joists for a floor on rigid supports was calculated to be 7 (4) and (5).

4. The maximum force occurring at the support for the I beam

resulting from the heel drop load was 429 lbs. A series of
modal response analyses as presented in (1) was performed on a
simply supported beam. Different load distributions were
assumed to be acting on the beam to model different load
transfer mechanisms between the joist floor and the I beam.
Each of the loadings vary with time according to the idealized
triangular loading applied to the floor (Fig. 3). The assumed
loading distribution and the calculated maximum load at the
support are shown in Fig. 5.

Figure 5   Modal Analyses for various load
distributions and resulting
maximum end reactions.

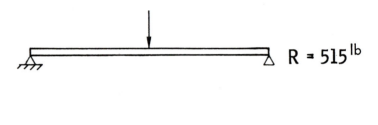

R = 515$^{lb}$

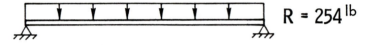

R = 254$^{lb}$

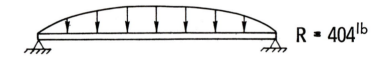

R = 404$^{lb}$

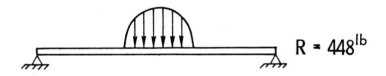

R = 448$^{lb}$

## V   CONCLUSIONS

The principal conclusions made following this investigation are
listed

1   The load cell data indicates that only those joists within the
effective width of the floor (4) transmit load to the supporting I beams. The load distribution varies in a sinusoidal
fashion along the portion of the I beam corresponding to the
effective width of the floor.

2   The emperical untested relationship for determining the
frequency of a joist-concrete floor on flexible supports as
presented in (4) yields satisfacotry results (7.7-8.3 cps
versus 8.22 cps).

3       Dynamic modal response methods in (1) may be used to predict the
        load transferred to the supports below the I beams. A sinu-
        soidal load distribution acting over the effective joists gave
        the best results (429 lb measured versus 448 lb calculated).

4       The equations for floor displacements for a rigid foundation
        floor in (4) yielded accurate results for the deflection of the
        joist relative to the supporting I beam. However, it did not
        correctly predict the total deflection of the floor since the
        total deflection is the sum of the joist deflection plus the
        I beam deflection.

5       Human sensitivity to vibration depends on the degree of
        acceleration experienced and is a function of the floor's
        frequency and amplitude of motion. This relationship is
        discussed by Goldman (3). Since the vibration experienced in
        floor systems is transient in nature, Goldman's perception
        scale was modified for use in floor systems (Fig. 1). Based
        on this criteria for human perception to vibration and the
        vibrational analysis of joist-concrete floor systems supported
        by flexible I beams as developed in this study. The degree of
        preception to vibration may be predicted.

REFERENCES

(1)     Biggs, John M., Introduction to Structural Dynamics, 1969,
        McGraw-Hill Book Company, New York, New York.

(2)     English, L. A., A Report on the Vibrational Characteristics of
        the Open Web Steel Floor System at the Metcalf South Shopping
        Center, Overland Park, Kansas, 1961.

(3)     Goldman, D. E. A., A Review of Subjective Responses to Vibratory
        Motion of the Human Body in the Frequency Range 1 to 70 cps,
        Naval Medical Research Institute, Report NW-004-001, Washington
        1948.

(4)     Moderow, Richard, R., Vibration Characteristics of Steel Joist-
        Concrete Slab Floor Systems, M.S. Thesis, University of Kansas,
        1970.

(5)     Vrooman, Douglas K., Load Transfer and Vibrational Properties
        of a Simply Supported Steel Joint-Concrete Slab Floor System.
        M.S. Thesis, University of Kansas, 1975.

## 37

# The effect of a support pedestal/foundation discontinuity on the dynamics of a rotating system

M Simon and G R Tomlinson
*Simon Engineering Laboratories, Manchester University, UK*

SUMMARY

A rotating machine and its supporting foundation is modelled as a coupled three degree-of-freedom system and the effect on the system's dynamic characteristics of a clearance between the foundation and one of the bearing pedestals, considered as a non-linear stiffness element, using modal analysis methods, is investigated. The investigation employs a simulation computer program which provides the frequency response data from the non-linear model which in turn is analysed using modal analysis curve fitting methods. It is shown that the effect of a discontinuity between the rotating machine and its supporting structure is to modify the mode shapes, resonances and effective damping within the system.

INTRODUCTION

Relatively simple rotating machinery systems, such as large centrifugal fans, often have their operating speeds below the first critical speed of the shaft. The mathematical model of such a system can be related to the coupled foundation translation/rocking modes and the shaft's first critical speed. In this speed zone the governing factors concerning vibration levels are the principal stiffnesses such as the rotor shaft and hub bending stiffness, foundation stiffness, interface stiffness between the bearing pedestals and the foundation and the damping levels associated with these. Frequently abnormal vibration levels are blamed upon such factors as the running speed being near the critical speed or the presence of a foundation resonance at the running speed. It is not uncommon in practice, however, to find excessive vibration levels due to looseness (i.e. clearances) between the holding-down bolts and the foundation. This paper presents some results from an initial study into the effect of such discontinuities between a rotating machine and its foundation using modal analysis methods. The results are purely theoretical although an attempt has been made to use realistic data for the system description. This data has been obtained through collaboration with a major manufacturer of large industrial centrifugal fans.

The particular discontinuity investigated in this initial study is a clearance which is represented as a localised non-linearity in the stiffness connecting the shaft and rotor coordinate to the foundation coordinate. Physically this can be interpreted as a looseness between the bearing pedestal holding-down bolts and the foundation platform. The work shows the application of modal analysis techniques based upon frequency response data to a system with a variable excitation force (mass imbalance) and indicates the changes that occur in the system's dynamic characteristics with a clearance discontinuity.

## MODAL ANALYSIS OF ROTATING SYSTEMS

Modal analysis techniques are well established and have been applied in the identification of structural systems (1). The attraction of this technique is that it is possible to analyse an existing system by using only measured frequency response data without recourse to the possible form of the system's governing differential equations. Briefly, the procedure for obtaining a model of a system from measured frequency response data begins with the familiar matrix equation:

$$[M]\{\ddot{q}\} + [K + jH]\{q\} = \{F\}e^{j\omega t} \tag{1}$$

where $[M]$, $[K]$ and $[H]$ are the mass, stiffness and hysteretic damping matrices of order $n \times n$, and $\{q\}$, $\{F\}$ are the response and excitation vectors respectively of order $n \times 1$. Assuming a solution of the form $\{q\} = \{\hat{q}\}e^{j\omega t}$ and employing the fact that the response vector $\{\hat{q}\}$ can be expressed as a linear sum of the $n$ independent eigenvectors $\{\phi\}^{(r)}$, where the superscript $(r)$ is the $r^{th}$ mode, we can write:

$$\{\hat{q}\} = \sum_{r=1}^{n} \{\phi\}^{(r)} \hat{p}_r \tag{2}$$

where the $\hat{q}$ coordinates have been transformed to the $\hat{p}$ coordinates. Substitution of $\{q\} = \{\hat{q}\}e^{j\omega t}$ and employing the orthogonality principle gives:

$$\{\hat{q}\} = \sum_{r=1}^{n} \frac{\{\phi\}^{(r)^t}\{F\}\{\phi\}^{(r)}}{m_r\omega_r^2[1 - (\omega/\omega_r)^2 + j\delta_r]} \tag{3}$$

where $\delta_r = {}^{h_r}/_{k_r}$ is the hysteretic damping factor.

Since we are usually concerned with the response at some station $j$ due to an excitation at station $k$, equation (3) reduces to:

$$\hat{q}_j = \sum_{r=1}^{n} \frac{\phi_j^{(r)}F_k\phi_k^{(r)}}{m_r\omega_r^2[1 - (\omega/\omega_r)^2 + j\delta_r]} \tag{4}$$

which, in terms of the velocity response, for rotating systems with an excitation force proportional to the square of the speed can be written as:

$$\dot{\hat{q}}_j = \sum_{r=1}^{n} \frac{j\omega^3\psi_j^{(r)}\psi_k^{(r)}me}{\omega_r^2(1 + j\delta_r) - \omega^2} \tag{5}$$

where $m$ is the mass imbalance at a radius $e$ and running speed $\omega$. The velocity response is used for the simple reason that it provides the minimum signal to noise ratio over a wide frequency range.

Each term in the series expression of equation (5) contains complex eigenvector elements ($\psi_{j,k}^{(r)}$) and complex eigenvalues $\omega_r^2(1 + j\delta_r)$. It is these constants which govern the behaviour of the model and which may be found by means of non-linear least squares curve fitting to measured data. The computer program written to carry out the curve fitting to the expressions given in equation (5) was based upon a technique detailed in reference (3) whereby the error function, $\epsilon_k$,

relating the fitted parameters $f(j\omega_k^3, a_1, a_i)$ to the measured
values $H(j\omega^3)$ is expressed as:

$$E_k = f(j\omega_k^3, a_i) - H(j\omega_k^3) \tag{6}$$

where the function $f$ is given by equation (5) at each value $\omega_k$, i.e.

$$f(j\omega_k^3, a_i) = \sum_r \frac{j\omega_k^3 X^{(r)}}{\omega^{(r)2} - \omega_k^2} \tag{7}$$

where $X^{(r)} = \psi_j^{(r)} \psi_k^{(r)} me$

$$\omega^{(r)2} = \omega_r^2 (1 + j\delta).$$

The computer program provides the mode shape parameter X, damping, $\delta$,
and the resonant frequency, $\omega_r$, for each mode.  This information can
then be used to reconstruct a model in terms of a set of mass, stiffness and damping matrices (2).

An example of the application of these techniques is shown in Fig. 2
which shows the curve fit obtained in terms of amplitude and phase
from frequency response data derived using the linear model of the
machine and foundation system shown in Fig. 1 (i.e. D = 0).

## MODEL OF THE SHAFT/FOUNDATION SYSTEM

The simplified three degree-of-freedom model is represented in Fig.1
together with the values of the parameters used in the analysis.  The
model was based upon an actual system for which it was considered
that the most important modes were the coupled foundation translation
and rocking modes and the shaft first bending mode (in the transverse
plane).  Hysteretic damping was employed in the model for both the
shaft and foundation.  The bearings in the actual system were spherical rollers and thus offered no viscous effects and the medium around
the concrete foundation was relatively dry, sandy soil for which the
damping is considered to be strain- rather than frequency-dependent
(4).

### Representation of a "Looseness" Discontinuity

The non-linearity, which represents a looseness or clearance effect
between the shaft and foundation, was modelled as a complex, spring
element combination of pure clearance and linear stiffness, i.e. for
the shaft/pedestal stiffness ks of Fig. 1, the model was:

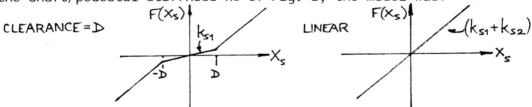

In the above the relative amplitude of vibration of the shaft with
respect to its foundation is defined as:

$$X_s = X - Y - Z \tag{8}$$

### Analysis of the Machine/Foundation System

Two computer programs were used for the analysis: a simulation
program to provide frequency response data of the non-linear system

and the curve-fitting program to produce the modal properties. The equations of motion representing the system shown in Fig. 1 were programmed on a computer simulation package DARE P which is described in reference (5). The DARE P simulation language can be used to integrate the results of time domain analyses to produce frequency spectra and hence permits the harmonic analysis of system response. DARE P employs a Runge-Kutta-Merson variable step integration procedure which has been tested for numerical accuracy by analysing a single degree-of-freedom system with Coulomb Friction and comparing the frequency response of this non-linear system with that derived from an exact analysis (6).

The procedure involved in analysing the machine/foundation model was to obtain the fundamental component of vibration velocity as a function of increasing speed at three coordinate positions X, Y, Z, representing the shaft translation, translation of the centroid of the foundation, and the rocking of the foundation about this centroid respectively. Thus for each computer simulation run three frequency response curves were obtained. The clearance effect could be studied by changing the term representing the discontinuity between runs. The vibration velocity at defined frequency intervals for each co-ordinate was then used as input data to the modal analysis curve fitting program. This program identified the mode shapes, modal dampings and resonant frequencies.

## EFFECT OF CLEARANCE ON SYSTEM DYNAMICS

Three conditions were considered in the investigation: zero clearance (i.e. a linear system) and clearances of 0.025 and 0.05 mm which represent 0, 1.5% and 3% of shaft resonant amplitude respectively. The mass imbalance of the machine was 0.16 kgmm, giving rise to foundation vibration levels of 4.7 $mms^{-1}$ at the nominal running speed of 77 $rads^{-1}$ (735 r.p.m.). Fig. 3 shows a typical set of mode shapes for the machine and foundation system.

The clearance causes a drop in the natural frequency of all three modes and significant changes in the modal damping (Table 1). In particular, the damping of the second mode which coincides with running speed decreases by more than 10%.

Fig. 4 shows Bode plots of vibration velocity for the shaft (co-ordinate X) and the foundation upper edge (co-ordinate (Y + Z)). The changes in natural frequency and the increased vibration level at running speed can be seen as the clearance increases.

Tables 2 and 3 give the complex mode shapes derived from the fits for zero and maximum clearance. The increase in the X-co-ordinate of mode 2 is again significant; there is also a general increase in the complex components which implies that the degree of coupling between the modes is increased.

## CONCLUSIONS

An initial study into the effect of a clearance discontinuity on the modal characteristics of a rotating machine and its foundation shows decreases in system natural frequencies corresponding to the reduced effective stiffness at the discontinuity. It also illustrates the substantial increases in vibration level at some frequencies that can be caused by the presence of a weakness in the support structure of a rotating machine.

It must be stressed that the results are unique to the system modelled; however, the system was modelled from a real industrial situation. The techniques of modal analysis can thus be used to identify system characteristics.

In the above work Coulomb damping effects, which are likely to arise in combination with movements due to clearances, were ignored and these combined non-linearities are to be included in the next stage of the investigation.

## REFERENCES

(1)    KLOSTERMAN, A, and ZIMMERMAN, R, 'Modal survey activity via frequency response functions', Society of Automobile Engineers, Paper No. 751068 (November 1975).

(2)    EWINS, D J, 'The whys and wherefores of modal analysis', Imperial College, London - Dynamics Section Internal Report (1977).

(3)    GOYDER, H G D, 'Methods and application of structural modelling from measured structural frequency response data', Journal of Sound and Vibration, 68 (2), 209-230 (1980).

(4)    CHOW, Y K, and SMITH, I M, 'Static and periodic infinite solid elements', International Journal for Numerical Methods in Engineering, 17, 503-526 (1981).

(5)    WAIT, J V, and CLARKE, D, DARE P, continuous system simulation language, Report No. 299, Computer Research Science Laboratory, University of Arizona (1976).

(6)    TOMLINSON, G R, 'An analysis of the distortion effects of Coulomb damping on the vector plots of lightly damped systems', Journal of Sound and Vibration, 71 (3), 443-451 (1980).

Table 1    Modal frequencies and damping

| Mode | D = 0 | | D = 0.05 mm | |
|------|-------|--------|-------|--------|
| | Frequency | Damping($\delta$) | Frequency | Damping($\delta$) |
| 1 | 21.1 | 0.099 | 20.3 | 0.124 |
| 2 | 77.0 | 0.097 | 75.8 | 0.087 |
| 3 | 143.0 | 0.024 | 141.5 | 0.020 |

Table 2    Complex mode shapes for linear (D = 0) system

| Coordinate | Mode 1 | Mode 2 | Mode 3 |
|------------|--------|--------|--------|
| X | 15.50 - j0.01 | 5.63 + j0.12 | 15.60 - j0.09 |
| Y | 9.10 - j0.04 | -9.06 - j0.03 | -5.62 + j0.15 |
| Z | 5.70 - j0.03 | 11.50 - j0.18 | -9.83 - j0.03 |

Table 3    Complex mode shapes for clearance D = 0.05 mm

| Coordinate | Mode 1 | Mode 2 | Mode 3 |
|------------|--------|--------|--------|
| X | 15.40 + j0.13 | 7.47 + j1.36 | 13.80 - j0.10 |
| Y | 9.84 - j1.05 | -9.32 + j2.14 | -4.97 + j0.07 |
| Z | 5.97 - j0.94 | 8.61 - j6.35 | -8.72 + j0.04 |

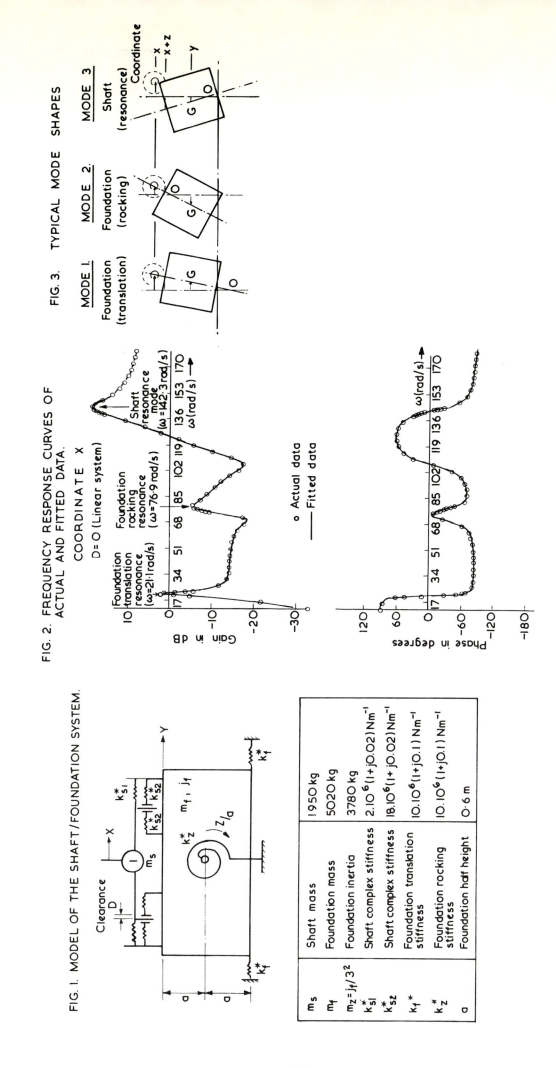

FIG. 3. TYPICAL MODE SHAPES

MODE 1.
Foundation
(translation)

MODE 2.
Foundation
(rocking)

MODE 3
Shaft
(resonance)

x
x + z
Coordinate

FIG. 2. FREQUENCY RESPONSE CURVES OF ACTUAL AND FITTED DATA.

COORDINATE X

D = O (Linear system)

o Actual data
— Fitted data

Shaft resonance mode (ω = 142·3 rad/s)

Foundation rocking resonance (ω = 76·9 rad/s)

Foundation translation resonance (ω = 21·1 rad/s)

ω(rad/s)

Gain in dB

Phase in degrees

ω(rad/s)

FIG. 1. MODEL OF THE SHAFT/FOUNDATION SYSTEM.

| $m_s$ | Shaft mass | 1950 kg |
|---|---|---|
| $m_f$ | Foundation mass | 5020 kg |
| $m_z = j_f/3^2$ | Foundation inertia | 3780 kg |
| $k_{s1}^*$ | Shaft complex stiffness | $2.10^6(1+j0.02)$ Nm$^{-1}$ |
| $k_{s2}^*$ | Shaft complex stiffness | $18.10^6(1+j0.02)$ Nm$^{-1}$ |
| $k_f^*$ | Foundation translation stiffness | $10.10^6(1+j0.1)$ Nm$^{-1}$ |
| $k_z^*$ | Foundation rocking stiffness | $10.10^6(1+j0.1)$ Nm$^{-1}$ |
| $a$ | Foundation half height | 0·6 m |

355

**Figure 4**          **Changes in Shaft and Foundation Vibration Levels**

KEY
———— D=0 (LINEAR)
— — · — D=0.025 MM
— — — — D=0.05 MM

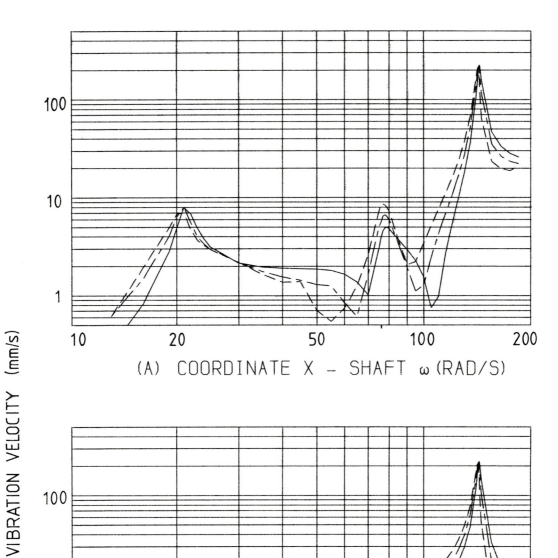

(A) COORDINATE X — SHAFT ω (RAD/S)

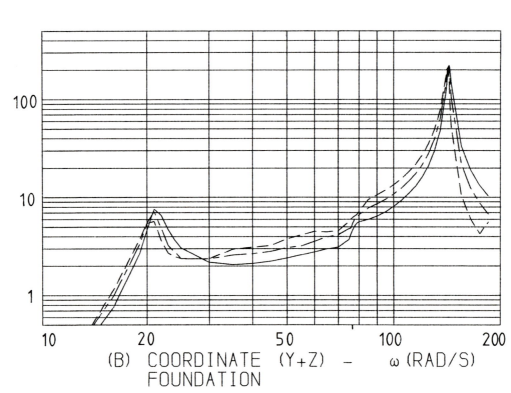

(B) COORDINATE (Y+Z) — ω (RAD/S)
FOUNDATION

VIBRATION VELOCITY (mm/s)

# Chairman's summing up

J B Menzies
*Head of Structural Integrity Division, Building Research Establishment, UK*

This afternoon five very interesting and wide-ranging papers have been presented and discussed. Each paper has contained theoretical analyses aimed at predicting the behaviour of the dynamically loaded structure under consideration. In most cases the structure has been a part, such as a beam or shell, of a complete structure, the analyses being carried out on the basis of assumed boundary conditions of the part and idealisations of the material behaviour. There have been some comparisons of the predictions from different analyses but generally comparisons have been made with the results from tests on small models although in the floor system investigation the model was in fact a section of full-scale floor.

The aim is of course to predict the behaviour of the structure at full scale with its actual boundary conditions, as constructed materials and in-service loading, i.e. incorporating all the inevitable deviations from design assumptions. To model these fully is seldom possible and modelling, analytical or experimental, is usually based on a compromise between expense and the accuracy of prediction. For structures loaded dynamically an approximate indication of global response may well suffice and a relatively simple model can be used. Alternatively study of the detailed behaviour may be essential. A decision should always be made concerning the accuracy necessary for the model results to be useful. The possibilities are many and we have heard some of these discussed this afternoon. The point which strikes me most from our discussion is that there has been little reference to the sensitivity of the results from models to differences between the model and full-scale structure. Perhaps the accuracy of prediction of models would be a good subject for the next International Seminar organised by the Institution of Structural Engineers Informal Study Group.

# Index

Model materials, cont.

   plastic roof cladding, 269

   polyurethane foam, 73

   prestressing wire, 254

   PVC sheet, 321

   reinforcement, 87, 105, 128-9, 169, 243

   sand, 142

   stainless steel, 78

   steel, 172-3, 202, 211, 238

   threaded rod reinforcement, 105

Model structures

   beams, 234-41, 242-53, 260-8, 291-301

   circular pretensioned cable roofs, 269-75

   concrete shelters, 198-219, 276-82

   cooling towers, 63-70

   damped, 291-301

   guyed masts, 78-83

   impulsively loaded, 234-41

   multi-storey reinforced concrete, 82-92, 93-102, 103-25, 126-40,
      162-70

   off-shore floating structures, 42-4, 57-62

   off-shore gravity structures, 4, 7-20, 21-6, 42-3, 151-4

   open spherical shells, 320-8

   piles, 3

   precast concrete large panel, 103-25

   prestressed concrete slab, 254-9

   projectiles, 3, 205, 210, 217

   reactor core, 329-42

   reactor building, 189-96

   reinforced concrete panels, 220-33

   steel joist-concrete slab floor, 343-49

   steel structures, 21-6, 78-83, 162-70, 171-9, 269-75

   structural walls, 155-61

   tall buildings, 86-92, 93-102, 103-25, 162-70, 171-7, 283-91,
      302-11

   see also Rigid-plastic models

Modelling

   analytical, 21-6, 202-3; see also Numerical analysis

   dimensional analysis, see scale

   engineering, 2, 7-21, 71-8, 86-92, 93-102, 126-40, 141-7,
      155-61, 162-70, 171-7, 254-9, 276-83, 291-301, 350-6